IEEE Recommended Practice for
Protection and Coordination of
Industrial and Commercial Power Systems

Published by
The Institute of Electrical and Electronics Engineers, Inc

Acknowledgement

Appreciation is expressed to members of the Power Systems Relaying Committee of the IEEE Power Engineering Society for their review and valuable suggestions in the preparation of this standards document.

Recognized as an
American National Standard (ANSI)

IEEE
Std 242-1986
(Revision of IEEE
Std 242-1975)

IEEE Recommended Practice for Protection and Coordination of Industrial and Commercial Power Systems

Sponsor

**Industrial and Commercial Power Systems Committee
of the
IEEE Industry Applications Society**

Approved September 19, 1985
Reaffirmed June 27, 1991

IEEE Standards Board

Approved February 28, 1986
Reaffirmed December 9, 1991

American National Standards Institute

Corrected Edition

NOTE: Corrections have been made on pages 64, 72, 75, 76, 80, 81, 82, 83, 84, 85, 86, 94, 104, 106, 379, and 381. A black bar has been added opposite each correction to aid in identifying the changes made to this printing.

Seventh Printing
February 1996

ISBN 0-471-85392-5

Library of Congress Catalog Number 86-81948

September 9, 1986

SH10702

Foreword

(This Foreword is not part of ANSI/IEEE Std 242-1986, IEEE Recommended Practice for Protection and Coordination of Industrial and Commercial Power Systems.)

At the October 1965 IEEE Industrial and Commercial Power Systems Conference in Buffalo, a paper was presented to the attendees by Tom Higgins and Norman Peach. That paper, resulting from a suggestion by A. C. Friel (deceased), proposed a new IEEE publication covering protection and coordination for industrial and commercial power systems. The first sentence of that paper states, "The proposed publication is intended for the plant electrical engineer with broad responsibilities for the plant electric system. He is familiar with the general principles of industrial system design, but has not had an opportunity to specialize in system protection." The Power Systems Protection Subcommittee (now Committee) of the Industrial and Commercial Power Systems (ICPS) Committee (now Department) accepted the responsibility of preparing the new book.

During the next seven years, thirteen of the fifteen chapter subjects were presented formally at sessions of IEEE conferences in which the ICPS Department participated. The last three years were occupied in reviewing and updating the papers and obtaining the necessary approvals. Copies of the manuscript were forwarded to each of the 74 ICPS Department members and approved through the usual IEEE Standards procedures.

The content of this standards document is evident from an inspection of the title of each chapter. The value of proper and adequate protection in electric power distribution systems has been emphasized by the provisions of the Occupational Safety and Health Act of 1969. Good coordination helps to ensure the availablity of electric energy, a requirement of steadily increasing economic importance.

The ICPS Department will maintain this document current with the state of developing technology, and revisions of this book will be issued at reasonable intervals in order to maintain all chapters current with the rapidly changing techniques of circuit protection. Suggestions for improvements are welcomed. These should be directed to:

> Secretary, IEEE Standards Board
> 345 East 47th Street
> New York, NY 10017
> USA

Recommended Practice for Electric Power Distribution for Industrial Plants (IEEE Red Book), ANSI/IEEE Std 141-1986.

Recommended Practice for Grounding of Industrial and Commercial Power Systems (IEEE Green Book), ANSI/IEEE Std 142-1982.

Recommended Practice for Electric Power Systems in Commercial Buildings (IEEE Gray Book), ANSI/IEEE Std 241-1983.

Recommended Practice for Industrial and Commercial Power System Analysis (IEEE Brown Book), ANSI/IEEE Std 399-1980.

Recommended Practice for Emergency and Standby Power for Industrial and Commercial Applications (IEEE Orange Book), ANSI/IEEE Std 446-1980.

Recommended Practice for the Design of Reliable Industrial and Commercial Power Systems (IEEE Gold Book), ANSI/IEEE Std 493-1980.

Recommended Practice for Electric Systems in Health Care Facilities (IEEE White Book), ANSI/IEEE Std 602-1986.

Recommended Practice for Energy Conservation and Cost-Effective Planning in Industrial Facilities (IEEE Bronze Book), ANSI/IEEE Std 739-1984.

The following persons were on the balloting committee that approved this document for submission to the IEEE Standards Board:

James D. Bailey
David S. Baker
Donald A. Bly
William Burt
Rene Castenschiold
Walt Chumakov
Keith Cooper
Jerome Frank
Stephen W. Herholtz

Charles James
Robert Jones
Clif LaPlatney
Ralph Lee
Daniel J. Love
Bal K. Mathur
Harvey E. Meisel
Russell O. Ohlson

Norman Peach
John F. Perkins
Dorn F. Pettit
William J. Rooney
Vince Saporita
Steve Schaeffer
John Scott
Robert L. Simpson
James C. Wilson

When the IEEE Standards Board approved this standard on September 19, 1985, it had the following membership:

John E. May, *Chairman*   **John P. Riganati,** *Vice Chairman*

Sava I. Sherr, *Secretary*

James H. Beall
Fletcher J. Buckley
Rene Castenschiold
Edward Chelotti
Edward J. Cohen
Paul G. Cummings
Donald C. Fleckenstein

Jay Forster
Daniel L. Goldberg
Kenneth D. Hendrix
Irvin N. Howell
Jack Kinn
Joseph L. Koepfinger*
Irving Kolodny
R. F. Lawrence

Lawrence V. McCall
Donald T. Michael*
Frank L. Rose
Clifford O. Swanson
J. Richard Weger
W. B. Wilkens
Charles J. Wylie

*Member emeritus

Protection and Coordination of Industrial and Commercial Power Systems

Second Edition

Protection and Coordination of Industrial and Commercial Power Systems

Working Group Members and Contributors

John Cooper, *Chairman*
Jerome Frank, *Co-Chairman*

Chapter 1—First Principles: Norman Peach, *Chairman*

Chapter 2—Calculation of Short-Circuit Currents: Russell O. Ohlson, *Chairman;* M. Wayne Puckett

Chapter 3—Instrument Transformers: John V. Scott, *Chairman;* Ron T. Carter, Ermal Curd, Dal Dalasta, Frank A. Denbrock, Harold H. Fahnoe, Daniel J. Love, Louis J. Powell, Ralph Stetson, Edwin F. Troy

Chapter 4—Selection and Application of Protective Relays: Keith Cooper, *Chairman;* David S. Baker, Louis J. Powell, Robert L. Simpson

Chapter 5—Fuses: Charles James, *Chairman;* S. Doebele, Harold Fahnoe, John V. Scott, Kenneth Swain

Chapter 6—Low-Voltage Circuit Breakers: Harvey E. Meisel, *Chairman;* Ray Hanson, Dan Kelly

Chapter 7—Ground-Fault Protection: Rene Castenschiold, *Chairman;* Walter V. Chumakov, Arthur Freund, Russell O. Ohlson, Norman Peach, Clarence Tsung

Chapter 8—Conductor Protection: Ralph H. Lee, *Chairman;* Keith M. Grimm, Timothy T. Ho, Richard H. Kaufman

Chapter 9—Motor Protection: Dorn L. Pettit (deceased), *Chairman;* Jerome Frank

Chapter 10—Transformer Protection: Jerome Frank, *Chairman;* Carey J. Cook, John Cooper, Harold H. Fahnoe, Daniel J. Love, Bal K. Mathur, R. L. Smith, Candido Soares, Roy Uptegraff, Jr

Chapter 11—Generator Protection: David S. Baker, *Chairman;* Barry L. Christen, Thomas D. Higgins

Chapter 12—Bus and Switchgear Protection: William Burt, *Chairman;* Samual P. Axe, S. Doebele, Leon E. Goff

Chapter 13—Service Supply Line Protection: Bal K. Mathur, *Chairman;* S. Doebele

Chapter 14—Overcurrent Coordination: Daniel J. Love, *Chairman;* Keith Cooper, S. Doebele

Chapter 15—Maintenance, Testing, and Calibration: James C. Wilson, *Chairman;* John Cooper, G. Shliapnikoff, R. L. Smith, Jr, Edwin F. Troy

CONTENTS

FIGURES

IEEE Recommended Practice for
Protection and Coordination of
Industrial and Commercial Power Systems

1. First Principles

1.1 Objectives. The objectives of electrical system protection and coordination are to prevent injury to personnel, to minimize damage to the system components, and to limit the extent and duration of service interruption whenever equipment failure, human error, or adverse natural events occur on any portion of the system. The circumstances causing system malfunction are usually unpredictable, though sound design and preventive maintenance can reduce the likelihood of their happening. The electrical system, therefore, should be designed and maintained in such a way as to protect itself automatically.

1.1.1 Safety. Prevention of human injury is the most important objective of electrical system protection. Interrupting devices should have adequate interrupting capability and energized parts should be sufficiently enclosed or isolated so as not to expose personnel to explosion, fire, arcing, or shock. Safety has priority over service continuity, equipment damage, or economics.

These fundamental principles of safety have always been adhered to by responsible engineers engaged in the design and operation of electrical systems. ANSI/NFPA 70-1984 [4][1] (National Electrical Code [NEC]) and state and local codes have prescribed practices intended to enhance the safety of electrical systems. In recent years an increased concern about safety has led to many studies resulting in detailed recommendations and regulations relating to electrical systems. Prominent among these are the regulations of the Occupational Safety and Health Administration (OSHA) of the United States Department of Labor. Engineers engaged in the design and operation of electrical system protection should familiarize themselves with the most recent OSHA regulations and all other applicable codes and regulations relating to human safety.

1.1.2 Equipment Damage Versus Service Continuity. Whether minimizing the risk of equipment damage or preserving service continuity is the more important objective depends on the operating philosophy of the particular plant or business. Some operations can afford limited service interruptions to minimize the possibility of equipment repair or replacement costs, while others will regard such an expense as small compared with even a brief interruption of service. The latter attitude is dominant in the process industries that have developed, from experience, practices in system design, operation, and maintenance that make it possible to operate with sustained overloads and reduce the likelihood of minor

---

[1]The numbers in brackets correspond to those of the references listed at the end of this chapter.

faults rapidly turning into major ones. A case in point is ungrounded systems that require special regard to maintenance, fault detection, and fault location to clear the first ground fault before a second ground fault occurs. In such installations this basic approach is applied throughout the electric system and all other systems impinging on the critical process. While generally considered obsolete, ungrounded systems are still preferred by some designers of electrical systems where process interruption is critical.

In industries where the process is not highly critical, electrical protection should be designed for the best compromise between equipment damage and service continuity. Keeping in mind that a prime objective is to obtain selectivity so as to minimize the extent of equipment shutdown in case of a fault, most operations probably would prefer that faulted equipment be deenergized as soon as the fault is detected.

1.1.3 Economic Considerations. The cost of system protection can never be ignored, and it will determine the degree of system protection that can be feasibly designed into a system. Many features can be added that will improve system performance, reliability, and flexibility, but at increased initial cost. On the other hand, failure to design into a system at least the minimum safety and reliability requirements will inevitably result in unsatisfactory performance with the probability of expensive downtime. Modifying a system that proves inadequate will be more expensive and in most cases less satisfactory than designing these features into a system at the outset.

1.2 Scope. This publication presents in a step-by-step, simplified, yet comprehensive, form the principles of system protection and the proper application and coordination of those components that may be required to protect industrial and commercial power systems against any abnormalities that could reasonably be expected to occur in the course of system operation.

Coverage is limited to system protection and coordination as it pertains to system design treated in ANSI/IEEE Std 141-1986 [1][1] and ANSI/IEEE Std 241-1983 [3]. No attempt is made to cover utility systems or residential systems, though much of the material presented is applicable to these systems.

1.3 Planning System Protection. The designer of electric power systems has available several techniques to minimize the effects of abnormalities occurring on the system itself or on the utilization equipment that it supplies. One can design into the electric system features that will:

(1) Quickly isolate the affected portion of the system and in this manner maintain normal service for as much of the system as possible and minimize damage to the affected portion of the system

(2) Minimize the magnitude of the available short-circuit current and in this manner minimize potential damage to the system, its components, and the utilization equipment it supplies

(3) Provide alternate circuits, automatic throwovers or automatic reclosing devices, or both, where applicable, and in this manner minimize the duration or the extent, or both, of supply and utilization equipment outages.

System protection encompasses all of the above techniques; however, this text deals mainly with the prompt isolation of the affected portion of the system. Accordingly, the function of system protection may be defined as "the detection and prompt isolation of the affected portion of the system whenever a short circuit or other abnormality occurs that might cause damage to, or adversely affect, the operation of any portion of the system or the load that it supplies."

Coordination is the selection or setting, or both, of protective devices so as to isolate only that portion of the system where the abnormality occurs. It is a basic ingredient of well-designed electric distribution system protection and is mandatory in certain health care and continuous process industrial systems.

System protection is one of the most essential features of an electrical system and must be considered concurrently with all other essential features. There is too often a tendency to consider system protection after all other design features have been determined and the basic system design established. Such an approach can result in an unsatisfactory system that cannot be adequately protected, except by a disproportionately high expenditure. System protection is so basic to the safety of personnel and the reliability of electrical supply, and can have such profound influence on the economics of system design, that examining system protection needs only after all other design features have been determined is a completely unrealistic approach. It would, therefore, seem that any competent designer should thoroughly examine the question of system protection at each stage of the planning and incorporate in the final system a fully integrated protection plan that is capable of being coordinated and is flexible enough to grow with the system.

In laying out electric power systems the designer should endeavor to keep the final design as simple as would be compatible with safety, reliability, maintainability, and economic considerations. Designing additional reliability or flexibility, or both, into a system may lead to a more complex system requiring more complex coordination of the protective system. Such additional complexity should be avoided except where the requisite personnel, equipment, and know-how are known to be available to adequately service and maintain a complex electric power system.

1.4 Abnormalities to Protect Against. The principal abnormalities to protect against are short circuits and overloads. Short circuits may be caused in many ways, including failure of insulation due to excessive moisture, mechanical damage to electrical distribution equipment, and failure of utilization equipment as a result of overloading or other abuse. Circuits may become overloaded simply by connecting larger or additional utilization equipment to the circuit. Overloads may be caused by improper installation and maintenance, such as misaligned shafts and worn bearings. Improper operating procedures are also a cause of equipment overload or damage. These include too frequent starting, extended accelerating periods, and obstructed ventilation.

Short circuits may occur between two phase conductors, between all phases of a polyphase system, or between one or more phases and ground. The short circuit may be "solid" (bolted) or welded, in which case the short circuit is permanent and has a relatively low impedance. Or (under 250 V) the short circuit may burn itself clear, probably opening one or more conductors in the process. The short circuit may involve an arc having relatively high impedance. Such an arcing short circuit can do extensive damage without producing exceptionally high current. An arcing short circuit may or may not extinguish itself. Another type of short circuit is one with a high impedance path, such as the dirt accumulated on an insulator, in which a flashover occurs. The flashover may be harmlessly extinguished or the ionization produced may lead to a more extensive short circuit. These different types of short circuits produce somewhat different conditions in the system. Electric systems should be protected against the highest current short circuits that can occur, keeping in mind, however, that this maximum fault protection may not simultaneously provide adequate protection against lower current faults that may involve an arc and that are potentially very destructive.

Other sources of abnormality, such as lightning, load surges, and loss of synchronism, usually have little or no effect on system overcurrent selectivity, but should not be ignored. They usually can be best handled on an individual basis for the specific equipment involved, such as transformers, motors, generators, etc.

Reference should be made to texts dealing with these problems.

1.5 Basic Protective Equipment. The isolation of short circuits and overloads requires the application of protective equipment that will both sense that an abnormal current flow exists and then remove the affected portion from the system. In some types the sensing device and the interrupting device are completely separate, interconnected only through external control wiring. In other types the sensing and interrupting functions are combined in the same device. In still other types, the sensing and interrupting devices, although actually separate, are included in the same equipment and mechanically coupled so as to function as a single device.

A fuse is both a sensing and interrupting device. It is connected in series with the circuit and responds to thermal effects produced by the current flow through it. Its fusible element is designed to open at a predetermined time depending on the amount of current that flows. Different types are available having the time-current characteristics required for the proper protection of the circuit components. Fuses may be noncurrent-limiting or current-limiting, depending upon their design and construction. Fuses are not resettable since their fusible elements are consumed in the process of interrupting current flow. Fuses, their characteristics, applications, and limitations are described in detail in Chapter 5.

Circuit breakers are interrupting devices only and must be used in conjunction with sensing devices to fulfill the detection function. In the case of medium-voltage (1–72.5 kV) circuit breakers, the sensing devices are separate protective relays or combinations of relays, covered in Chapter 4. In the case of low-voltage (under 1000 V) circuit breakers, sensing devices may be external protection relays or combinations of relays. In most applications, either molded-case circuit breakers or other low-voltage circuit breakers have sensing devices built into the equipment. These sensors may be thermal or magnetic series devices, or they may be integrally mounted but otherwise separate protective relays. Low-voltage circuit breakers, their applications, characteristics, and limitations are covered in Chapter 6.

Overcurrent relays used in conjunction with medium-voltage circuit breakers are available with different functional characteristics. They may be either directional or nondirectional in their action. They may be instantaneous or time-delay in response. Various time-current characteristics, such as inverse time, very inverse time, extremely inverse time, and definite minimum time, are available over a wide range of current settings. For specific applications, various types of differential overcurrent relays are available. Overcurrent relays and their selection, application, and settings are covered in detail in Chapter 4. Such relays generally are used in conjunction with instrument transformers, which are covered in Chapter 3.

1.6 Preliminary Design. The designer of an electric power system should first determine the load requirement, including the sizes and types of loads, and any special requirements. He should also determine the available short-circuit current at the point of delivery, the time-current curves and settings of the nearest utility company protective device, and any contract restrictions on ratings and settings of protective relays in the user's system. (See Fig 1.)

The designer can then proceed with a preliminary system design. Chapter 2 covers the fundamentals of short-circuit behavior and the calculation of short-circuit duty requirements that will permit evaluation of the preliminary design for compatibility with available ratings of circuit breakers and fuses. At this point some modification of the design may be necessary because of economic considerations or equipment availability. Preliminary design should be evaluated from the standpoint of system coordination, as covered in Chapter 14. If the protection provided in the preliminary design cannot be selectively coordinated with utility company settings and contractual restrictions on protective device settings, the design should be modified to provide proper selective coordination.

Ground-fault protection is an essential part of system protection and is given detailed coverage in Chapter 7 for two reasons. First, although many of the devices used to obtain ground-fault protection are similar to those covered in Chapters 3 through 6, the need for protection and the potential problems of proper applications are frequently not fully appreciated. Second, proper selective coordination of ground-fault protection seldom has any significant effect on overall system selectivity, although its effect should be taken into consideration in the same general manner covered in Chapter 14.

1.7 Special Protection. In addition to developing a basic protection design, the designer may also need to develop protective schemes for specific pieces of equipment or for specific portions of the system. Such specialized protection should be coordinated with the basic system protection. Specialized protection applications include:

Chapter 8 Conductor Protection

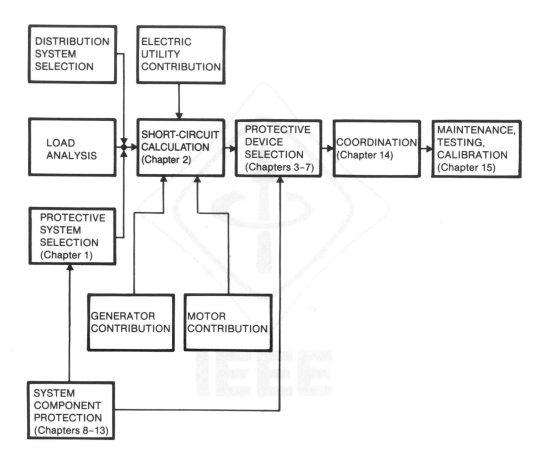

Fig 1
Sequence of Steps in System Protection and Coordination

1.8 Field Follow-up. Proper application of the principles covered in the first 14 chapters of this book will result in the installation of system protection capable of coordinated selective isolation of system faults. But this capability will be to no avail if the proper field follow-up is not planned and executed. This followup has three elements: proper installation, including testing and calibration of all protective devices; proper operation of the system and its components; and a proper preventive maintenance program, including periodic retesting and recalibration of all protective devices. A separate chapter, Chapter 15, has been included to cover testing and maintenance.

1.9 References. The following publications shall be used in conjunction with this chapter.

[1] ANSI/IEEE Std 141-1986, IEEE Recommended Practice for Electric Power Distribution for Industrial Plants.[2]

[2] ANSI/IEEE Std 142-1982, IEEE Recommended Practice for Grounding of Industrial and Commercial Power Systems.

[3] ANSI/IEEE Std 241-1983, IEEE Recommended Practice for Electric Power Systems in Commercial Buildings.

[4] ANSI/NFPA 70-1984, National Electrical Code.[3]

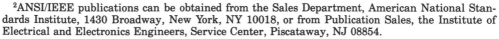

---

[2]ANSI/IEEE publications can be obtained from the Sales Department, American National Standards Institute, 1430 Broadway, New York, NY 10018, or from Publication Sales, the Institute of Electrical and Electronics Engineers, Service Center, Piscataway, NJ 08854.

[3]ANSI/NFPA publications can be obtained from the Sales Department, American National Standards Institute, 1430 Broadway, New York, NY 10018, or from Publication Sales, National Fire Protection Association, Batterymarch Park, Quincy, MA 02269.

2. Calculation of Short-Circuit Currents

2.1 General Discussion. Short-circuit currents introduce large amounts of destructive energy in the forms of heat and magnetic force into a power system. Calculations should be made to ensure that the short-circuit ratings of the equipment are adequate to handle the currents available at their locations. In general, the procedure is to (1) develop a graphical representation of the system with symbolic voltage sources and circuit impedances, then (2) determine the total equivalent impedance from the source to designated points, and (3) at each point divide the voltage by the total impedance to that point to derive the short-circuit current.

2.1.1 Basic Assumptions. Certain simplifying assumptions are customarily made when calculating fault current. An important assumption is that the fault is "bolted," that is, it has zero impedance. This assumption simplifies calculation, since the resulting calculated values are a maximum, and equipment selected on this basis will always have an adequate rating. Furthermore, a three-phase fault is customarily assumed, because this type of fault generally results in the maximum short-circuit current available in a system. In most systems the three-phase fault is frequently the only one calculated.

The actual fault current is usually less than the calculated bolted three-phase value because most faults involve arcing resistance or other undefined impedances, or both, not used in the calculations. Faults which are not three-phase will usually be less. Bolted line-to-line currents are about 87% of the three-phase value, while bolted line-to-ground currents can range from about 25–125% of the three-phase value, depending on system parameters. However, line-to-ground currents of more than 100% of the three-phase value rarely occur in industrial and commercial systems. Line-to-ground arcing fault currents in low-voltage systems are sometimes less than normal load currents and yet can be extremely destructive. This has led to techniques specifically directed to ground-fault protection (see Chapter 8).

Several assumptions which simplify the calculations are usually made. First, load currents are ignored. Second, the voltages of power company and generator sources are assumed to be equal to their nominal values at no load, although the actual voltages may be in a range of ± 5% of the nominal values. Third, motors are assumed to be running with their rated voltage at the terminals when a fault occurs. Fourth, the transformer percent impedance values used may be actual values or are often nominal values possibly subject to their ± 7.5% tolerance to anticipate the worst case. Fifth, when source X/R ratios are unknown, a relatively high value is generally assumed. This normally results in calculated values of short-circuit current that are slightly high. Sixth, switchboard and panelboard bus impedances can be ignored. These impedance values are usually small enough to only slightly increase the calculated values and are difficult to determine with any reasonable degree of accuracy. Furthermore, integrated equipment short-circuit ratings of panelboards and switchboards already involve consideration of their bus impedances. Other assumptions about source contributions and circuit element impedances are discussed later in this chapter.

2.1.2 Sources of Fault Current. The basic sources of fault current are the power company system, generators, synchronous motors, and induction motors.

(1) A typical modern electric utility represents a large and complex interconnection of generating plants. In a typical system, the individual generators are not affected by a maximum short circuit in an industrial plant. Transmission lines, distribution lines, and transformers introduce impedances between the utility generators and the industrial customer. Were it not for such impedances, the utility system would be an infinite source of fault current. Before performing calculations, accurate values of present and projected available short-circuit currents and the X/R ratio, or the $R + jX$ source impedance, at the delivery point must be obtained from the supplying utility.

(2) In-plant generators react to system short circuits in a predictable way. Fault current from a generator decreases exponentially from a high initial value to a lower steady-state value some time after the initiation of the fault. Since a generator continues to be driven by its prime mover, and to have its field energized from its separate exciter, the steady-state fault current will persist unless interrupted by some circuit interrupter.

For purposes of fault current calculation, the variable generator reactance can be represented by three values. X_d'', the direct-axis subtransient reactance, usually determines the current magnitude during the first cycle after the fault occurs. In some cases, the time constant is so short that the magnitude of the first peak may actually be determined by the direct-axis transient reactance. X_d', the direct-axis transient reactance, determines the current magnitude in the range up to 1 or 2 s. X_d, the direct-axis synchronous reactance, determines the current flow after a steady-state condition is reached. Most fault protective devices, such as circuit breakers or fuses, operate before steady-state conditions are reached. Therefore, generator synchronous reactance is seldom used in calculating fault currents for the application of these devices.

(3) Synchronous motors supply current to a fault in much the same manner as do synchronous generators. The drop in system voltage due to a fault causes the synchronous motor to receive less power from the system for driving its load. The inertia of the motor and its load acts as a prime mover, and with field excitation maintained, the motor acts as a generator. This fault current diminishes as the motor slows down and the motor field excitation decays.

The variable reactance of a synchronous motor is described with the same designation as for a generator. However, numerical values of the three reactances X_d'', X_d', and X_d will often be different for motors than for generators.

(4) The fault current contribution of an induction motor results from generator

action produced by inertia driving the motor after the fault occurs. In contrast to the synchronous motor, the field flux of the induction motor is produced by induction from the stator rather than from a direct-current field winding. This flux decays on removal of source voltage resulting from a fault, and so the contribution of an induction motor drops off at a rapid exponential rate and soon disappears. As a consequence, induction motors are assigned only a reactance that is about equivalent to the synchronous machine subtransient reactance X_d''. This value will be about equal to the locked-rotor reactance, and hence the initial fault current contribution will be about equal to the full-voltage starting current of the machine. However, the resistance in small motors may be large enough to cause significant decay in their fault current contribution before the first peak of fault current is experienced.

Wound-rotor induction motors normally operate with their rotor rings short circuited and will contribute fault current in the same manner as a squirrel-cage induction motor. Occasionally, large wound-rotor motors are operated with external resistance maintained in their rotor circuits. This gives them short-circuit time constants that are so low that their fault contribution is insignificant. However, a specific investigation should be made before neglecting the contribution from a wound-rotor motor.

Capacitor discharge currents, because of their very short-time constant, can be neglected in most cases. However, there are some applications in which very high transitory currents can be developed when a short circuit occurs close to a bank of energized capacitors. These currents, generally of much higher frequency than normal operating frequency, may exceed in magnitude the power-frequency short-circuit currents and persist long enough to impose severe duty on the circuit components that carry this current. If a problem is anticipated, the capacitor currents could be calculated the same way as for back-to-back switching and compared to the short-circuit ratings of the devices [B5][4].

2.1.3 Total Short-Circuit Current as a Function of Time. When a short circuit occurs, a new circuit is established with lower impedance, most of which is inductance, and the current consequently increases. In the case of a bolted short circuit the impedance is drastically reduced, and the current increases to a very high value in a fraction of a cycle. Figure 2 represents a symmetrical short-circuit current, that is, a short-circuit current that has the same axis as the normal current that was flowing before the fault occurred. To produce a symmetrical short-circuit current under the assumption that the short-circuit power factor is zero, the fault should occur exactly when the normal voltage is maximum. In Fig 2 the system voltage is assumed to remain constant although the current increases.

The total short-circuit current is made up of components from all sources connected to the circuit. The contributions from rotating machinery decrease at various rates, so that the symmetrical current is initially at a maximum, then

---

[4] The numbers in brackets correspond to those of the references listed at the end of this chapter. The numbers in brackets with the prefix "B" corresponds to those of the bibliography listed at the end of this chapter.

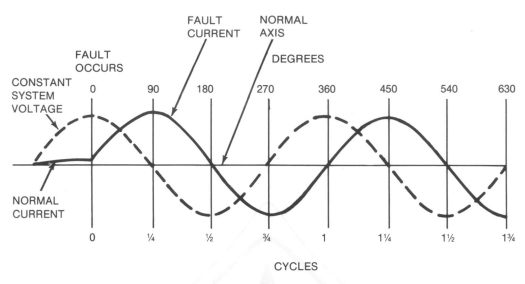

Fig 2
A Symmetrical Short-Circuit Current Wave for a Totally Reactive System

decreases until a steady-state value is reached. This decrease is known as the alternating-current decrement of the short-circuit current. Figure 3 shows a decreasing symmetrical short-circuit current.

Most short-circuit currents are not symmetrical; they are offset from the normal-current axis for a period of several cycles. If the power factor is essentially zero until a steady-state value is reached and the short circuit occurs at a zero point on the voltage wave, the current starts to build up from zero, but cannot follow the normal-current axis because the current should lag behind the voltage by 90°. Although the current is symmetrical with respect to a new axis, it is asymmetrical with respect to the original axis. Figure 4 illustrates the case where a short circuit with low power factor produces the highest first peak of short-circuit current. The magnitude of current offset for a typical fault will be between the two extremes of complete symmetry and complete asymmetry because the odds are against the fault occurring exactly at a voltage peak or a voltage zero.

An analysis of a typical asymmetrical current wave is made in Fig 4. The offset of the asymmetrical current wave from a symmetrical wave having equal peak-to-peak displacement is a positive value of current that may be considered as a direct current. The asymmetrical current, therefore, may be thought of as the sum of an alternating-current component b and a direct-current component a. At the instant of fault initiation (0 cycles in Fig 4), b is negative and $a + b = 0$. Just before quarter-cycle the symmetrical alternating-current component is zero, and the total current is equal to the direct-current component. At near half-cycle the total current is maximum, being the sum of the maximum positive alternating-current component and the direct-current component.

ANSI/IEEE
Std 242-1986

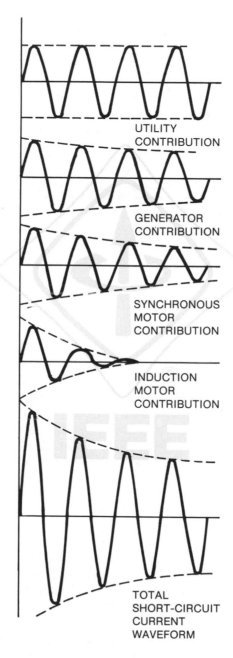

**Fig 3
Decreasing Symmetrical Short-Circuit Current Waveform**

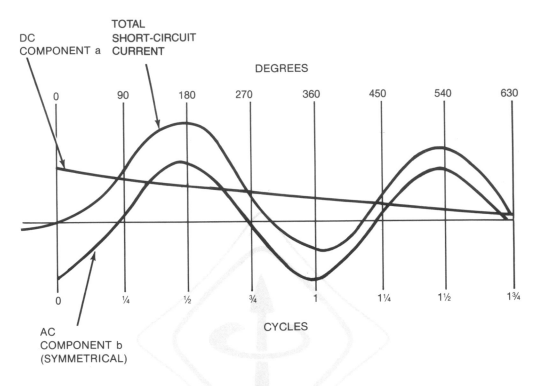

Fig 4
Analysis of Asymmetrical Current Waveform

The direct-current component decreases eventually to zero as the stored energy it represents is expended in the form of I^2R losses in the resistance of the system. The initial rate of decay of the direct-current component is inversely proportional to the X/R ratio of the system from the source to the fault. The lower the X/R ratio, the more rapid is the decay. This decay is called the "direct-current decrement." The total short-circuit current is thus affected by both an alternating-current decrement and a direct-current decrement before reaching its steady-state value.

2.1.4 Short-Circuit Ratings of Equipment. From the foregoing it can be seen that short-circuit currents may act differently in the first few cycles than later, if allowed to persist. Former practice was to establish rms asymmetrical short-circuit ratings of equipment. Calculations for symmetrical values of short-circuit currents were made, and asymmetrical values were then determined by applying simple multipliers to the symmetrical values. The trend in recent years to rate protective equipment on an rms symmetrical basis is now nearly complete. Under the new rating basis, asymmetry is accounted for by various application formulas depending on the class of equipment. Some of these application formulas, embodied in national standards, are quite involved [6].

Short-circuit current calculation concepts are also examined from another point of view. While the symmetrical equipment rating concept was gaining acceptance, the operating times of interrupting devices were being reduced. The interrupting period of protective devices became crowded into the first few cycles and even into the first half-cycle for current-limiting fuses. To an increasing number of engineers, therefore, maximum symmetrical values of short-circuit current have not appeared fully adequate to define short-circuit conditions.

The concept of I^2t has been introduced to supplement the symmetrical current concept because it represents the actual thermal stresses imposed on equipment carrying short-circuit current in the first few cycles [11]. The quantity I^2t represents $\int i^2 dt$, the time integral of the current squared for the time under consideration (Fig 5). It is quite possible that in the future some pieces of protective equipment may be time-current coordinated and rated on an I^2t basis rather than a maximum symmetrical current basis.

2.2 Calculation Preliminaries

2.2.1 One-Line Diagram. A one-line diagram is a graphical representation of the power system and should be prepared as the first step in making a short-circuit current study. This diagram should show all sources of short-circuit current and all significant circuit elements. Reactance and resistance values of all these elements should be included in the diagram. Reactance and resistance data can be obtained from Tables 1–14, or preferably from the equipment supplier. In any case, the calculations cannot be more accurate than the data used.

Fig 5
Finding I^2t in Fault-Current Waveforms
(a) Current-Limiting Fuse   (b) 1.5 Cycle Circuit Breaker

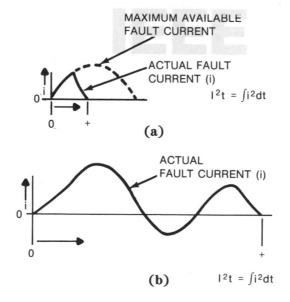

Table 1
Impedance Data for Three-Phase Transformers With Primaries of Up to 15 000 V and Secondaries of 600 V or Less

kVA 3Θ	Suggested X/R Ratio for Calculation	Normal Range of Percent Impedance (see notes)
112.5	3.0	1.6 − 2 Min − 6.2
150.0	3.5	1.5 − 2 Min − 6.4
225.0	4.0	2.0 − 2 Min − 6.6
300.0	4.5	2.0 − 4.5 Min − 6.0
500.0	5.0	2.1 − 4.5 Min − 6.1
750.0	6.0	3.2 − 5.75 − 6.75 − 6.8
1000.0	7.0	3.2 − 5.75 − 6.75 − 8.0
1500.0	7.0	3.5 − 5.75 − 6.75 − 6.8
2000.0	8.0	3.5 − 5.75 − 6.75 − 6.8
2500.0	9.0	3.5 − 5.75 − 6.75 − 6.8

NOTES: (1) Underlined values are from ANSI C57.12.10-1977 [1], ANSI C57.12.22-1980 [2], and NEMA 210-1976 [10].
(2) Network transformers (with three-position swithches) have 5.0 %Z for 300–1000 kVA, 7.0 % Z for 1500–2500 kVA, (with two-position switches) 4.0 %Z for 500–750 kVA. See ANSI C57.12-40-1982 [3].
(3) Three-phase banks with three single-phase transformers may have values as low as 1.2%.

Table 2
Data for Three-Phase Transformers With Secondaries of 2400 V or More (750–60 000 kVA)

Primary kV	Primary kV BIL	Standard Percent Impedance (see notes)
2.4 − 22.9	60 − 150	5.5 or 6.5
− 34.4	−200	6.0 or 7.0
− 43.8	−250	6.5 or 7.5
− 67.0	−350	7.0 or 8.0
− 115.0	−450	7.5 or 8.5
− 138.0	−550	8.0 or 9.0

NOTES: (1) Actual values are generally within ± 7.5% of the standard values [1].
(2) Add 0.5% for load tap changing [8].
(3) Lower values are usually for OA 55 °C or OA 55/65 °C rise transformers.
(4) Higher values are usually for OA 65 °C rise transformers.
(5) X/R values are similar to those in Table 1. Consult manufacturer or use the values in [4] for transformers rated over 2500 kVA.

Table 3
Constants of Medium-Voltage Copper Conductors for 1 Ft Delta Spacing

Conductor Size		Resistance R at 25 °C, 50 Hz (mΩ/conductor/100 ft)	Reactance X_A at 1 ft Spacing, 60 Hz (mΩ/conductor/100 ft)
(MCM or kcmil)	(AWG No)		
1000.00		1.19	7.58
900.00		1.30	7.69
800.00		1.45	7.82
750.00		1.53	7.90
700.00		1.63	8.00
600.00		1.88	8.18
500.00		2.24	8.39
450.00		2.49	8.54
400.00		2.80	8.67
350.00		3.17	8.83
300.00		3.71	9.02
250.00		4.44	9.22
211.60	4/0	5.24	9.53
167.80	3/0	6.60	9.81
133.10	2/0	8.31	10.10
105.50	1/0	10.49	10.30
83.69	1	13.22	10.60
66.37	2	16.51	10.80
52.63	3	20.70	11.10
41.74	4	26.27	11.30
33.10	5	33.01	11.60
26.25	6	41.31	12.10
20.80	7	51.98	12.30
16.51	8	65.66	12.60

NOTE: For a three-phase circuit the total impedance, line-to-neutral, is
$$Z = R + j (X_A + X_B) \cdot M$$
To determine X_B, use Tables 5–8. For overhead lines, $M = 1$.
For cable in conduit, determine M from Table 9.

Table 4
Constants of Medium-Voltage Aluminum Cable
Steel Reinforced for 1 Ft Delta Spacing

Conductor Size (MCM or kcmil)	(AWG No)	Resistance R at 25 °C, 60 Hz (mΩ/conductor/100 ft)	Reactance X_A at 1 ft Spacing, 60 Hz (mΩ/conductor/100 ft)
1590.0		1.18	6.79
1431.0		1.31	6.92
1272.0		1.47	7.04
1192.5		1.56	7.12
1113.0		1.67	7.19
954.0		1.94	7.38
795.0		2.22	7.44
715.5		2.49	7.56
636.0		2.80	7.68
556.5		3.21	7.86
477.0		3.38	8.02
397.5		4.06	8.24
336.4		4.80	8.43
266.8		6.04	11.45
	4/0	7.62	10.99
	3/0	9.59	11.75
	2/0	12.13	12.12
	1/0	15.27	12.42
	1	19.33	12.59
	2	24.35	12.15
	3	30.73	12.51
	4	38.67	12.40
	5	48.70	12.59
	6	61.47	12.73

NOTE: For a three-phase circuit the total impedance, line-to-neutral, is
$Z = R + j (X_A + X_B) \cdot M$
To determine X_B, use Tables 5–8. For overhead lines, $M = 1$.
For cable in conduit, determine M from Table 9.

Table 5
60 Hz Reactance Spacing Factor X_B, in Milliohms per Conductor per 100 Ft

(feet)	Separation (inches) 0	1	2	3	4	5	6	7	8	9	10	11
0	–	−5.71	−4.12	−3.19	−2.52	−2.01	−1.59	−1.24	−0.93	−0.66	−0.42	−0.20
1	0.00	0.18	0.35	0.51	0.61	0.80	0.93	1.06	1.17	1.29	1.39	1.49
2	1.59	1.69	1.78	1.86	1.95	2.03	2.11	2.18	2.55	2.32	2.39	2.46
3	2.52	2.59	2.65	2.71	2.77	2.82	2.88	2.93	2.99	3.04	3.09	3.14
4	3.19	3.23	3.28	3.33	3.37	3.41	3.46	3.50	3.54	3.58	3.62	3.66
5	3.70	3.74	3.77	3.81	3.85	3.88	3.92	3.95	3.99	4.02	4.05	4.09
6	4.12	4.15	4.18	4.21	4.24	4.27	4.30	4.33	4.36	4.39	4.42	4.45
7	4.47	4.50	4.53	4.55	4.58	4.60	4.63	4.66	4.68	4.71	4.73	4.76
8	4.78											

Table 6
60 Hz Reactance Spacing Factor X_B, in Milliohms per Conductor per 100 Ft

Separation (inches)	Factor X_B	(inches)	Separation (quarter inches)			
			0	$\frac{1}{4}$	$\frac{2}{4}$	$\frac{3}{4}$
0.40	−7.95	2	−4.12	−3.840	−3.590	−3.390
0.50	−7.29	3	−3.19	−3.010	−2.820	−2.670
0.60	−6.95	4	−2.52	−2.380	−2.250	−2.120
0.70	−6.58	5	−2.01	−1.795	−1.795	−1.684
0.75	−6.36	6	−1.59	−1.494	−1.399	−1.323
0.80	−6.22	7	−1.24	−1.152	−1.078	−1.002
0.90	−5.95	8	−0.93	−0.852	−0.794	−0.719
1.00	−5.71	9	−0.66	−0.605	−0.529	−0.474
1.10	−5.48	10	−0.42	—	—	—
1.20	−5.25	11	−0.20	—	—	—
1.25	−5.19	12	—	—	—	—
1.30	−5.08					
1.40	−4.90					
1.50	−4.77					
1.60	−4.62					
1.70	−4.50					
1.75	−4.43					
1.80	−4.37					
1.90	−4.28					
2.00	−4.12					

Table 7
Typical Conductor Spacings for Overhead Lines

Nominal System Voltage (volts)	Equivalent Delta Spacing (inches)
120	12
240	12
480	18
600	18
2400	30
4160	30
6900	36
13 800	42
23 000	48
34 500	54
69 000	96
115 000	204

NOTE: When conductors are not arranged in a delta, the following formula may be used to determine the equivalent delta:

$d = 3\sqrt{A \cdot B \cdot C}$

When the conductors are located in one plane and the outside conductors are equally spaced from the middle conductor, the equivalent is 1.26 times the distance between the middle conductor and an outside conductor. For example,

equivalent delta spacing $= 3\sqrt{A \cdot A \cdot 2A}$
$= 1.26A$

O–A–O–A–O
——2A——

55

Table 8
Approximate Minimum Conductor Spacings for Medium-Voltage Cable
(Copper and Aluminum) in Conduit*

Conductor Size (MCM or kcmil)	(AWG No)	Spacings for 5 kV Single-Conductor Cable in Conduit** (inches)	Spacings for 5 kV Multiconductor Conduit† (inches)
1000		1.4	1.0
750		1.3	0.9
500		1.1	0.8
350		0.9	0.7
250		0.8	0.6
	4/0	0.8	0.6
	3/0	0.7	0.6
	2/0	0.7	0.6
	1/0	0.6	0.5
	1	0.6	0.5
	2	0.5	0.5
	4	0.5	0.5
	6	0.4	0.4
	8	0.4	0.4

* Use actual cable outside diameter, if known.
**Add 0.3 for 8 kV, 0.5 for 15 kV.
† Add 0.1 for 8 kV, 0.2 for 15 kV.

Table 9
Medium-Voltage Cable in Conduit—Reactance Factor *M* for Various Constructions and Installations

| Conductor Size (MCM or kcmil) | Multiple-Conductor Cable in Conduit or Armor | | | | Three Single-Conductor in Conduit with Random Lay | | | |
| | Nonshielded | | Shielded | | Nonshielded | | Shielded | |
	Non-magnetic	Magnetic	Non-magnetic	Magnetic	Non-magnetic	Magnetic	Non-magnetic	Magnetic
250 and Smaller	1.0	1.149	0.945	1.086	1.2	1.5	1.320	1.650
300	1.0	1.146	0.946	1.084	1.2	1.5	1.302	1.628
350	1.0	1.140	0.947	1.080	1.2	1.5	1.298	1.623
400	1.0	1.134	0.949	1.076	1.2	1.5	1.295	1.619
450	1.0	1.128	0.951	1.073	1.2	1.5	1.292	1.616
500	1.0	1.122	0.952	1.068	1.2	1.5	1.290	1.613
600	1.0	1.111	0.955	1.061	1.2	1.5	1.285	1.607
700	1.0	1.100	0.957	1.053	1.2	1.5	1.282	1.602
750	1.0	1.095	0.959	1.050	1.2	1.5	1.278	1.598
1000	1.0	1.070	0.962	1.029	1.2	1.5	1.271	1.589

Table 10
60 Hz Low-Voltage Cable in Conduit-Resistance (R) and Reactance (X) Data for Insulation Types THW, RHW, RHH, and Use in Milliohms per Conductor per 100 Ft at 25 °C Copper Conductor

Cable Size	Several Single-Conductor Cables (1/C)						One Multiple-Conductor Cable (3/C)					
	Steel		Aluminum		Plastic		Steel		Aluminum		Plastic	
	R	X	R	X	R	X	R	X	R	X	R	X
14	257.00	5.60	257.00	4.48	257.00	4.48	257.00	4.29	257.00	3.73	257.00	3.73
12	162.00	5.23	162.00	4.18	162.00	4.18	162.00	4.01	162.00	3.49	162.00	3.49
10	101.80	4.90	101.80	3.92	101.80	3.92	101.80	3.76	101.80	3.27	101.80	3.27
8	64.04	5.14	64.04	4.12	64.04	4.12	64.04	3.94	64.04	3.43	64.04	3.43
6	41.00	5.04	41.00	4.03	41.00	4.03	41.00	3.86	41.00	3.36	41.00	3.36
4	25.90	4.77	25.90	3.82	25.90	3.82	25.90	3.65	25.90	3.18	25.90	3.18
3	20.50	4.58	20.50	3.66	20.50	3.66	20.50	3.50	20.50	3.05	20.50	3.05
2	16.40	4.49	16.40	3.59	16.20	3.59	16.40	3.44	16.40	2.99	16.20	2.99
1	13.03	4.58	13.03	3.66	12.90	3.66	13.03	3.50	13.03	3.05	12.90	3.05
1/0	10.40	4.46	10.40	3.56	10.20	3.56	10.40	3.41	10.40	2.97	10.20	2.97
2/0	8.35	4.35	8.35	3.48	8.12	3.48	8.35	3.33	8.35	2.90	8.12	2.90
3/0	6.68	4.22	6.68	3.37	6.43	3.37	6.68	3.23	6.68	2.81	6.43	2.81
4/0	5.34	4.14	5.34	3.31	5.11	3.31	5.34	3.17	5.34	2.76	5.11	2.76
250	4.57	4.23	4.57	3.38	4.33	3.38	4.57	3.24	4.57	2.82	4.33	2.82
300	3.85	4.14	3.85	3.31	3.62	3.31	3.85	3.16	3.85	2.76	3.62	2.76
350	3.33	4.07	3.33	3.25	3.11	3.25	3.33	3.09	3.33	2.71	3.11	2.71
400	2.97	4.04	2.97	3.23	2.73	3.23	2.97	3.05	2.97	2.69	2.73	2.69
500	2.44	3.96	2.44	3.17	2.20	3.17	2.44	2.96	2.44	2.64	2.20	2.64
600	2.09	4.01	2.09	3.21	1.85	3.21	2.09	2.97	2.09	2.68	1.85	2.68
750	1.74	3.94	1.74	3.15	1.50	3.15	1.74	2.88	1.74	2.63	1.50	2.63
1000	1.40	3.86	1.40	3.09	1.15	3.09	1.40	2.74	1.40	2.57	1.15	2.57

Aluminum Conductor

Cable Size	Several Single-Conductor Cables (1/C)						One Multiple-Conductor Cable (3/C)					
	Steel		Aluminum		Plastic		Steel		Aluminum		Plastic	
	R	X	R	X	R	X	R	X	R	X	R	X
14*	422.00	5.60	422.00	4.48	422.00	4.48	422.00	4.29	422.00	3.73	422.00	3.73
12	266.00	5.23	266.00	4.18	266.00	4.18	266.00	4.01	266.00	3.49	266.00	3.49
10	167.00	4.90	167.00	3.92	167.00	3.92	167.00	3.76	167.00	3.27	167.00	3.27
8	105.00	5.14	105.00	4.12	105.00	4.12	105.00	3.94	105.00	3.43	105.00	3.43
6	67.40	5.04	67.40	4.03	67.40	4.03	67.40	3.86	67.40	3.36	67.40	3.36
4	42.40	4.77	42.40	3.82	42.40	3.82	42.40	3.65	42.40	3.18	42.40	3.18
3	33.60	4.58	33.60	3.66	33.60	3.66	33.60	3.50	33.60	3.05	33.60	3.05
2	26.60	4.49	26.60	3.59	26.60	3.59	26.60	3.44	26.60	2.99	26.60	2.99
1	21.10	4.58	21.10	3.66	21.10	3.66	21.10	3.50	21.10	3.05	21.10	3.05
1/0	16.80	4.46	16.80	3.56	16.80	3.56	16.80	3.41	16.80	2.97	16.80	2.97
2/0	13.30	4.35	13.30	3.48	13.30	3.48	13.30	3.33	13.30	2.90	13.30	2.90
3/0	10.60	4.22	10.60	3.37	10.50	3.37	10.60	3.23	10.60	2.81	10.50	2.81
4/0	8.44	4.14	8.44	3.31	8.38	3.31	8.44	3.17	8.44	2.76	8.38	2.76
250	7.22	4.23	7.22	3.38	7.09	3.38	7.22	3.24	7.22	2.82	7.09	2.82
300	6.02	4.14	6.02	3.31	5.92	3.31	6.02	3.16	6.02	2.76	5.92	2.76
350	5.20	4.07	5.20	3.25	5.07	3.25	5.20	3.09	5.20	2.71	5.07	2.71
400	4.60	4.04	4.60	3.23	4.44	3.23	4.60	3.05	4.60	2.69	4.44	2.69
500	3.75	3.96	3.75	3.17	3.56	3.17	3.75	2.96	3.75	2.64	3.56	2.64
600	3.19	4.01	3.19	3.21	2.98	3.21	3.19	2.97	3.19	2.68	2.98	2.68
750	2.64	3.94	2.64	3.15	2.40	3.15	2.64	2.88	2.64	2.63	2.40	2.63
1000	2.11	3.86	2.11	3.09	1.82	3.09	2.11	2.74	2.11	2.57	1.82	2.57

* Size 14 for type "use" not listed in 1984 NEC.

Table 11
60 Hz Low-Voltage Cable in Conduit-Resistance (R) and Reactance (X) Data for Insulation Types THWN and THHN in Milliohms per Conductor per 100 Ft at 25 °C Copper Conductor

| Cable Size | Several Single-Conductor Cables (1/C) | | | | | | One Multiple-Conductor Cable (3/C) | | | | | |
| | Steel | | Aluminum | | Plastic | | Steel | | Aluminum | | Plastic | |
	R	X	R	X	R	X	R	X	R	X	R	X
14	257.00	4.93	257.00	3.94	257.00	3.94	257.00	3.51	257.00	3.05	257.00	3.05
12	162.00	4.68	162.00	3.74	162.00	3.74	162.00	3.33	162.00	2.90	162.00	2.90
10	101.80	4.63	101.80	3.71	101.80	3.71	101.80	3.37	101.80	2.93	101.80	2.93
8	64.04	4.75	64.04	3.80	64.04	3.80	64.04	3.51	64.04	3.05	64.04	3.05
6	41.00	4.37	41.00	3.49	41.00	3.49	41.00	3.24	41.00	2.82	41.00	2.82
4	25.90	4.41	25.90	3.53	25.90	3.53	25.90	3.28	25.90	2.85	25.90	2.35
3	20.50	4.30	20.50	3.44	20.50	3.44	20.50	3.20	20.50	2.79	20.50	2.79
2	16.40	4.20	16.40	3.36	16.20	3.36	16.40	3.13	16.40	2.73	16.20	2.73
1	13.03	4.27	13.03	3.42	12.90	3.42	13.03	3.19	13.03	2.77	12.90	2.77
1/0	10.40	4.17	10.40	3.34	10.20	3.34	10.40	3.12	10.40	2.72	10.20	2.72
2/0	8.35	4.09	8.35	3.27	8.12	3.27	8.35	3.06	8.35	2.66	8.12	2.66
3/0	6.68	4.00	6.68	3.20	6.43	3.20	6.68	3.00	6.68	2.61	6.43	2.61
4/0	5.34	3.93	5.34	3.14	5.11	3.14	5.34	2.95	5.34	2.57	5.11	2.57
250	4.57	3.99	4.57	3.19	4.33	3.19	4.57	2.99	4.57	2.61	4.33	2.61
300	3.85	3.93	3.85	3.14	3.62	3.14	3.85	2.95	3.85	2.57	3.62	2.57
350	3.33	3.83	3.33	3.11	3.11	3.11	3.33	2.90	3.33	2.54	3.11	2.54
400	2.97	3.85	2.97	3.08	2.73	3.08	2.97	2.86	2.97	2.52	2.73	2.52
500	2.44	3.79	2.44	3.03	2.20	3.03	2.44	2.79	2.44	2.49	2.20	2.49
600	2.09	3.82	2.09	3.05	1.85	3.05	2.09	2.78	2.09	2.50	1.85	2.50
750	1.74	3.76	1.74	3.01	1.50	3.01	1.74	2.71	1.74	2.47	1.50	2.47
1000	1.40	3.70	1.40	2.96	1.15	2.96	1.40	2.60	1.40	2.43	1.15	2.43

Aluminum Conductor

| Cable Size | Several Single-Conductor Cables (1/C) | | | | | | One Multiple-Conductor Cable (3/C) | | | | | |
| | Steel | | Aluminum | | Plastic | | Steel | | Aluminum | | Plastic | |
	R	X	R	X	R	X	R	X	R	X	R	X
14	422.00	4.93	422.00	3.94	422.00	3.94	422.00	3.51	422.00	3.05	422.00	3.05
12	266.00	4.68	266.00	3.74	266.00	3.74	266.00	3.33	266.00	2.90	266.00	2.90
10	167.00	4.63	167.00	3.71	167.00	3.71	167.00	3.37	167.00	2.93	167.00	2.93
8	105.00	4.75	105.00	3.80	105.00	3.80	105.00	3.51	105.00	3.05	105.00	3.05
6	67.40	4.37	67.40	3.49	67.40	3.49	67.40	3.24	67.40	2.82	67.40	2.82
4	42.40	4.41	42.40	3.53	42.40	3.53	42.40	3.28	42.40	2.85	42.40	2.85
3	33.60	4.30	33.60	3.44	33.60	3.44	33.60	3.20	33.60	2.79	33.60	2.79
2	26.60	4.20	26.60	3.36	26.60	3.36	26.60	3.13	26.60	2.73	26.60	2.73
1	21.10	4.27	21.10	3.42	21.10	3.42	21.10	3.19	21.10	2.77	21.10	2.77
1/0	16.80	4.17	16.80	3.34	16.80	3.34	16.80	3.12	16.80	2.72	16.80	2.72
2/0	13.30	4.09	13.30	3.27	13.30	3.27	13.30	3.06	13.30	2.66	13.30	2.66
3/0	10.60	4.00	10.60	3.20	10.50	3.20	10.60	3.00	10.60	2.61	10.50	2.61
4/0	8.44	3.93	8.44	3.14	8.38	3.14	8.44	2.95	8.44	2.57	8.38	2.57
250	7.22	3.99	7.22	3.19	7.09	3.19	7.22	2.99	7.22	2.61	7.09	2.61
300	6.02	3.93	6.02	3.14	5.92	3.14	6.02	2.95	6.02	2.57	5.92	2.57
350	5.20	3.88	5.20	3.11	5.07	3.11	5.20	2.90	5.20	2.54	5.07	2.54
400	4.60	3.85	4.60	3.08	4.44	3.08	4.60	2.86	4.60	2.52	4.44	2.52
500	3.75	3.79	3.75	3.03	3.56	3.03	3.75	2.79	3.75	2.49	3.56	2.49
600	3.19	3.82	3.19	3.05	2.98	3.05	3.19	2.78	3.19	2.50	2.98	2.50
750	2.64	3.76	2.64	3.01	2.40	3.01	2.64	2.71	2.64	2.47	2.40	2.47
1000	2.11	3.70	2.11	2.96	1.82	2.96	2.11	2.60	2.11	2.43	1.82	2.43

Table 12
60 Hz Low-Voltage Busway-Resistance *(R)* and Reactance *(X)* Data in Milliohms per 100 Ft at 25 °C

Ampere Rating	Feeder				Plug-In			
	Copper		Aluminum		Copper		Aluminum	
	R	X	R	X	R	X	R	X
100	—	—	—	—	11.82	4.00	21.96	4.00
225	—	—	—	—	7.15	3.42	6.12	3.42
400	—	—	—	—	1.78	2.30	3.11	2.60
600	—	—	2.56	0.99	1.78	2.30	1.71	1.59
800	1.26	0.99	1.78	0.81	1.05	2.17	1.80	2.17
1000	1.05	0.82	1.59	0.50	1.05	2.17	1.20	1.43
1350	0.76	0.65	1.06	0.44	0.71	1.43	0.84	1.00
1600	0.70	0.53	0.89	0.40	0.49	1.00	0.75	0.90
2000	0.52	0.41	0.63	0.31	0.43	0.90	0.60	0.72
2500	0.38	0.32	0.47	0.25	0.35	0.72	0.42	0.50
3000	0.35	0.30	0.46	0.20	0.24	0.50	0.40	0.48
4000	0.25	0.20	0.31	0.16	—	—	—	—
5000	0.19	0.15	—	—	—	—	—	—

Table 13
Multipliers for Source Short-Circuit Current Contributions

Type of Source(s)	Type of Calculation					
	First-Cycle		Interrupting		Medium-Voltage Circuit Breaker Close and Latch*	
	Multiply SCA or SCA_M by	Multiply Xd'' by	Multiply SCA or SCA_M by	Multiply Xd'' by	Multiply SCA or SCA_M by	Multiply Xd'' by
Utility or Power Company	1.0	1.0	1.0	1.0	1.0	1.0
Generators**	1.0	1.0	1.0	1.0	1.0	1.0
Synchronous Motors	1.0	1.0	0.667	1.5	1.0	1.0
Induction motors Above 1000 hp at 1800 rpm	1.0	1.0	0.667	1.5	1.0	1.0
Above 250 hp at 3600 rpm	1.0	1.0	0.667	1.5	1.0	1.0
All others, 50 hp and above	1.0	1.0	0.333	3.0	0.833	1.2
All smaller than 50 hp	1.0	1.0	NEGLECT	NEGLECT	NEGLECT	NEGLECT

* Refers to calculations for medium-voltage circuit breakers as developed in national standards (see [5] and [4]).
** Use $0.75Xd'$ for hydrogenerators without amortisseur windings.

Table 14
Impedance Data for Single-Phase Transformers

kVA 1ϕ	Suggested X/R Ratio for Calculation	Normal Range of Percent Impedance (%Z)*	Impedance Multipliers** for Line-to-Neutral Fault	
			for %X	for %R
25.0	1.1	1.2−6.0	0.6	0.75
37.5	1.4	1.2−6.5	0.6	0.75
50.0	1.6	1.2−6.4	0.6	0.75
75.0	1.8	1.2−6.6	0.6	0.75
100.0	2.0	1.3−5.7	0.6	0.75
167.0	2.5	1.4−6.1	1.0	0.75
250.0	3.6	1.9−6.8	1.0	0.75
333.0	4.7	2.4−6.0	1.0	0.75
500.0	5.5	2.2−5.4	1.0	0.75

* National standards do not specify %Z for single-phase transformers. Consult manufacturer for values to use in calculation.
** Based on rated current of the winding (one-half nameplate kVA divided by secondary line-to-neutral voltage).

2.2.2 Calculation Method. Two methods of making short-circuit calculations are presented. They are identified as the *direct method* and the *per-unit method*. While they represent different calculation concepts, if both use the same data, they produce results of the same degree of accuracy.

The direct method of short-circuit calculation presented here is so designated because, for the most part, it uses the system one-line diagram directly, uses system and equipment data such as volts, amperes, and ohms directly, and uses basic electrical equations and relationships directly without utilizing special diagrams, abstract units, or mathematical techniques.

The per-unit method is more representative of conventional electrical circuit analysis. It involves converting the system one-line diagram into an equivalent impedance diagram and reducing this to a single impedance value. This is best accomplished especially when several voltage levels are involved by using a special mathematical technique that establishes base (or reference) values for volts, current, kVA, and ohms and then refers the actual parameters to these bases in special equations to derive per-unit (or sometimes percent) values. Applying these values in special equations produces the short-circuit values.

For those who do not specialize in short-circuit calculation work, the direct method is usually easier to comprehend since it uses the familiar system one-line diagram and is compatible with the intuitive analogy of water flowing from sources to point of fault. Also, it instills confidence since system and equipment data are applied directly to familiar electrical equations producing recognizable values that can be immediately appraised.

The direct method is particularly adapted to progressive analysis of a whole or portion of a system starting at the source, considering each eschelon step by step, and determining short-circuit values at each location out to the end of the various circuits. Such an analysis is typically associated with planning an entire power system of a new building or facility where short-circuit values at all points must be determined before proper equipment can be selected.

The per-unit method with its special mathematical technique is particularly adapted to calculating short-circuit values at one or more specific points in the system, especially when several voltage levels exist between the source and the short-circuit point. In the per-unit method each point of fault is considered separately. A system equivalent impedance diagram is developed using those parameters that will have an effect on short-circuit current at that point. The diagram is reduced by delta and wye conversion equations to a single impedance, which applied to the appropriate equation produces the short-circuit value. Each fault location requires its own separate equivalent impedance diagram, subsequent reduction, and calculation. It is not dependent on short-circuit values obtained from a proceeding circuit section. Since each point is considered separately, the per-unit method may be more expedient when a single specific remote location is being analyzed (perhaps in an existing system) since progressive calculation at interim points would not be involved.

In the evolution of short-circuit calculation techniques, most of the major short-circuit studies have pertained to systems where the characteristics of the per-unit method were of an advantage—typically, existing medium- and high-voltage systems with many voltage levels, such as power company systems. As a result, the per-unit method is generally considered the official standard method of calculation, as presented in great detail in many references, including ANSI standards and other IEEE publications and books. It is not the intent of this discussion to duplicate the voluminous material presented in these references. Instead, a brief discussion of the per-unit method is presented here in comparison with the direct method and the reader is encouraged to study these other references if the per-unit method is chosen for calculation.

2.2.3 Calculation Times. Because some protective devices operate after a few cycles and others after a time delay, short-circuit currents may need to be calculated at the following recommended times.

(1) The First-Cycle Considerations. Maximum symmetrical values immediately after fault initiation are always required and are often the only values needed. In this chapter these values are referred to as first-cycle current and are symmetrical rms short-circuit amperes of alternating current unless otherwise noted.

These values are used in selecting proper short-circuit ratings for low-voltage equipment and when converted to asymmetrical values are the basis for selecting circuit medium-voltage switch and fuse ratings and circuit breaker close and latch ratings. They often are used in the process of selecting medium-voltage circuit breaker interrupting ratings.

The first-cycle values are also required for coordinating protective devices according to their time-current characteristics. Even if a device does not interrupt until several cycles after fault initiation, thus allowing the fault current to decay, the protective devices and all series devices should withstand the maximum current as well as the total energy. A device that interrupts in less than one-half cycle (and before the first current peak is attained) reduces the withstand requirements of series devices. The let-through current for such a device can be determined from the maximum symmetrical values in the first one-half cycle usually referred to as prospective short-circuit current.

(2) After 1.5–8 Cycles (Interrupting Considerations). Maximum values after a few cycles are required for comparison with the interrupting ratings of medium-voltage circuit breakers. Only the basic considerations and calculations are discussed in this chapter. (They are discussed fully in [6] and [7] and referred to in the example.)

(3) About 30 Cycles. These reduced fault currents are sometimes needed for estimating the performance of time-delay relays and fuses. Often, minimum values should be calculated to determine whether sufficient current is available to open the protective devices within a satisfactory time.

2.2.4 Impedance Tables. Tables 1–12 give impedance data for three-phase transformers, medium-voltage cable, low-voltage cable, and low-voltage busway. When specific data cannot be obtained from the manufacturer, these tables can be used and should provide satisfactory results.

It should be noted that transformers usually provide the most critical impedances in any system. A small difference between an assumed transformer impedance and the actual impedance can significantly alter the calculated fault current and consequently the equipment selected. Therefore, percent impedance values for transformers being built should be obtained from the supplier (manufacturer or power company).

2.3 Calculating Three-Phase Short-Circuit Currents by the Direct Method. With the one-line diagram of the actual power system, refer to Figs 6–16, which present the direct method of calculation. The pictorial representation in each figure clearly shows where in the one-line diagram it applies. The format shows what data must be acquired and how they are used to calculate the results. Supplemental comments and explanations appear on the right side of the figures. The method is based on familiar concepts. Visualize current flow as if it were water flowing from sources to fault. Fundamental ohms law is the basis for most calculations, but the concept of *percent impedance* is used when data warrant. Percent impedance represents the percent of normal voltage that, applied to the primary of a transformer, etc, will cause full-load current to flow in a short-circuited secondary. Maximum short-circuit current occurs when full (100%) voltage is applied and is equal to full-load current times the ratio of 100% divided by the total percent impedance value. Throughout the procedure, resistances are totaled separately, reactances are totaled separately, and these used to determine resulting impedance; mathematical conversion of one to the other (vectorially) is done with simple trigonometric equations easily performed on popular small inexpensive calculators.

2.3.1 Description of Procedure. Fault calculation points on the system one-line diagram are marked with *bus* numbers. The various sources are next analyzed to determine the total source short-circuit current available at the various busses. Each segment from one bus to another is then handled separately to determine the short-circuit current at the *start* and *finish* of that segment, beginning with the service point and working through the system until every segment has been analyzed. Each circuit segment is handled separately according to one of the following sections.

2.3.2 Basic Equations. To calculate the available short-circuit current at a given point in the system, the following fundamental equations may be used:

$$kVA_{SC} = \frac{\sqrt{3} \cdot V_{LL} \cdot I_{SC}}{1000} = \sqrt{3} \cdot kV_{LL} \cdot I_{SC}$$

$$I_{SC} = \frac{V_{LN}}{Z}$$

from the right triangle relationship

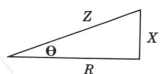

$$Z^2 = R^2 + X^2$$

$$X = \frac{Z\,(X/R)}{\sqrt{(X/R)^2 + 1}}$$

$$\frac{X}{R}\ \text{ratio} = \frac{X}{R} = \tan \Theta$$

$$\Theta = \arctan \frac{X}{R}$$

$$R = Z \cos \Theta$$

$$X = Z \sin \Theta$$

$$X = Z \sin \left(\arctan \frac{X}{R}\right)$$

$$Z = \frac{X}{\sin \left(\arctan \dfrac{X}{R}\right)}$$

where

kVA_{SC} = three-phase first-cycle short-circuit rms symmetrical kilovolt-amperes (also denoted as SCKVA; see 2.3.3)

I_{SC} = three-phase first-cycle short-circuit rms symmetrical current amperes (also denoted as SCA; see 2.3.3)

V_{LL} = line-to-line volts

kV_{LL} = line-to-line kilovolts

V_{LN} = line-to-neutral volts

R = resistance in ohms, on a line-to-neutral basis

X = reactance in ohms, on a line-to-neutral basis

Z = impedance in ohms, on a line-to-neutral basis

$\frac{X}{R}$ratio = reactance divided by resistance

2.3.3 Determining Source Contributions for First-Cycle Calculations. In using the direct method of short-circuit calculation, the total source current at each bus should be determined before the short-circuit currents can be calculated. This procedure for first-cycle calculations is detailed in the next several paragraphs. Note that certain abbreviations are used for simplicity and to avoid excessive subscripting:

SCKVA = Short-Circuit kilovolt-Amperes = kVA_{SC}

SCA = Short-Circuit Amperes = I_{SC}

2.3.4 Determining Source Contributions for Medium-Voltage Circuit Breaker Interrupting and Close and Latch Capacity Calculations. Many factors have an influence on proper application of medium-voltage circuit breakers. Persons involved in such applications are advised to study ANSI/IEEE C37.010-1979 [5] and other appropriate references. The great detail presented in these standards is not repeated in this text, but portions are briefly discussed to provide some understanding of the procedure relating medium-voltage circuit breaker application to system short-circuit calculations.

Devices that interrupt after a few cycles, such as medium-voltage circuit breakers, may be required to interrupt less current than the available first-cycle short circuit. This is due to the rapid decrease of some source short-circuit contributions by the time the circuit breaker interrupts. Data in Table 13 are used in these calculations.

Both the short-circuit current and X/R ratio are important in medium-voltage circuit breaker application. However, if the available symmetrical short-circuit current at time of interruption (or the higher first-cycle symmetrical short-circuit current) is not more than 80% of the circuit breaker symmetrical current interrupting rating, any X/R ratio will be satisfactory and therefore need not be calculated. This 80% condition is often sufficient to justify the circuit breaker interrupting rating. If this does not suffice, the available short-circuit current

Fig 6
Power Company Source

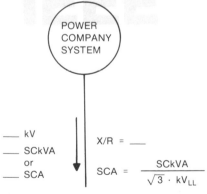

The Power Company will advise the maximum available symmetrical short-circuit kVA or amperes that its system can produce at the user's service connection. This is the maximum fault possible at that point expressed in terms of normal voltage and steady-state short-circuit current. It influences the size of short circuits in the user's distribution system. If this information is not available, be guided by the interrupting rating of the nearest circuit breaker or fuse intended to clear this fault.

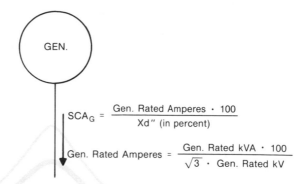

$$SCA_G = \frac{\text{Gen. Rated Amperes} \cdot 100}{Xd'' \text{ (in percent)}}$$

$$\text{Gen. Rated Amperes} = \frac{\text{Gen. Rated kVA} \cdot 100}{\sqrt{3} \cdot \text{Gen. Rated kV}}$$

Values vary widely, depending on design. Consult generator manufacturer for data. Xd'' is the subtransient reactance. If Xd'' is given in per-unit, multiply by 100 to yield percent.

**Fig 7
Generator**

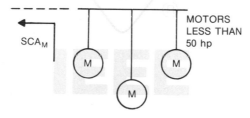

$$SCA_M = 4 \cdot \text{Sum of Motor Full-Load Amperes}$$

It is not practical to consider each small motor separately, and so it is usual practice to combine them all at each location. Consider all motors at each location which may be running even though at partial load. ANSI standards permit neglecting motors less than 50 hp when considering medium-voltage circuit breaker applications—see Table 13.

A multiplier of 4 is used throughout the industry. This is generally lower than for large motors to account for the rapid decay of their short-circuit contribution.

**Fig 8
Small Motors (Less Than 50 hp)**

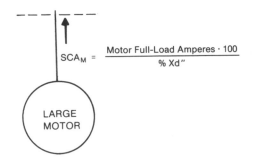

$$\text{SCA}_M = \frac{\text{Motor Full-Load Amperes} \cdot 100}{\% \ Xd''}$$

Handle each motor separately, unless several of identical type are connected to the same bus.

The exact short-circuit contribution can be determined from the subtransient reactance (Xd''), if known. If Xd'' is given in per-unit, multiply by 100 to obtain percent.

Motor Rated kVA $\cong$ Motor Rated hp (for 80% power factor)

Motor Rated kVA $\cong$ 0.8 $\cdot$ Motor Rated hp (for 100% power factor)

$$\text{Motor Full-Load Amperes} = \frac{\text{Motor Rated kVA}}{\sqrt{3} \cdot \text{Motor Rated kV}}$$

1 hp = .746 kW but .8 usually used in determining kVA

New individual motors are rated at 200, 230, 460, 575, 2300, 4000, 6600 and 13 200 volts. Older motors may be rated lower, and new motors included in a complete unit (such as a chiller unit) may be rated at the system voltage.

APPROXIMATIONS

(a) Large Induction Motors

$$\text{SCA}_M \cong \text{Locked-Rotor Amperes}$$

or

$$\text{SCA}_M \cong \frac{\text{NEC (and NEMA)} \ \text{Coded Letter Multiple} \cdot \text{hp}}{\sqrt{3} \cdot \text{Motor Rated kV}}$$

or

If subtransient reactance cannot be determined, the approximations will be close.

Locked-rotor current is nearly the same as short-circuit contribution.

Code letter on motor nameplate identifies range of ratio of locked-rotor kVA ÷ hp. Maximum values are as follows:

Letter	Multiple	Letter	Multiple
A	3.15	L	10.0
B	3.55	M	11.2
C	4.0	N	12.5
D	4.5	P	14.0
E	5.0	R	16.0
F	5.6	S	18.0
G	6.3	T	20.0
H	7.1	U	22.4
J	8.0	V	Over 22.4
K	9.0		

$$\text{SCA}_M \cong 5 \cdot \text{Full-Load Amperes}$$

(b) Large Synchronous Motors

$$\text{SCA}_M \cong 6.7 \cdot \text{Motor Full-Load Amperes for 1200 rpm}$$
$$\cong 5 \cdot \text{Motor Full-Load Amperes for 514-900 rpm}$$
$$\cong 3.6 \cdot \text{Motor Full-Load Amperes for 450 rpm or less}$$

Assumes motor impedance is 20% (usual range 15-25% with 20% more popular).

Assumes $X''d$ = 15%

Assumes $X''d$ = 20%

Assumes $X''d$ = 28%

If rpm is unknown, assume 514–900 rpm, since 1200 rpm is uncommon.

Fig 9
Large Motors (50 hp or More)

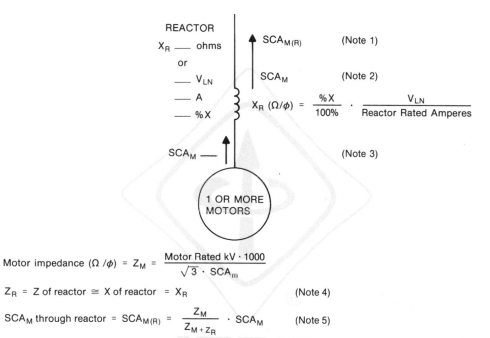

Motor impedance $(\Omega/\phi) = Z_M = \dfrac{\text{Motor Rated kV} \cdot 1000}{\sqrt{3} \cdot \text{SCA}_m}$

$Z_R = Z$ of reactor $\cong X$ of reactor $= X_R$ (Note 4)

SCA_M through reactor $= \text{SCA}_{M(R)} = \dfrac{Z_M}{Z_{M+Z_R}} \cdot \text{SCA}_M$ (Note 5)

NOTES
(1) Reactor impedance significantly reduces motor short-circuit contribution.
(2) If system has no neutral, use V_{LN} equal to V_{LL} divided by $\sqrt{3}$.
(3) SCA_M is total short-circuit amperes from motors (obtained from Figs 8 and 9).
(4) Impedance is virtually reactance, very high X/R ratio.
(5) Usually can add these impedances arithmetically instead of vectorially since both reactor and motor will have high X/R ratio.

Fig 10
Motor Contribution Through Reactor

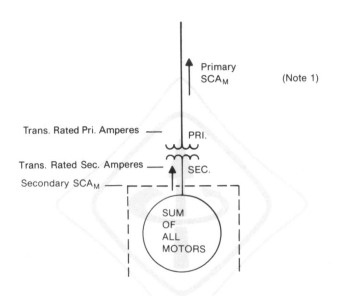

APPROXIMATION   (Note 2)

$$\text{Primary SCA}_M = \frac{\text{Trans. Rated Pri. Amperes}}{\text{Trans. Rated Sec. Amperes}} \cdot \text{Secondary SCA}_M$$

When L.V. is 240 V or higher Pri. SCA$_M$ = 3.5 · transformer primary A   (Note 3)
When L.V. is 208 V or lower Pri. SCA$_M$ = 2 · transformer primary A

MORE EXACT DETERMINATION   (Note 4)

$$\text{Primary SCA}_M = \frac{\text{Trans. Rated Pri. Amperes}}{\dfrac{\text{Trans. \%Z}}{100\%} + \dfrac{\text{Trans. Rated Sec. Amperes}}{\text{Secondary SCA}_M}}$$

NOTES

(1) Motors generate short circuit current which can be transmitted through transformer to primary system.
(2) Simple estimate yielding somewhat high results by ignoring transformer impedance.
(3) Based on only small motors (see Fig 8) normal transformer impedance, all transformer load being motors if L.V. is 240 volts and above and ½ transformer load being motors if voltage is 208 volts or lower.
(4) Yields more accurate results when transformer impedance is known.

Fig 11
Transformer Secondary Motor Contribution to Primary System

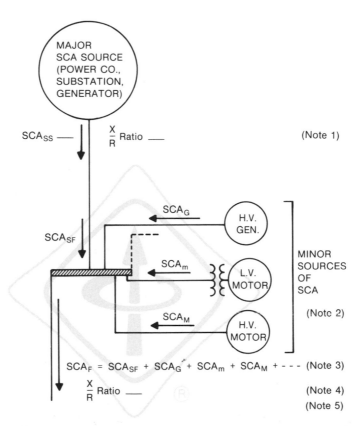

$$SCA_F = SCA_{SF} + SCA_G + SCA_m + SCA_M + - - - \quad \text{(Note 3)}$$

$\frac{X}{R}$ Ratio ____ (Note 4)

(Note 5)

NOTES

(1) The SCA_M at the start of the feeder is the combination of the SCA from the various sources which supply this feeder.

(2) The SCA values from the various sources (as determined in Figs 6 through 11) will be reduced by the impedance of the conductors from these sources. However, the impedance of some and perhaps all of these conductors is often ignored (considered zero) to make calculation easier and data may not be available. This produces calculated SCA values which are higher than the actual values and are safe but may require larger capacity equipment than necessary. Generally, the impedance of short conductor segments and conductors from minor sources are often ignored. When the conductor impedance is to be considered the calculations are as indicated in Figs 14, 15, 16, and perhaps 13.

(3) Add arithmetically.

(4) The X/R ratio at the start of the feeder is determined mostly by the X/R ratio of the major sources.

(5) Short circuit ampere values from the various sources should, for maximum accuracy, be added vectorily and can be, using the X/R ratios. However, since the SCA values from the major sources are usually influenced by high X/R ratios and minor sources contributions are usually estimates anyway, adding these arithmetically produces sufficiently accurate results. This is much easier and safe since arithmetic addition will not produce SCA values lower than more precise vector addition.

**Fig 12
Total Short-Circuit Available at Beginning/Start of A Feeder**

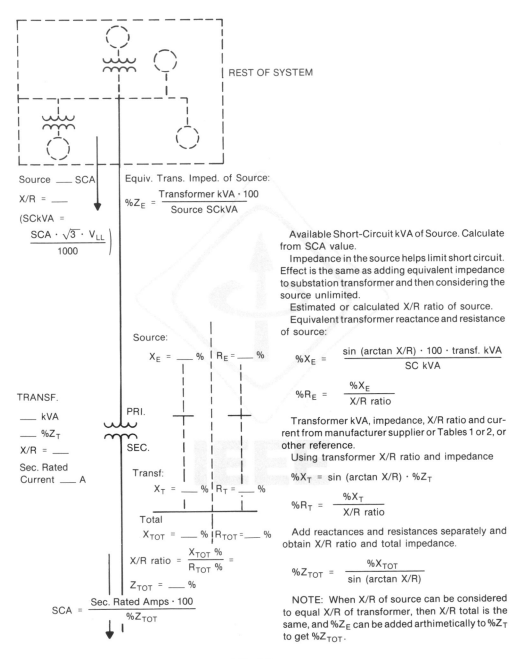

Source ___ SCA

$X/R = $ ___

$\left(\text{SCkVA} = \dfrac{\text{SCA} \cdot \sqrt{3} \cdot V_{LL}}{1000}\right)$

Equiv. Trans. Imped. of Source:

$\%Z_E = \dfrac{\text{Transformer kVA} \cdot 100}{\text{Source SCkVA}}$

Source:

$X_E = $ ___ % $R_E = $ ___ %

TRANSF.

___ kVA

___ $\%Z_T$

$X/R = $ ___

Sec. Rated
Current ___ A

PRI.

SEC.

Transf:

$X_T = $ ___ % $R_T = $ ___ %

Total

$X_{TOT} = $ ___ % $R_{TOT} = $ ___ %

$X/R \text{ ratio} = \dfrac{X_{TOT}\,\%}{R_{TOT}\,\%} = $

$Z_{TOT} = $ ___ %

$\text{SCA} = \dfrac{\text{Sec. Rated Amps} \cdot 100}{\%Z_{TOT}}$

Available Short-Circuit kVA of Source. Calculate from SCA value.

Impedance in the source helps limit short circuit. Effect is the same as adding equivalent impedance to substation transformer and then considering the source unlimited.

Estimated or calculated X/R ratio of source.

Equivalent transformer reactance and resistance of source:

$\%X_E = \dfrac{\sin(\arctan X/R) \cdot 100 \cdot \text{transf. kVA}}{\text{SC kVA}}$

$\%R_E = \dfrac{\%X_E}{X/R \text{ ratio}}$

Transformer kVA, impedance, X/R ratio and current from manufacturer supplier or Tables 1 or 2, or other reference.

Using transformer X/R ratio and impedance

$\%X_T = \sin(\arctan X/R) \cdot \%Z_T$

$\%R_T = \dfrac{\%X_T}{X/R \text{ ratio}}$

Add reactances and resistances separately and obtain X/R ratio and total impedance.

$\%Z_{TOT} = \dfrac{\%X_{TOT}}{\sin(\arctan X/R)}$

NOTE: When X/R of source can be considered to equal X/R of transformer, then X/R total is the same, and $\%Z_E$ can be added arthimetically to $\%Z_T$ to get $\%Z_{TOT}$.

Fig 13
Transformer Secondary Short-Circuit Current

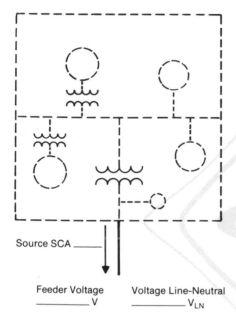

CONSOLIDATED
SOURCE TO
FEEDER

Source SCA _____

Available Short-Circuit Amperes from Source.

Feeder Voltage
_____ V

Voltage Line-Neutral
_____ V_{LN}

Short-circuit analysis is performed on a per-phase basis. Necessary to use line-to-neutral voltage. Even if system has no neutral use V_{LN} equal to voltage line-to-line divided by $\sqrt{3}$.

Z = Source Impedance $(m\Omega/\phi)$

$$Z = \frac{V_{LN} \cdot 1000}{SCA}$$

Ohm's Law. Effective impedance is obtained by dividing known voltage by available short-circuit current.

Impedance, reactance, and resistance are expressed in milliohms, instead of ohms, for greater convenience. One milliohm = $1\,m\Omega$ = 1/1000 ohm = .001 ohm. This accounts for 1000 factor in equation.

Source X/R ratio _____

X/R ratio value determined from previous calculations or estimate of supply. (In low voltage [600 V max] systems X/R = 12 is often used if more specific information is not available.)

Reactance | Resistance

Source: X = ___ mΩ | R = ___ mΩ

$$X = Z \cdot \sin \left(\arctan \frac{X}{R} \text{ ratio} \right)$$

$$R = \frac{X}{\frac{X}{R} \text{ ratio}}$$

Fig 14
Effective Impedance (Reactance and Resistance) of System Supplying Feeder

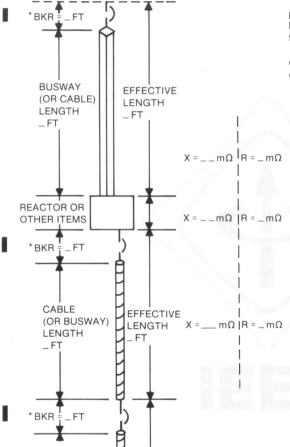

Determine milliohms reactance and resistance per phase of each circuit section by considering length and reactance and resistance data in manufacturer's literature or accompanying tables.

For parallel conductors per phase divide milliohms values of one conductor by number of identical conductors.

Short-circuit analysis is performed on a per-phase (line-to-neutral) basis. Circuit length is overall three-phase length, not the sum of the individual conductor lengths.

Reactance (X) and resistance (R) values are expressed in milliohms (mΩ) for convenience.

Busway data in Table 12.

Cable data in Tables 3 through 11.

Reactance and resistance milliohms for any miscellaneous devices, components, switchboard bussing, etc, may be considered if available.

Use reactor actual milliohms for X and X/80 for R (assumes X/R = 80).

Circuit breaker impedances are usually ignored for ease of calculation. When calculated current values slightly exceed equipment rating it may be beneficial to consider circuit breaker impedance. Circuit breaker reactance and resistance values depend on breaker size, design, and manufacturer. However, these values usually are similar to those of the circuit supplied by the breaker (feeder breaker) or supplying the breaker (main breaker). From this and for convenience, circuit breakers are represented here as equivalent lengths of cable or busway. A similar relationship for switch and fuse devices has not yet been established.

*If circuit breaker impedances are considered, then treat each circuit breaker other than at point of assumed fault as equal to —

or 5 ft of cable or plug-in busway
10 ft of feeder busway

Use load side cable/busway for feeder bkrs; line side cable/busway for equipment main bkrs.

NOTE: The reactance and resistance of a circuit breaker help determine its interrupting rating. Therefore, circuit breaker reactance and resistance cannot be used in determining the system short-circuit current at the point where that breaker is considered for use. Circuit breakers at other points can be considered.

Fig 15
Reactance and Resistance of Circuit Segments

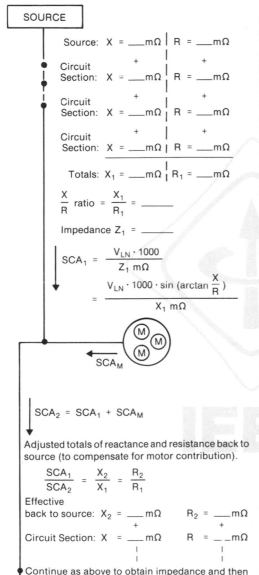

Often, medium-voltage cable impedance is ignored, especially since it has very little effect on calculations in the subsequent low-voltage system. This is easier and faster, and usually sufficiently precise, and produces safe values. However, if the short-circuit values obtained at remote locations on medium-voltage feeders are higher than the ratings of the equipment proposed for use there, it may be desirable to consider the medium-voltage cable impedance. Values can be obtained from the cable manufacturer or from Tables 3 through 9.

Low-voltage conductor impedance has significant effect and should be considered unless the conductor segment is very short.

Reactance and resistance milliohm values from Figs 14 and 15.

Add arithmetically all reactance values, then all resistance values — keeping reactances separate from resistances. X/R ratio is total reactance divided by total resistance. Obtain resultant impedance Z by using total reactance and X/R ratio.

$$Z_1 = \frac{X_1}{\sin (\arctan X/R)}$$

Ohm's Law: Phase voltage divided by phase impedance equals available short-circuit amperes in each phase.

Motor contribution explained in Figs 8 and 9.

Add short-circuit currents from source and motors arithmetically (assumes X/R ratio of supply feeder and motor circuit are equal). This is a safe approximation and easier than more rigorous vector addition.

With voltage considered a constant, the increased short-circuit (due to addition from motors) can be considered a result of reducing the impedance of the supply feeder circuit, then ignoring the motors. In Ohm's Law, impedance (also reactance and resistance) varies inversely as the short-circuit current, permitting the proportion equations at left. Knowing four values, the other two are easily found.

To the new reactance and resistance milliohm values representing effective totals back to the source add X and R values of subsequent circuit sections. Procedure is same as above; totaling reactances, resistances; determining impedance; applying Ohm's Law to get available short-circuit amperes.

Fig 16
Determination of Short-Circuit Current on Feeder

may be as great as 100% of the circuit breaker interrupting rating as long as the X/R ratio does not exceed $X/R = 15$ for 5- and 3-cycle circuit breakers and 10 for 8-cycle circuit breakers. But if the X/R ratio is greater than these limits, multipliers from curves in ANSI/IEEE C37.010-1979 [5] should be applied. These evaluations require calculating or conservatively estimating the system X/R ratio at point of fault.

A satisfactory, safe X/R ratio can be obtained from vector analysis of the short-circuit currents and X/R ratios of the various sources. One way is to factor the various source short-circuit currents into effective reactance and resistance components by considering: reactance component = short-circuit current · sin (arctan X/R ratio); the resistance component = reactance component divided by X/R ratio. The arithmetic total of reactance components divided by the arithmetic total of resistance components will produce an acceptable X/R ratio for use with the total short-circuit current. Simplifying estimates tending toward a higher X/R ratio will be on the safe side in the medium-voltage circuit breaker applications.

In some cases circuit breakers with adequate interrupting capacity may have inadequate close and latch ratings. ANSI/IEEE C37.010-1979 [5] discusses this case. In essence, the calculated first-cycle short-circuit symmetrical rms current should be multiplied by a factor of 1.6 to obtain an asymmetrical rms current value to be compared with the close and latch rating of the circuit breaker expressed in asymmetrical rms amperes. If the circuit breaker close and latch rating is inadequate, a second evaluation can be made by reducing the induction motor short-circuit contribution, as shown in Table 13.

2.3.5 Calculating Short-Circuit Currents Through Impedances. After the short-circuit currents produced by the various sources throughout the system have been determined, short-circuit values at various locations can be calculated. This involves considering the impedance of each circuit segment (from start at one bus to finish at another bus) until every pertinent bus or eschelon of the system has been considered. The procedure is detailed in the following Figs 13–16. Note that since resistance, reactance, and impedance values are very small they are usually expressed in milliohms and a factor of 1000 appears in the ohms law equations.

2.3.6 Current-Limiting Protective Devices. In industrial and commercial power systems, it is common practice to use current-limiting devices to protect other devices against high values of short-circuit current, which exceed their short-circuit ratings. Short-circuit calculations should be made without considering the effects of current-limiting devices to determine the "prospective" short-circuit current, or that which would flow if no current-limiting device were present. This "prospective" short-circuit current is used in applying the devices. The manufacturer of the device to be protected should be consulted to determine the suitability of the application for the level of short-circuit current available at the line-side terminals of the protected device. At this time the best practice is to determine the suitability of various applications by performance testing in accordance with recognized test procedures of qualifying organizations, such as Underwriters Laboratories. No general agreement has yet been reached on how to use the I^2t and peak current let-through values of a current-limiting device to establish its capability to protect another device.

It should be noted that a current-limiting device cannot be used to protect another current-limiting device. Also, current-limiting devices in series do not work together to compound the current-limiting effect. The faster acting of the two, which is usually the one to be protected, will try to interrupt the full short-circuit current before the other device is able to operate.

2.3.7 Short-Circuit Current Tables and Curves. These are often found in application literature as a quick means for determining short-circuit values. However, these should be used with caution. They usually apply to a very limited portion of a power system and are based on assumptions that may not be representative of the actual system being analyzed. The tables and curves usually refer to a transformer supplying a feeder and short-circuit current values are presented for various feeder lengths. To enable this simplicity, assumptions are made for supply system capacity, transformer impedance and X/R ratio, motor short-circuit contribution, and feeder cable electrical parameters. Error can be very significant if the assumptions, particularly the transformer imped-ance, do not represent the actual conditions. In view of this, the reader may find it more reliable to perform the few calculations described here using the actual data pertinent to his/her system or by making and appraising his/her own assump-tions.

2.4 Example of Direct Method Short-Circuit Current Calculations. The direct method is used in a brief short-circuit study of a power system. Figure 17 is a one-line diagram representing a comprehensive medium voltage and low voltage power system. Data from this diagram are used to illustrate the calculation techniques. Locations in the system where a short-circuit current value is desired are designated with bus numbers. The direct method is used to determine the short-circuit current at fourteen bus locations. (In 2.6 the per-unit method is used to determine the short-circuit current at three of these bus locations. The results from the two methods are essentially the same.)

The general procedure involves three steps. First, the total motor and genera-tor first-cycle short-circuit contribution flowing toward the source at each bus is calculated working from the loads up to the service point. Second, the total first-cycle symmetrical short-circuit current is calculated at the start and finish of each circuit segment (from one bus to another) by working from the service point down to the loads. Third, any necessary medium-voltage circuit breaker interrupting and close and latch short-circuit current values are calculated by refiguring the various source contributions.

Much of the direct method of short-circuit calculation is condensed into a short-circuit calculation worksheet, which is used in this example (Figs 18, 19, 20). This worksheet offers a concise procedure and a convenient way of storing the data for future reference. Filling in the columns from top to bottom indicates what data should be obtained and how they are to be used. When calculations are necessary the equations are shown. Reactance and resistance values are used for accurate vector addition of impedances as well as X/R ratios. The relation of reactance, resistance, and impedance is conveniently handled with the single trigonometric expression "sin arctan X/R ratio" (sine of the angle whose tangent is the X/R ratio). This is easily calculated with even very inexpensive readily available calculators.

A blank copy of this worksheet is included, which the reader may find useful (Fig 18). Also the format and sequence of calculation may be a helpful guide to those who would like to develop a simple computer program for short-circuit calculations.

2.4.1 First-Cycle Motor and Generator Short-Circuit Contributions. The total motor and generator short-circuit contribution flowing toward the source must be determined first. The actual calculations for each bus are shown.

(1) Busses 14 and 15. No motor contribution.
(2) Bus 13. 100 kVA for 100 hp induction motor at 80% power factor.

$$\text{SCA}_M = 5 \cdot I_{FL}$$
$$= 5 \cdot \frac{100 \text{ kVA}}{\sqrt{3} \cdot 0.46 \text{ kV}} \quad \text{(see Fig 9)}$$
$$= 5 \cdot 125.5 \text{ A}$$
$$= 628 \text{ SCA}$$

(3) Busses 11 and 12.

$$\text{SCA}_M = 2 \cdot 628 \text{ A} = 1256 \text{ SCA}$$

(4) Bus 10.

$$\text{transformer primary amperes} = \frac{750 \text{ kVA}}{\sqrt{3} \cdot 4.16}$$
$$= 104.09 \text{ A}$$

$$\text{transformer rated secondary amperes} = \frac{750 \text{ kVA}}{\sqrt{3} \cdot 0.480}$$
$$= 902.11 \text{ A}$$

$$\text{SCA}_M = \frac{104.09 \text{ A}}{\dfrac{5.5\%}{100\%} + \dfrac{902.11 \text{ A}}{1256 \text{ A}}} \quad \text{(see Fig 11)}$$
$$= 135 \text{ SCA}$$

(5) Bus 9. No motor contribution.
(6) Busses 7 and 8. Generator contribution.

$$\text{generator rated amperes} = \frac{625 \text{ kVA}}{\sqrt{3} \cdot 4.16 \text{ kV}}$$
$$= 86.74 \text{ A}$$

$$\text{SCA}_G = \frac{86.74 \text{ A} \cdot 100\%}{9\%} \quad \text{(see Fig 7)}$$
$$= 964 \text{ SCA}$$

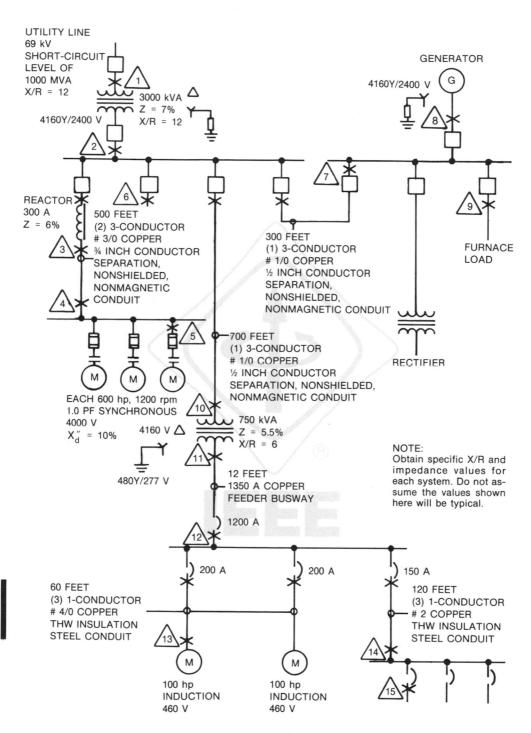

**Fig 17
One-Line Diagram for Typical Industrial Distribution System**

Fig 18
Sample Worksheet for Short-Circuit Calculations by Direct Method

Short-Circuit Calculation Sheet
(Three-Phase Fault)

A step-by-step procedure for calculating maximum short-circuit amperes (SCA) rms sym. in 3φ electrical systems. Applicable to medium-voltage source, transformer, and/or low-voltage circuits.

Simply fill in columns. "•" designates data obtained from system. "*" designates values calculated using functional "item letter" equations which are derived from more complex electrical equations. A calculator with sin, tan, arc/INV functions is required.

TRANSFORMER CIRCUIT SEGMENT (See Fig 13)	Busses →
Source:	
• SCkVA	$c \cdot d \div 577.35 = a$
• X/R Ratio	b
SCA rms sym.	c
• Voltage (line-to-line)	d
Transformer:	
• kVA 3φ	e
• % Impedance	f
• X/R Ratio	g
• Secondary Voltage (line-to-line)	h
Source:	
* Equivalent % X	$\sin(\arctan b) \cdot 100 \cdot e \div a = j$
* Equivalent % R	$j \div b = k$
Transformer:	
* % X	$\sin(\arctan g) \cdot f = m$
* % R	$m \div g = n$
At Transformer Secondary:	
* Total % X [Back	$j + m = p$
* Total % R to	$k + n = q$
* X/R Ratio Source]	$p \div q = r$
* SCA from Transformer	$\sin(\arctan r) \cdot e \cdot 57\,735 \div h \div p = s$

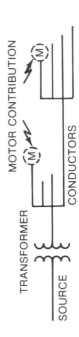

MOTOR CONTRIBUTION

TRANSFORMER

SOURCE

CONDUCTORS

CONDUCTOR CIRCUIT SEGMENT (See Figs 14, 15, and 16)

Busses →	a								
• Voltage (line-to-line)	a								
• Length in feet	b								
• Number of conductors/ϕ	c								
• X mΩ/conductor/100 ft	d								
• R mΩ/conductor/100 ft	e								

At finish of preceding upstream segment

• SCA from preceding upstream segment	f								
• X/R Ratio back to Source	g								
* X mΩ back to Source	① (s←) h								
* R mΩ back to Source	② (t←) j								

At start of this segment

• SCA motor contribution	k								
* SCA Total	f + k = m								
* X mΩ new — Back to Source	h · f ÷ m = n								
* R mΩ new — Source	j · f ÷ m = p								

Conductor segment

* X mΩ	b · d ÷ c · 100 = q								
* R mΩ	b · e ÷ c · 100 = r								

At finish of this segment

* Total X mΩ — Back to	n + q = s								
* Total R mΩ — Source	p + r = t								
* X/R Ratio	s ÷ t = u								
* SCA from this segment	sin (arctan u) · a · 577.35 ÷ s = v								

① h = sin (arctan g) · a · 577.35 ÷ f
h = s of preceding upstream conductor segment (s ←)

② j = h ÷ g
j = t of preceding upstream conductor segment (t ←)

Fig 19
First of Two Pages showing Example Problem Solved By Direct Method

Short-Circuit Calculation Sheet
(Three-Phase Fault)

A step-by-step procedure for calculating maximum short-circuit amperes (SCA) rms sym. in 3ϕ electrical systems. Applicable to medium-voltage source, transformer, and/or low-voltage circuits.

Simply fill in columns. "●" designates data obtained from system. "*" designates values calculated using functional "item letter" equations which are derived from more complex electrical equations. A calculator with sin, tan, arc/INV functions is required.

MV Cable

Busses	Calculations
3-4	$3.45 = (9.81 - 6.36) \cdot 1$
2-7	$3.01 = (10.30 - 7.29) \cdot 1$

TRANSFORMER CIRCUIT SEGMENT (See Fig 13)

		Busses →	1-2
Source:			
● SCkVA	$c \cdot d \div 577.35 =$	a	1 000 000
● X/R Ratio		b	12
● SCA rms sym.		c	8367.4
● Voltage (line-to-line)		d	69 000
Transformer:			
● kVA 3ϕ		e	3000
● % Impedance		f	7
● X/R Ratio		g	12
● Secondary Voltage (line-to-line)		h	4160
Source:			
* Equivalent % X	$\sin(\arctan b) \cdot 100 \cdot e \div a =$	i	0.299
* Equivalent % R	$j \div b =$	k	0.025
Transformer:			
* % X	$\sin(\arctan g) \cdot f =$	$.m$	6.976
* % R	$m \div g =$	n	0.581
At Transformer Secondary:			
* Total % X	Back	$j + m = p$	7.275
* Total % R	to	$k + n = q$	0.606
* X/R Ratio	Source	$p \div q = r$	12
* SCA from Transformer	$\sin(\arctan r) \cdot e \cdot 57\ 735 \div h \div p = s$		5703

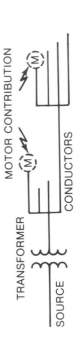

SOURCE — TRANSFORMER — CONDUCTORS — MOTOR CONTRIBUTION

CONDUCTOR CIRCUIT SEGMENT

Busses →		2-3	3-4	4-5	2-6	2-7	7-8
• Voltage (line-to-line)	a	4160	4160	4160	4160	4160	4160
• Length in feet	b		500	0	0	300	0
• Number of conductors/ϕ	c		2			1	
• X mΩ/conductor/100 ft	d		3.45			3.01	
• R mΩ/conductor/100 ft	e		6.60			10.49	
At finish of preceding upstream segment							
• SCA from preceding upstream segment	f	5703	2885	2852	5703	5703	7014
• X/R Ratio back to Source	g	12	23.553	16.220	12	12	5.734
* X mΩ back to Source ① (s ←) h	h	419.69	831.88			419.69	
* R mΩ back to Source ② (t ←) j	j	34.97	35.32			34.97	
At start of this segment							
• SCA motor contribution	k	1099	0	1386	2551	1587	0
* SCA Total f + k = m	m	6802	2885	4238	8254	7290	7014
* X mΩ new Back to h · f ÷ m = n	n	351.88	831.88			328.33	
* R mΩ new Source j · f ÷ m = p	p	29.32	35.32			27.36	
Conductor segment							
* X mΩ b · d ÷ c ÷ 100 = q	q	430.00	8.63	0	0	9.03	
* R mΩ b · e ÷ c ÷ 100 = r	r	6.00	16.50	0	0	31.47	
At finish of this segment							
* Total X mΩ Back n + q = s	s	831.88	840.51			337.36	
* Total R mΩ to p + r = t	t	35.32	51.82			58.83	
* X/R Ratio Source s ÷ t = u	u	23.553	16.220	16.220	12	5.735	
* SCA from this segment sin (arctan u) · a · 577.35 ÷ s = v	v	2885	2852	4238	8254	7014	7014

① h = sin (arctan g) · a · 577.35 ÷ f
  h = s of preceding upstream conductor segment (s ←)

② j = h ÷ g
  j = t of preceding upstream conductor segment (t ←)

Fig 20
Second of Two Pages Showing Example Problem Solved by Direct Method

TRANSFORMER CIRCUIT SEGMENT (See Fig 13)		
Busses →		10-11
Source:		
• SCkVA	$c \cdot d \div 577.35 = a$	52 325
• X/R Ratio	b	3.223
• SCA rms sym.	c	7262
• Voltage (line-to-line)	d	4160
Transformer:		
• kVA 3ϕ	e	750
• % Impedance	f	5.5
• X/R Ratio	g	6
• Secondary Voltage (line-to-line)	h	480
Source:		
* Equivalent % X	$\sin(\arctan b) \cdot 100 \cdot e \div a = j$	1.369
* Equivalent % R	$j \div b = k$	0.425
Transformer:		
* % X	$\sin(\arctan g) \cdot f = m$	5.425
* % R	$m \div g = n$	0.904
At Transformer Secondary:		
* Total % X	Back to Source $\quad j + m = p$	6.794
* Total % R	$k + n = q$	1.329
* X/R Ratio	$p \div q = r$	5.112
* SCA from Transformer	$\sin(\arctan r) \cdot e \cdot 57\ 735 \div h \div p = s$	13 031

Short-Circuit Calculation Sheet
(Three-Phase Fault)

A step-by-step procedure for calculating maximum short-circuit amperes (SCA) rms sym. in 3ϕ electrical systems. Applicable to medium-voltage source, transformer, and/or low-voltage circuits.

Simply fill in columns. "•" designates data obtained from system. "*" designates values calculated using functional "item letter" equations which are derived from more complex electrical equations. A calculator with sin, tan, arc/INV functions is required.

MV Cable

Busses	Calculation
2-10	$3.01 = (10.30 - 7.29) \cdot 1$

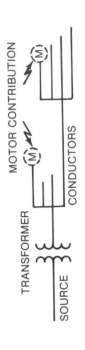

MOTOR CONTRIBUTION

TRANSFORMER

SOURCE   CONDUCTORS

(See 10-11 above)

CONDUCTOR CIRCUIT SEGMENT (See Figs 14, 15 and 16)

	Busses →	7-9	2-10	11-12	12-13	12-14	14-15
• Voltage (line-to-line)	a	4160	4160	480	480	480	480
• Length in feet	b	0	700	12	60	120	0
• Number of conductors/ϕ	c		1	1	1	1	
• X mΩ/conductor/100 ft	d		3.01	0.65	4.14	4.49	
• R mΩ/conductor/100 ft	e		10.49	0.76	5.34	16.40	
At finish of preceding upstream segment							
• SCA from preceding upstream segment	f	7014	5703	13 031	12 974	12 974	8168
• X/R Ratio back to Source	g	5.734	12	5.112	5.024	5.024	1.043
* X mΩ back to Source	① (s ←) h		419.69	20.87	20.95	20.95	
* R mΩ back to Source	② (t ←) j		34.97	4.08	4.17	4.17	
At start of this segment							
• SCA motor contribution	k	964	2416	0	628	1256	0
* SCA Total	f + k = m	7978	8119		13 602	14 230	
* X mΩ new Back to	h · f ÷ m = n		294.80		19.98	19.10	
* R mΩ new Source	j · f ÷ m = p		24.56		3.98	3.80	
Conductor segment							
* X mΩ	b · d ÷ c ÷ 100 = q	0	21.07	0.08	2.48	5.39	0
* R mΩ	b · e ÷ c ÷ 100 = r	0	73.43	0.09	3.20	19.68	0
At finish of this segment							
* Total X mΩ Back	n + q = s		315.87	20.95	22.46	24.49	
* Total R mΩ to	p + r = t		97.99	4.17	7.18	23.48	
* X/R Ratio Source	s ÷ t = u	5.734	3.223	5.024	3.128	1.043	1.043
* SCA from this segment	sin (arctan u) · a · 577.35 ÷ s = v	7978	7262	12 974	11 753	8168	8168

① h = sin (arctan g) · f
    h = s of preceding upstream conductor segment (s ←)

② j = h ÷ g
    j = t of preceding upstream conductor segment (t ←)

(7) Bus 6. No motor contribution.

(8) Bus 5. 480 kVA (0.8 · 600) for 600 hp synchronous motor at 100% power factor.

$$\text{SCA}_M = \frac{100\%}{10\%} \cdot I_{FL} \qquad \text{(Fig 9)}$$

$$= 10 \cdot \frac{480 \text{ kVA}}{\sqrt{3} \cdot 4.000 \text{ kV}}$$

$$= 693 \text{ SCA}$$

(9) Busses 3 and 4. 3 · 693 A = 2079 A

(10) Bus 2. Sum of feeder motor and generator contributions.

From Bus 3 through reactor:

$$Z_M = \frac{4.000 \cdot 1000}{\sqrt{3} \cdot 2079} \qquad \text{(Fig 10)}$$

$$= 1.111 \ \Omega$$

$$Z_R = \frac{6\%}{100\%} \cdot \frac{2400 \text{ V}}{300 \text{ A}}$$

$$= 0.480 \ \Omega$$

SCA_M through reactor =

$$\frac{1.111}{1.111 + 0.480} \cdot 2079 \text{ SCA} = 1452 \text{ SCA}$$

Total SCA_M = 1452 SCA + 135 SCA + 964 SCA = 2551 SCA

(11) Bus 1. Calculation unnecessary.

2.4.2 First-Cycle Symmetrical Short-Circuit Calculations for Each Bus. The next step is to begin at the service point and make first-cycle symmetrical short-circuit calculations for each bus in the system. For convenience, the impedance of circuit breakers is neglected. The calculated values and equations are shown on the completed worksheets in Figs 19 and 20.

Note that total motor and generator short-circuit contribution at the start of a segment does not include load-side contribution. For example, the motor and generator contribution at the start of the segment from Bus 2 to Bus 10 is 2551 SCA − 135 SCA = 2416 SCA.

Additional calculations must also be made for the reactor as follows:

$Z_R = 0.48000 \ \Omega$ (see 2.4.1 item 10)

   $= 480.00 \text{ m}\Omega$

Assuming $X/R = 80$.

$Z \cong Z_R = 480.00 \text{ m}\Omega$

$R = \dfrac{480.00}{80} = 6.00 \text{ m}\Omega$

2.4.3 Medium-Voltage Circuit Breaker Interrupting and Close and Latch Calculations. The final step is to make any necessary medium-voltage circuit breaker interrupting and close and latch calculations to verify the adequacy of medium-voltage circuit breakers. The feeder circuit breaker at Bus 6 will have the highest calculated short-circuit current at 4160 V. Therefore, all of the medium-voltage circuit breakers need only have ratings higher than or equal to the calculated current values for that bus.

Refer to 2.3.4 and ANSI/IEEE C37.010-1979 [5].

In a first evaluation, if the calculated system first-cycle short-circuit symmetrical rms current is not more than 80% of the circuit breaker symmetrical rms interrupting current rating, the circuit breaker is adequate from an interrupting capacity standpoint (irrespective of the X/R ratio) and no further calculations are required. This situation occurs frequently and is the case in this example since the smallest medium-voltage circuit breaker likely to be applied would be rated at least 12 000 A interrupting capacity and the 8254 first-cycle SCA is much less than 80% of this circuit breaker interrupting rating.

However, if the calculated first-cycle SCA were greater than 80% of the circuit breaker interrupting rating, a further calculation refinement could be used. Since at the time of circuit breaker interruption motor contribution will have decreased, a new short-circuit current value may be calculated using interrupting multipliers shown in Table 13. In this example, the source contributions back to Bus 2 may be recalculated as follows:

(1) Utility Source. Same as before at Bus 2, which was 5703 SCA.

(2) Synchronous Motors Through Reactor. Use 0.667 of first-cycle value:

(Table 13)

$$SCA_M = 0.667 \cdot 2079 \text{ A} = 1386 \text{ SCA}$$
$$Z_M = \frac{4.000 \cdot 1000}{\sqrt{3} \cdot 1386} \qquad \text{(Fig 10)}$$
$$= 1.666 \ \Omega$$
$$Z_R = 0.480 \ \Omega$$

(see item 10 of 2.4.1)

$$SCA_M \text{ through reactor} = \frac{1.666}{1.666 + 0.480} \cdot 1386 \text{ SCA}$$
$$= 1076 \text{ SCA}$$

(3) 100 hp Induction Motors Through Transformer. Use 0.333 of first-cycle value. (Table 13.)

$$\text{Sec. } SCA_M = 0.333 \cdot 1256 \text{ SCA}$$
$$= 419 \text{ SCA}$$
$$SCA_M = \frac{104.09 \text{ A}}{\dfrac{5.5\%}{100\%} + \dfrac{902.11 \text{ A}}{419 \text{ A}}}$$
$$= 47 \text{ SCA}$$

(4) 625 kVA Generator. Assume no exponential decay in short-circuit contribution.

$$SCA_G = 964 \text{ SCA}$$

(5) Total at Bus 6.

$$\begin{aligned} SCA &= 5703 \text{ SCA} + 1076 \text{ SCA} + 47 \text{ SCA} + 964 \text{ SCA} \\ &= 7790 \text{ SCA} \end{aligned}$$

If this recalculated short-circuit current (7790 SCA) is not more than 80% of the circuit breaker interrupting current rating, the circuit breaker will be adequate from an interrupting capacity standpoint (irrespective of the X/R ratio).

However, if after the previous two evaluations the circuit breakers were still not adequate from an interrupting capacity standpoint, a third evaluation determining that the system 7790 SCA at time of interruption was not more than 100% of the circuit breaker interrupting rating and the X/R ratio was not greater than 10 for 8-cycle circuit breakers or 15 for 5- and 3-cycle circuit breakers would justify adequate circuit breaker interrupting capacity. Obviously, if more convenient, the larger 8254 SCA value obtained from first-cycle calculation may also be used instead of 7790 SCA to justify a sufficient interrupting capacity rating, possibly avoiding the need to calculate the 7790 SCA value. A satisfactory X/R ratio would have to be determined (see 2.3.4) and if it exceeded the 10 or 15 values above, multipliers from ANSI/IEEE C37.010-1979 [5] would be required.

The circuit breaker asymmetrical close and latch rating may be verified by being at least as large as the first-cycle 8254 SCA · 1.6 = 13 206 SCA asymmetrical rms. If it is not, a reduced current may be calculated based on less motor short-circuit contribution (see Table 13) as follows:

(1) Utility Source. Same as before at Bus 2, which was 5703 SCA.

(2) Synchronous Motors Through Reactor. Same as before at Bus 2, which was 1452 SCA.

(3) 100 hp Induction Motors Through Transformer. Use 0.833 of the usual first-cycle value. (Fig 11.)

$$\begin{aligned} \text{Sec. } SCA_M &= 0.833 \cdot 1256 \text{ SCA} \\ &= 1047 \text{ SCA} \end{aligned}$$

$$\begin{aligned} SCA_M &= \cfrac{104.09}{\dfrac{5.5\%}{100\%} + \dfrac{902.11 \text{ A}}{1047 \text{ A}}} \\ &= 114 \text{ SCA} \end{aligned}$$

(4) 625 kVA Generator. Same as before at Bus 2, which was 964 SCA.

(5) Total at Bus 6.

$$\begin{aligned} SCA &= 5703 \text{ SCA} + 1452 \text{ SCA} + 114 \text{ SCA} + 964 \text{ SCA} \\ &= 8233 \text{ SCA} \end{aligned}$$

The circuit breaker close and latch rating is adequate if it is not less than the calculated 8233 · 1.6 = 13 173 SCA asymmetrical rms.

2.5   Calculating Three-Phase Short-Circuit Currents by the Per-Unit Network Method. The basic concept in the per-unit method of calculation is to develop an equivalent impedance network diagram of the power system involved and resolve this down to a single impedance value through delta, wye, series, parallel conversion equations. While it is possible to do this with the actual system ohmic values, it becomes very complex when several voltage levels are involved since ohmic values at different voltages are not directly compatible. The per-unit mathematical technique was developed to facilitate this and while the equivalent diagram and the per-unit mathematics are separate considerations, they invariably are used together and referred to as the per-unit method of determining short-circuit values.

High-voltage and medium-voltage power systems are particularly attractive to the per-unit method since the analysis being undertaken usually involves a relatively few specific points often in an existing system. Also in these systems reactance usually far exceeds resistance (high X/R ratio), permitting resistance to be ignored, which greatly simplifies the mathematics.

2.5.1   Representation of Impedances. The principal problem in calculating short-circuit currents is assigning values of impedance to each circuit element and then converting to a per-unit base form to provide an easy way of combining values connected in series and parallel. If possible, obtain the impedance values directly from the equipment manufacturer. If not possible, the values included in Tables 1–12 are typical and will usually provide satisfactory answers. Additional values may also be found in other sources, such as ANSI/IEEE Std 141-1986 [4], or equipment nameplates.

The determination of short-circuit currents involves impedance from the source to the point of fault. Impedance consists of both reactance and resistance. Simplifying calculations by neglecting resistance is usually acceptable if the overall X/R ratio from the sources to the point of fault is 5 or greater (accuracy within 2% on the safe side). However, since the X/R ratios of some of the various circuit components may be considerably different, it may be difficult to estimate the overall X/R ratio at the fault point to decide whether ignoring resistance is prudent.

Many short-circuit calculations in low-voltage systems frequently involve low X/R ratios where resistance has a significant effect and should not be ignored.

Per-unit impedances, expressed on a chosen base, can be combined directly, regardless of how many voltage levels are encountered from source to fault. The per-unit system presented in this chapter is illustrated as it applies to balanced three-phase system faults.

2.5.2   Per-Unit Quantities on a Chosen Base. In the per-unit system there are four base quantities: base apparent power in kilovolt-amperes or megavolt-amperes, base voltage in volts or kilovolts, base impedance in ohms, and base current in amperes. Choosing any two at one voltage level automatically determines the rest. The relationship between base, per-unit, and actual quantities is

$$\text{per-unit quantity} = \frac{\text{actual quantity}}{\text{base quantity}}$$

actual quantity = per-unit quantity · base quantity

Normally, the base apparent power in kilovolt-amperes is selected first. Then, the voltage at one voltage level is chosen as the base voltage, which then determines the base voltages at the other levels with reference to the primary and secondary ratings of the transformers. The base apparent power may be any convenient value in kVA or MVA, often (but not necessarily) the rating of a transformer in the system, such as 10 000 kVA. Base voltages are usually line-to-line voltages in kilovolts. The formulas for deriving the bases are

$$I_b = \frac{kVA_b}{\sqrt{3} \cdot kV_b}$$

$$Z_b = \frac{kV_b \cdot 1000}{\sqrt{3} \cdot I_b} = \frac{(kV_b)^2 \cdot 1000}{kVA_b} = \frac{(kV_b)^2}{MVA_b}$$

where

I_b = base current, in amperes
kV_b = base voltage, in kilovolts, line-to-line
kVA_b = base apparent power, in kilovolt-amperes
MVA_b = base apparent power, in megavolt-amperes
Z_b = base impedance, in ohms/phase

Circuit element impedances are usually described in actual ohms or milliohms or in percent on an equipment rating as a base. Cable impedance is generally expressed in ohms and transformer impedance as percent on the self-cooled rating in kilovolt-amperes; for example, 5% on a 500 kVA transformer base. These actual element impedance values can be converted to per-unit by the formulas

$$Z_{pu} = \frac{Z_e \cdot kVA_b}{1000 \cdot kV_b{}^2}$$

$$Z_{pu} = \frac{Z_\% \cdot kVA_b}{100 \cdot kVA_e} \left(\frac{kV_e}{kV_b}\right)^2$$

$$Z_{pu} = \frac{Z_{pu\,e} \cdot kVA_b}{kVA_e} \left(\frac{kV_e}{kV_b}\right)^2$$

where

Z_{pu} = per-unit impedance on chosen kilovolt-ampere power base
$Z_{pu\,e}$ = per-unit impedance on kilovolt-ampere rating of element
Z_e = actual element impedance, in ohms
$Z_\%$ = percent element impedance based on element kilovolt-ampere rating
kVA_e = actual element apparent power rating, in kilovolt-amperes
kV_e = actual element voltage rating, in kilovolts

If resistance can be neglected, then these formulas can be used for reactance instead of impedance, by replacing impedance Z with reactance X.

2.5.3 Description of Procedure. A one-line diagram should be prepared showing all sources of short-circuit current and all significant circuit elements as illustrated in Fig 17. The next step is to construct from the one-line diagram an impedance diagram or diagrams with the significant fault points marked. Both reactance and resistance values should be shown. For faults at particular points, if it is obvious that the system X/R ratio is greater than 5, the resistance values can be neglected in the calculations.

For each fault point, it is necessary to combine the impedances to the point of fault into a single equivalent impedance and determine the short-circuit current. It is often suggested that the system X/R ratio to a fault point can be determined by treating reactances alone and resolving the network down to an equivalent reactance, and then the same procedure is followed treating resistances alone. The X/R ratio is determined from these two values. X and R values obtained in this way are generally usable but usually not completely accurate if many system parallel impedances are involved. It is obvious that the X/R ratio of the resultant impedance to the point of fault will always be greater than 5 if the X/R ratios of all the component impedances are greater than 5. However, if a major impedance has a lower X/R ratio, such as a large amount of cable, all calculations should be made using impedance values involving both reactance and resistance values. The effort to determine if the error will be tolerable when resistance is ignored may be better used in calculations involving resistance as well as reactance to produce more precise results.

Equations for determining equivalent series and parallel impedances are given in Fig 21. Figure 22 shows how to convert impedances in those systems in which three circuit elements form a wye or delta configuration.

2.6 Example of Per-Unit Network Method Short-Circuit Current Calculation. The per-unit method is used to calculate the short-circuit values at three locations in the medium-voltage and low-voltage power system represented by Fig 17. (These locations were also calculated by the direct method in 2.4.) The general procedure basically involves four steps. First, base values are assigned and all circuit element impedances are converted to per-unit values. Sources are treated as ideal voltage sources having a first-cycle series impedance. Second, an impedance diagram is constructed identifying significant fault points. Where possible, the series and parallel impedances are combined to show equivalent impedances. Third, for each fault point handled separately, the series and parallel impedances are resolved down to one equivalent impedance, and the short-circuit current is calculated. Fourth, the necessary interrupting and medium-voltage circuit breaker close and latch short-circuit calculations are made by repeating the first three steps, with impedance diagrams that have the required changes in source impedance values.

Both reactance and resistance are used in the impedance values to illustrate the full calculations for best accuracy and to be comparable with the direct method. If resistance or reactance were to be ignored, zero would be used for the ignored term in the various equations.

FOR COMBINATION OF BRANCHES IN SERIES

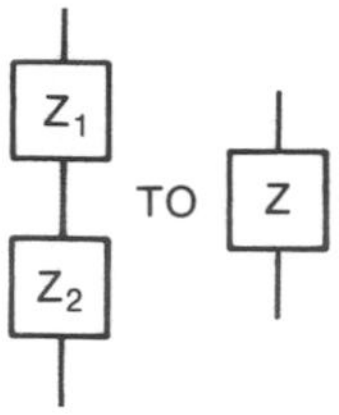

$$Z = Z_1 + Z_2$$
$$= R_1 + jX_1 + R_2 + jX_2$$
$$= (R_1 + R_2) + j(X_1 + X_2)$$

FOR COMBINATION OF BRANCHES IN PARALLEL

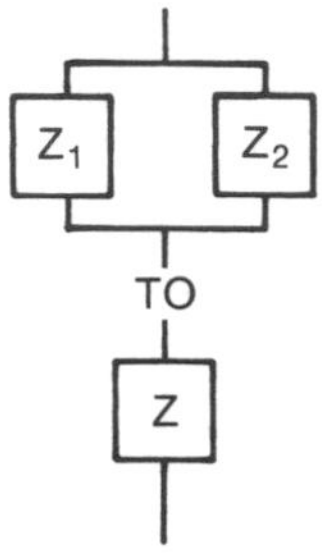

$$Z = \frac{Z_1 \cdot Z_2}{Z_1 + Z_2}$$

$$= \frac{(R_1 + jX_1)\ (R_2 + jX_2)}{(R_1 + R_2) + j(X_1 + X_2)}$$

$$= \frac{[(R_1R_2 - X_1X_2) + j(R_2X_1 + R_1X_2)]\ [(R_1 + R_2) - j(X_1 + X_2)]}{[(R_1 + R_2) + j(X_1 + X_2)]\ [(R_1 + R_2) - j(X_1 + X_2)]}$$

$$= \frac{R_1^2 R_2 + R_1R_2^2 + R_2X_1^2 + R_1X_2^2}{(R_1 + R_2)^2 + (X_1 + X_2)^2} + j\ \frac{R_1^2 X_2 + R_2^2 X_1 + X_1^2 X_2 + X_1X_2^2}{(R_1 + R_2)^2 + (X_1 + X_2)^2}$$

Fig 21
Combining Series and Parallel Impedances

2.6.1 Calculating Per-Unit Values. The base values are determined and then each per-unit value. They will be calculated to at least four decimal places to yield a high degree of accuracy.

(1) Base Values. The base apparent power will be chosen as 10 000 kVA. At 4160 V, base impedance will be

$$Z_b = \frac{(4.16)^2 \cdot 1000}{10\ 000} = 1.7306\ \Omega$$

At 480 V, base impedance will be

$$Z_b = \frac{(0.480)^2 \cdot 1000}{10\ 000} = 0.02304\ \Omega$$

(2) Utility Supply Equivalent Impedance.

$$Z_{pu} = \frac{1.0 \cdot 10\ 000}{1\ 000\ 000} \cdot \left(\frac{69}{69}\right)^2 = 0.01_{pu}$$

FOR TRANSFORMING WYE TO DELTA

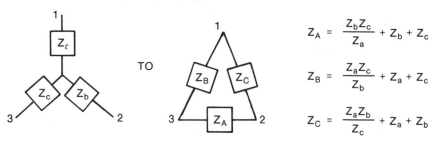

$$Z_A = \frac{Z_b Z_c}{Z_a} + Z_b + Z_c$$

$$Z_B = \frac{Z_a Z_c}{Z_b} + Z_a + Z_c$$

$$Z_C = \frac{Z_a Z_b}{Z_c} + Z_a + Z_b$$

FOR TRANSFORMING DELTA TO WYE

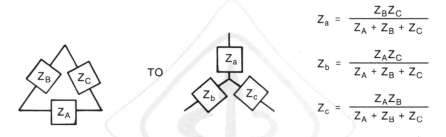

$$Z_a = \frac{Z_B Z_C}{Z_A + Z_B + Z_C}$$

$$Z_b = \frac{Z_A Z_C}{Z_A + Z_B + Z_C}$$

$$Z_c = \frac{Z_A Z_B}{Z_A + Z_B + Z_C}$$

TO TRANSFORM COMPLEX DENOMINATOR TO REAL

$$\left(\frac{1}{R + jX}\right)\left(\frac{R - jX}{R - jX}\right) = \frac{R}{R^2 + X^2} - j\,\frac{X}{R^2 + X^2}$$

Fig 22
Transforming Impedances Connected in Delta or Wye

For $X/R = 12$,

$X_{pu} = 0.997 \cdot Z_{pu} \cong 0.0100_{pu}$
$R_{pu} = X_{pu}/12 = 0.0008_{pu}$

(3) 3000 kVA Transformer. The per-unit impedance is 0.07 on the transformer rating kilovolt-ampere base.

$$Z_{pu} = \frac{0.07 \cdot 10\ 000}{3000}\left(\frac{4.16}{4.16}\right)^2 = 0.2333_{pu}$$

For $X/R = 12$ and from 2.3.2

$$X_{pu} = \frac{0.2333_{pu} \cdot 12}{\sqrt{12^2 + 1}} = 0.2325_{pu}$$
$$R_{pu} = 0.2325/12 = 0.0194_{pu}$$

(4) 625 kVA Generator. Given $Xd'' = 9\%$ impedance, the per-unit impedance is $9/100 = 0.09$ on the generator kilovolt-ampere base.

Assuming $R_{pu} \cong 0$, then $Z = X$

$$X''_{pu} = \frac{0.09 \cdot 10\ 000}{625} \left(\frac{4.16}{4.16}\right)^2 = 1.4400_{pu}$$

(5) 300 ft Tie Cable. From Table 3, $X_A = 10.30$ mΩ/100 ft and $R = 10.49$ mΩ/100 ft. From Table 6, $X_B = 7.29$ mΩ/100 ft for ½ in spacing. From Table 9, $M = 1$. Therefore,

$$X_{tot} = (X_A + X_B) \cdot M$$
$$= 10.30 - 7.29 = 3.01 \text{ m}\Omega/100 \text{ ft}$$

For 300 ft,

$$X_{tot} = \frac{3.0}{1000} \cdot \frac{300}{100} = 0.0093 \ \Omega$$
$$X_{pu} = \frac{0.00903}{X_b} = \frac{0.00903}{1.7306} \cong 0.0052_{pu}$$
$$R_{tot} = \frac{10.49}{1000} \cdot \frac{300}{100} = 0.03147 \ \Omega$$
$$R_{pu} = \frac{0.03147}{1.7306} = 0.0182_{pu}$$

(6) 500 ft Feeder Cable. $X_A = 9.81$ mΩ/100 ft, $R = 6.60$ mΩ/100 ft, $X_B = -6.36$ mΩ/100 ft for ¾ in spacing, $M = 1$.

The total reactance is

$$X_{tot} = (X_A + X_B) \cdot M$$
$$= 9.81 - 6.36 = 3.45 \text{ m}\Omega/100 \text{ ft}$$

For 500 ft,

$$X_{tot} = \frac{3.45}{1000} \cdot \frac{500}{100} = 0.01725 \ \Omega$$

For two parallel conductors per phase,

$$X_{tot} = \text{½} \cdot 0.01725 = 0.008625 \ \Omega$$
$$X_{pu} = \frac{0.008625}{X_b} = \frac{0.008625}{1.7306} = 0.0050_{pu}$$
$$R_{tot} = \text{½} \cdot \frac{6.60}{1000} \cdot \frac{500}{100} = 0.0165 \ \Omega$$
$$R_{pu} = \frac{0.0165}{1.7306} = 0.0095_{pu}$$

(7) 700 ft Feeder Cable. From the calculation for the 300 ft tie cable
$X_{tot} = 3.01$ mΩ/100 ft

For 700 ft,

$$X_{tot} = \frac{3.0}{1000} \cdot \frac{700}{100} = 0.02107 \ \Omega$$

$$X_{pu} = \frac{0.02107}{X_b} = \frac{0.02107}{1.7306} = 0.0122_{pu}$$

$$R_{tot} = \frac{10.49}{1000} \cdot \frac{700}{100} = 0.07343 \ \Omega$$

$$R_{pu} = \frac{0.07343}{1.7306} = 0.0424_{pu}$$

(8) *Rectifier and Furnace Loads.* These loads will neither contribute to nor limit the fault current in the system, and so they are neglected for the purpose of calculating fault currents.

(9) *Current-Limiting Reactor.* The reactor impedance from Fig 10 is

$$Z = \frac{6\%}{100} \cdot \frac{2400 \ V}{300 \ A}$$
$$= 0.480 \ \Omega$$

$$Z_{pu} = \frac{0.480}{1.7306} = 0.2774_{pu}$$

Assuming $X/R = 80$,

$$X_{pu} = 0.2774_{pu}$$

$$R_{pu} = \frac{0.2774}{80} = 0.0035_{pu}$$

(10) *750 kVA Transformer.* The per-unit impedance on the transformer kilovolt-ampere base = 0.055_{pu}

$$Z_{pu} = \frac{0.055 \cdot 10\ 000}{750} \left(\frac{0.48}{0.48}\right)^2 = 0.733_{pu}$$

$$X/R = 6$$

$$X_{pu} = \frac{0.7333 \cdot 6}{\sqrt{6^2 + 1}} = 0.7233_{pu}$$

$$R_{pu} = \frac{0.7233}{6} = 0.1206_{pu}$$

(11) *600 hp Synchronous Motors.* The motor's kilovolt-ampere base is 480 kVA ($0.8 \cdot 600$).

$$\text{motor full-load amperes} = \frac{480}{\sqrt{3} \cdot 4.000}$$ (see Fig 9)
$$= 69.3 \ A$$

$$SCA_M = \frac{69.3}{0.10} = 693 \ SCA$$

Assuming $R = 0$, then $Z = X$

$$X''_M = \frac{2400}{693} = 3.463 \ \Omega \qquad \text{(from 2.3.2 and 2.5.2)}$$

$$X''_{pu} = \frac{3.463}{Z_b} = \frac{3.463}{1.7306} = 2.0010_{pu}$$

(12) 1350 A Busway. $X = 0.65 \ \text{m}\Omega/100'$

$$X_{tot} = \frac{0.65}{1000} \cdot \frac{12}{100} = 0.000 \ 078 \ \Omega$$

$$X_{pu} = \frac{0.000 \ 078}{0.02304} = 0.0034_{pu}$$

$$R = 0.76 \ \text{m}\Omega/100',$$

$$R_{tot} = \frac{0.76}{1000} \cdot \frac{12}{100} = 0.000 \ 0912 \ \Omega$$

$$R_{pu} = \frac{0.000 \ 0912}{0.02304} = 0.0039_{pu}$$

(13) 60 ft Motor Feeder Cables. $X = 4.14 \ \text{m}\Omega/100'$,

$$X_{tot} = \frac{4.14}{1000} \cdot \frac{60}{100} = 0.002484 \ \Omega$$

$$X_{pu} = \frac{0.002484}{0.02304} = 0.1078_{pu}$$

$$R = 5.34 \ \text{m}\Omega/100'$$

$$R_{tot} = \frac{5.34}{1000} \cdot \frac{60}{100} = 0.003204 \ \Omega$$

$$R_{pu} = \frac{0.003204}{0.02304} = 0.1391_{pu}$$

(14) 100 hp Induction Motors. The motor's kilovolt-ampere base is 100 kVA,

$$\text{motor full-load amperes} = \frac{100}{\sqrt{3} \cdot 0.460}$$
$$= 125.5 \ \text{A}$$

$$\text{SCA}_M = 5 \cdot 125.5 \ \text{A} = 628 \ \text{A}$$

Assuming $R = 0$, then $Z = X$

$$X''_M = \frac{277}{628} = 0.441 \ \Omega$$

$$X''_{pu} = \frac{0.441}{Z_b} = \frac{0.441}{0.02304} = 19.1406_{pu}$$

(15) 120 ft Feeder Cable. X = 4.49 mΩ/100′,

$$X_{\text{tot}} = \frac{4.49}{1000} \cdot \frac{120}{100} = 0.005388 \ \Omega$$

$$X_{\text{pu}} = \frac{0.005388}{0.02304} = 0.2339_{\text{pu}}$$

$$R = 16.40 \ \text{m}\Omega/100′$$

$$R_{\text{tot}} = \frac{16.40}{1000} \cdot \frac{120}{100} = 0.01968 \ \Omega$$

$$R_{\text{pu}} = \frac{0.01968}{0.02304} = 0.8542_{\text{pu}}$$

(16) Circuit Breakers. Impedances for these will be neglected to simplify the calculations.

2.6.2 Constructing the Impedance Diagram. After the per-unit values are determined, a system impedance diagram is constructed. Figure 23 shows the first-cycle values arranged similarly to the circuits of Fig 17.

Many of the impedances can be combined with ease. Series impedances can be added numerically and represented as a single impedance. Parallel impedances can also be combined into a single value using the formulas in Fig 21. All rotating machine sources, including plant motors and generators plus utility generators, are represented by their per-unit impedances connected to an equivalent source bus. Thus, the utility supply equivalent impedance is in parallel with the motor and generator impedances.

The impedance diagram should be simplified as much as possible, retaining the points at which the fault current is to be calculated. Figure 24 illustrates a simplification of Fig 23 by combining series and parallel impedance values in connecting the equivalent source points. The dashed lines indicate busses of *equal potential* in so far as the short-circuit current calculations are concerned.

Further simplification of the impedance diagram can be made only for a specific fault location. For a fault at Bus 6, for example, it is no longer necessary to retain the other fault locations, and further simplifications of the impedance diagram can be made.

2.6.3 Performing the First-Cycle Symmetrical Short-Circuit Calculations. After the impedance diagram is simplified, first-cycle calculations should be made for each fault location. (The calculated values are only slightly different than those derived from the direct method, largely because of rounding errors and minor differences in the step-by-step process of each method.)

(1) Fault at Bus 6; 4160 V Fault on Main Switchgear Feeder. The simplification of the impedance diagram into a single equivalent impedance is shown in Fig 25. The first-cycle rms symmetrical fault current is

$$\begin{aligned}
\text{SCA} &= I_{\text{base}} \div Z_{\text{pu}} \\
&= \frac{10\ 000}{\sqrt{3} \cdot 4.160} \div 0.1681 \\
&= 8256 \ \text{SCA}
\end{aligned}$$

$$X/R = 16.1$$

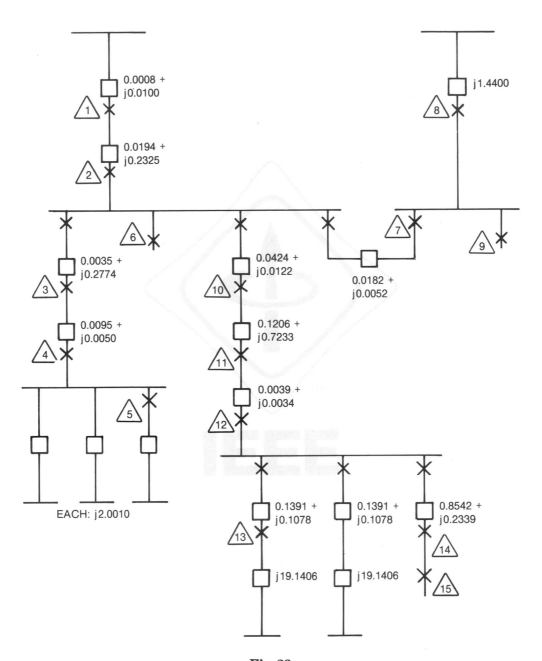

Fig 23
Impedance Diagram Constructed from One-Line Diagram
for First-Cycle Calculations

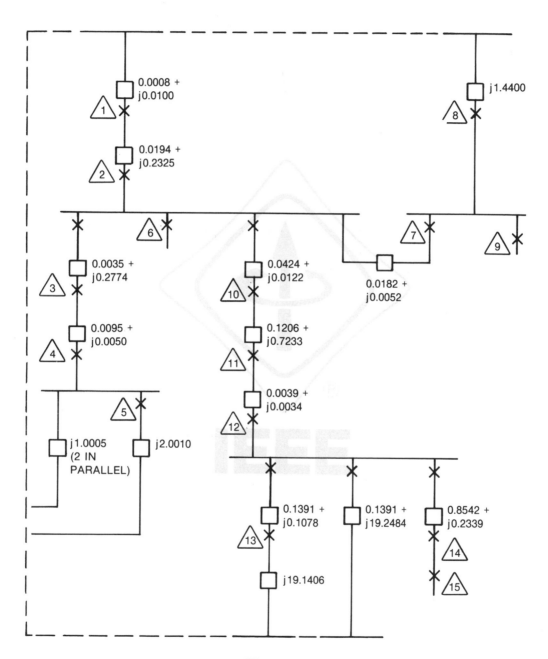

Fig 24
Impedance Diagram for First-Cycle Calculations
Simplified by Combining Impedances

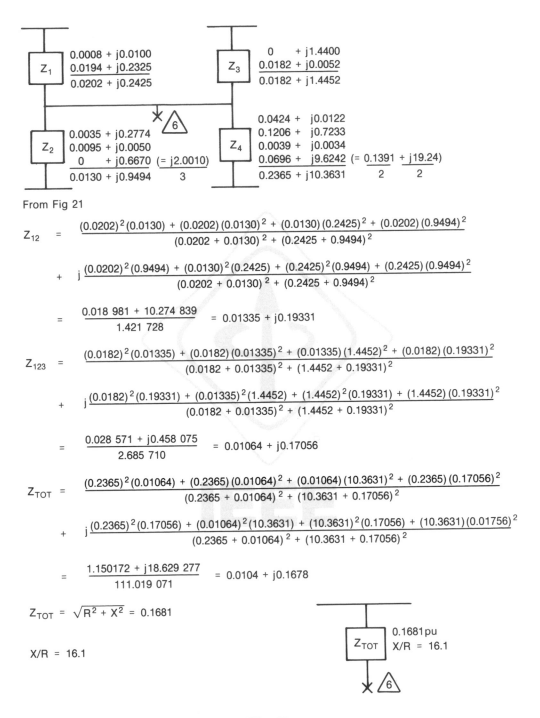

From Fig 21

$$Z_{12} = \frac{(0.0202)^2(0.0130) + (0.0202)(0.0130)^2 + (0.0130)(0.2425)^2 + (0.0202)(0.9494)^2}{(0.0202 + 0.0130)^2 + (0.2425 + 0.9494)^2}$$

$$+ \; j\frac{(0.0202)^2(0.9494) + (0.0130)^2(0.2425) + (0.2425)^2(0.9494) + (0.2425)(0.9494)^2}{(0.0202 + 0.0130)^2 + (0.2425 + 0.9494)^2}$$

$$= \frac{0.018\,981 + 10.274\,839}{1.421\,728} = 0.01335 + j0.19331$$

$$Z_{123} = \frac{(0.0182)^2(0.01335) + (0.0182)(0.01335)^2 + (0.01335)(1.4452)^2 + (0.0182)(0.19331)^2}{(0.0182 + 0.01335)^2 + (1.4452 + 0.19331)^2}$$

$$+ \; j\frac{(0.0182)^2(0.19331) + (0.01335)^2(1.4452) + (1.4452)^2(0.19331) + (1.4452)(0.19331)^2}{(0.0182 + 0.01335)^2 + (1.4452 + 0.19331)^2}$$

$$= \frac{0.028\,571 + j0.458\,075}{2.685\,710} = 0.01064 + j0.17056$$

$$Z_{TOT} = \frac{(0.2365)^2(0.01064) + (0.2365)(0.01064)^2 + (0.01064)(10.3631)^2 + (0.2365)(0.17056)^2}{(0.2365 + 0.01064)^2 + (10.3631 + 0.17056)^2}$$

$$+ \; j\frac{(0.2365)^2(0.17056) + (0.01064)^2(10.3631) + (10.3631)^2(0.17056) + (10.3631)(0.01756)^2}{(0.2365 + 0.01064)^2 + (10.3631 + 0.17056)^2}$$

$$= \frac{1.150172 + j18.629\,277}{111.019\,071} = 0.0104 + j0.1678$$

$$Z_{TOT} = \sqrt{R^2 + X^2} = 0.1681$$

$X/R = 16.1$

Fig 25
First-Cycle Impedance Calculations for Fault at Bus 6

(2) Fault at Start of Segment from Bus 12 to Bus 14; 480 V Fault on Switchboard Feeder. In low-voltage systems, only the first-cycle fault current values are of interest, since protective devices on these systems usually initiate operation within this time and are rated for those currents. Figure 26 shows the process of reducing the impedance diagram of Fig 24 to a single equivalent impedance. The first-cycle rms symmetrical fault current is

$$SCA = I_{base} \div Z_{pu}$$
$$SCA = \frac{10\ 000}{\sqrt{3} \cdot 0.480} \div 0.8465 = 14\ 209\ SCA \qquad\qquad X/R = 5.6$$

(3) Fault at Bus 15; 480 V Fault on Panelboard Feeder. Figure 27 shows the simplified impedance diagram for this fault location. The impedance of the 120 ft of cable need only be added to the equivalent impedance of the system for a fault at the start of the segment from Bus 12–Bus 14. The first-cycle rms symmetrical fault current is

$$SCA = \frac{10\ 000}{\sqrt{3} \cdot 0.480} \div 1.4645 = 8213\ SCA \qquad\qquad X/R = 1.06$$

(4) Faults on Other Busses. The calculations should be made for each fault location to insure that the equipment will be adequately rated for the available first-cycle fault currents. These further calculations will not be shown, since those already shown should clearly indicate the process by which further calculations would be made.

2.6.4 Medium-Voltage Circuit Breaker Interrupting and Close and Latch Calculations. After the usual first-cycle calculations are made, additional interrupting and close and latch calculations may be necessary for medium-voltage circuit breakers. To determine whether such calculations are necessary, the same line of reasoning as that in 2.4.3 is used. The circuit breakers should be rated for the calculated short-circuit current at Bus 6.

If *interrupting* calculations are found to be necessary, then the source impedances should be refigured as follows:

(1) Utility Supply Equivalent Impedance. Same as before, which was 0.0008 + j0.0100, in per-unit.

(2) 625 kVA Generator. Same as before, which was j1.4400, in per-unit.

(3) 600 hp Synchronous Motors. Use 1.5 times the first-cycle value.

(Table 13)

$$X_{pu} = 1.5 \cdot j2.0010$$
$$= j3.0015_{pu}$$

(4) 100 hp Induction Motors. From Table 13, use 3.0 times the first-cycle value.

$$X_{pu} = 3 \cdot j19.1406$$
$$= 57.4218_{pu}$$

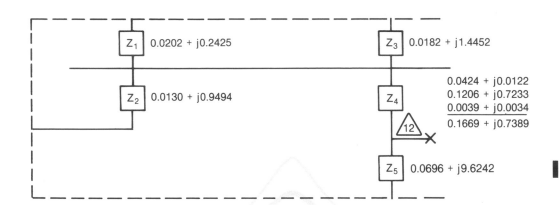

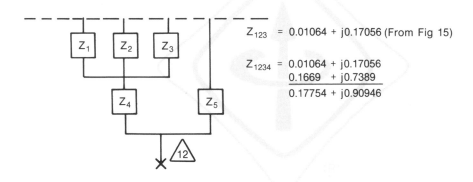

$$Z_{TOT} = \frac{(0.0696)^2(0.17754) + (0.0696)(0.17754)^2 + (0.17754)(9.6242)^2 + (0.0696)(0.90946)^2}{(0.0696 + 0.17754)^2 + (9.6242 + 0.90946)^2}$$

$$+ j\frac{(0.0696)^2(0.90946) + (0.17754)^2(9.6242) + (9.6242)^2(0.90946) + (9.6242)(0.90946)^2}{(0.0696 + 0.17754)^2 + (9.6242 + 0.90946)^2}$$

$$= \frac{16.505\ 304 + j92.507\ 047}{111.019\ 071} = 0.1487 + j0.8333$$

$$Z_{TOT} = \sqrt{R^2 + X^2} = 0.8465$$

$$X/R = 5.6$$

Z_{TOT}

0.8465_{pu}
$X/R = 5.6$

Fig 26
First-Cycle Impedance Calculations for
Fault at Start of Segment From Bus 12 to Bus 14

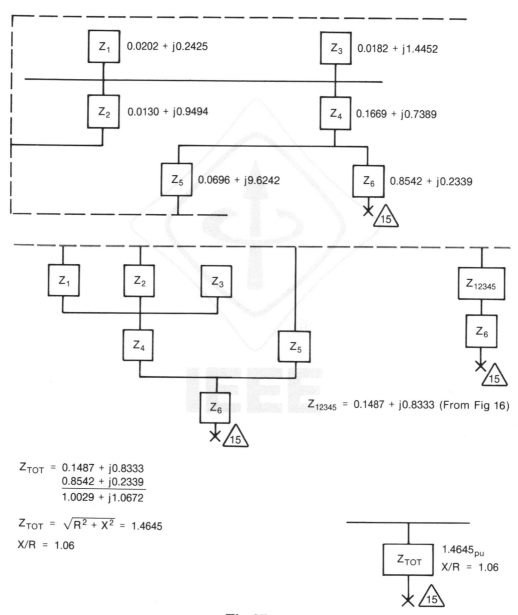

**Fig 27
First-Cycle Impedance Calculations for Fault at Bus 15**

Figure 28 shows the simplified impedance diagram with these new values for a fault at Bus 6. The interrupting rms symmetrical fault current is

$$\text{SCA} = \frac{10\ 000}{\sqrt{3} \cdot 4.160} \div 0.1782 = 7788\ \text{SCA} \qquad\qquad X/R = 15.6$$

If close and latch calculations for the medium-voltage circuit breakers are found to be necessary, then the source impedances may be refigured as follows:

(1) Utility Supply Equivalent Impedance. Same as before, which was 0.0008 + j0.0100, in per-unit.

(2) 625 kVA Generator. Same as before, which was j1.4400, in per-unit.

(3) 600 hp Synchronous Motors. Same as before, which was j2.0010, in per-unit.

(4) 100 hp Induction Motors. Use 1.2 times the usual first-cycle value.

$$X_{\text{pu}} = 1.2 \cdot \text{j}19.1406$$
$$= 22.9687_{\text{pu}}$$

Figure 29 shows the simplified impedance diagram with these new values for a fault at Bus 6. The calculated symmetrical rms current value should be multiplied by 1.6 to provide an asymmetrical rms value comparable with the circuit breaker close and latch rating.

$$\text{SCA} = \frac{10\ 000}{\sqrt{3} \cdot 4.160} \div 0.1685 \qquad\qquad X/R = 16.2$$
$$= 8237\ \text{SCA symmetrical rms}$$
$$= 8237 \cdot 1.6 = 13\ 179\ \text{asymmetrical rms}$$

The circuit breakers will be adquate if 7788 SCA symmetrical rms is not more than 80% of the circuit breaker interrupting rating (if it is more, see ANSI/IEEE Std C37.010-1979 [5] for multipliers) and 13 179 SCA asymmetrical rms is not more than the circuit breaker close and latch rating. (See 2.3.4.)

2.7 Single-Phase Short-Circuit Calculation

2.7.1 Single-Phase Taps from a Three-Phase System. Single-phase taps may be taken from a three-phase system as shown in Fig 30, where two-phase conductors and the three-phase neutral are extended from the three-phase four-wire system. A line-to-line fault on the single-phase system may be calculated using the same parameters as for three-phase fault calculations by the direct method.

The voltage driving the fault current (line-to-line voltage) equals $\sqrt{3}\ V_{\text{LN}}$, and the impedance to the fault is twice the total line-to-neutral impedance, provided $Z_{\text{L1}} = Z_{\text{L2}}$ and $Z_{\text{T1}} = Z_{\text{T2}}$.

Thus,

$$Z_{\text{FS}} = \frac{V_{\text{LN}}}{\text{SCA}_{\text{S}}}$$
$$\text{SCA}_{\text{LL}} = \frac{\sqrt{3}V_{\text{LN}}}{2\ Z_{\text{LN}}} = \frac{0.866\ V_{\text{LN}}}{Z_{\text{FS}} + Z_{\text{L1}} + Z_{\text{T1}}}$$

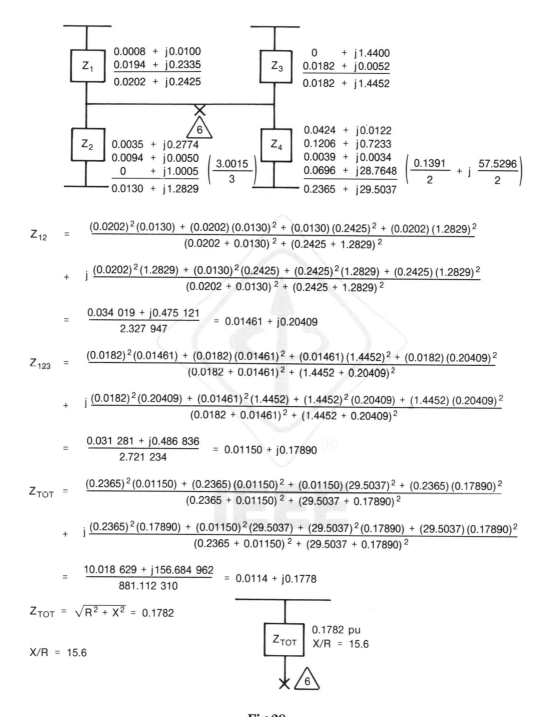

$$Z_{12} = \frac{(0.0202)^2(0.0130) + (0.0202)(0.0130)^2 + (0.0130)(0.2425)^2 + (0.0202)(1.2829)^2}{(0.0202 + 0.0130)^2 + (0.2425 + 1.2829)^2}$$

$$+ \; j \frac{(0.0202)^2(1.2829) + (0.0130)^2(0.2425) + (0.2425)^2(1.2829) + (0.2425)(1.2829)^2}{(0.0202 + 0.0130)^2 + (0.2425 + 1.2829)^2}$$

$$= \frac{0.034\ 019 + j0.475\ 121}{2.327\ 947} = 0.01461 + j0.20409$$

$$Z_{123} = \frac{(0.0182)^2(0.01461) + (0.0182)(0.01461)^2 + (0.01461)(1.4452)^2 + (0.0182)(0.20409)^2}{(0.0182 + 0.01461)^2 + (1.4452 + 0.20409)^2}$$

$$+ \; j \frac{(0.0182)^2(0.20409) + (0.01461)^2(1.4452) + (1.4452)^2(0.20409) + (1.4452)(0.20409)^2}{(0.0182 + 0.01461)^2 + (1.4452 + 0.20409)^2}$$

$$= \frac{0.031\ 281 + j0.486\ 836}{2.721\ 234} = 0.01150 + j0.17890$$

$$Z_{TOT} = \frac{(0.2365)^2(0.01150) + (0.2365)(0.01150)^2 + (0.01150)(29.5037)^2 + (0.2365)(0.17890)^2}{(0.2365 + 0.01150)^2 + (29.5037 + 0.17890)^2}$$

$$+ \; j \frac{(0.2365)^2(0.17890) + (0.01150)^2(29.5037) + (29.5037)^2(0.17890) + (29.5037)(0.17890)^2}{(0.2365 + 0.01150)^2 + (29.5037 + 0.17890)^2}$$

$$= \frac{10.018\ 629 + j156.684\ 962}{881.112\ 310} = 0.0114 + j0.1778$$

$$Z_{TOT} = \sqrt{R^2 + X^2} = 0.1782$$

$$X/R = 15.6$$

Z_{TOT} 0.1782 pu   X/R = 15.6

Fig 28
Impedance Calculations for Fault at Bus 6 at
Medium-Voltage Breaker Interrupting Time

107

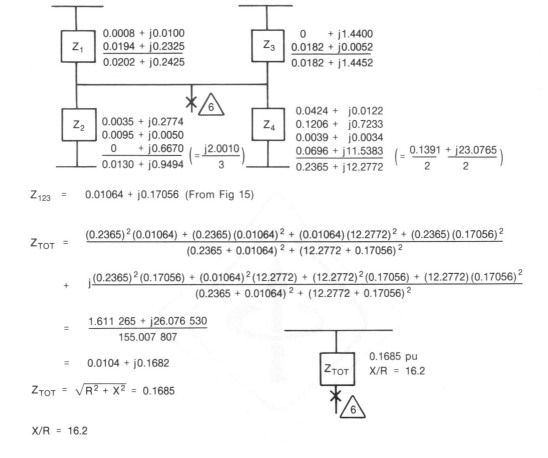

Fig 29
Impedance Calculations for Fault at Bus 6 at
Medium-Voltage Circuit Breaker Close and Latch Time

where

SCA_{LL} = line-to-line fault current of single-phase tap, in amperes

SCA_S = three-phase fault current available at transformer secondary terminals, in amperes

V_{LN} = line-to-neutral voltage of three-phase system, in volts

Z_{LN} = total line-to-neutral impedance, in ohms

Z_{FS} = line-to-neutral impedance for three-phase fault at transformer secondary terminals, in ohms

Z_{L1}, Z_{L2} = line impedances to point of tap, in ohms

Z_{T1}, Z_{T2} = line impedances from point of tap to fault point, in ohms

In the case of a line-to-neutral fault on a single-phase tap from a three-phase system (Fig 30), the first step is to determine the short-circuit current available

108

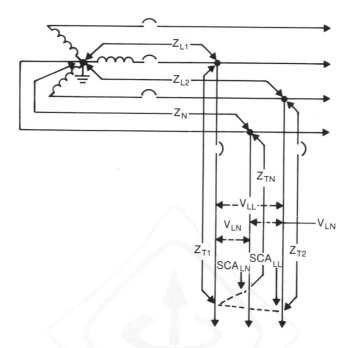

**Fig 30
Single-Phase Tap from a Three-Phase System**

for a line-to-neutral fault at the point of tap. This is difficult to determine, because the impedance of the neutral conductor depends upon the system ground(s). For practical purposes, however, the neutral impedance is often assumed to be equal to the line impedance. Therefore,

$$Z_{TN} = Z_{T1}$$
$$Z_N = Z_{L1}$$
$$SCA_{LN} = \frac{V_{LN}}{2\,Z_{LN}} = \frac{0.5\,V_{LN}}{Z_{FS} + Z_{L1} + Z_{T1}} = \frac{SCA_{LL}}{\sqrt{3}}$$

where SCA_{LL}, Z_{L1}, Z_{LN}, Z_{FS}, and Z_{T1} are as defined previously:

SCA_{LN} = single-phase line-to-neutral fault current at fault point, in amperes
Z_N = impedance of neutral conductor from transformer to point of tap, in ohms
Z_{TN} = impedance of neutral conductor from point of tap to fault point, in ohms

2.7.2 Single-Phase Tap Through a Single-Phase Transformer. In the case where a single-phase tap is made from a three-phase line through a single-phase transformer (Fig 31), the effect of the transformer should be taken into account. The transformer impedance is the critical determinant of the secondary short-circuit current, and so the actual impedance and X/R ratio of the transformer should be taken from the nameplate or supplied by the manufacturer. The typical X/R ratios in Table 14 may be used when necessary.

The short-circuit current of a line-to-line fault on the secondary circuit is found by a three-step process. The steps are as follows:

(1) Calculate the short-circuit current available at the transformer primary terminals as in 2.7.1. Also calculate the X/R ratio for this current.

(2) Calculate the short-circuit current and X/R ratio available at the secondary terminals from these formulas:

$$\%Z_E = \frac{kVA_T \cdot 100\%}{SCA_P \cdot kV_P}$$

$$\%X_E = \frac{\%Z_E \, (X_P/R_P)}{\sqrt{(X_P/R_P)^2 + 1}}$$

$$\%R_E = \%X_E/(X_P/R_P)$$

$$\%X_T = \%Z_T(X_T/R_T)/\sqrt{X_T R_T)^2 + 1}$$

$$\%R_T = \%X_T(X_T/R_T)$$

$$\%X_{tot} = \%X_E + \%X_T$$

$$\%R_{tot} = \%R_E + \%R_T$$

$$X/R = \%X_{tot}/\%R_{tot}$$

$$\%Z_{tot} = \sqrt{(\%R_{tot})^2 + (\%X_{tot})^2}$$

$$SCA_S = \frac{kVA_T}{kV_{LLS}} \cdot \frac{100}{\%Z_{tot}}$$

where

$\%Z_E, \%X_E, \%R_E, X_P/R_P$ = equivalent values for the short-circuit current available at the primary terminals

$\%Z_T, \%X_T, \%R_T, X_T/R_T$ = equivalent transformer values

$\%X_{tot}, \%R_{tot}, \%Z_{tot}$ are equivalent total values on secondary side

kVA_T = transformer rated kilovolt-amperes

SCA_P = short-circuit current available at transformer primary terminals in amperes (=SCA_{LL} or SCA_{LN} in 2.7.1)

kV_{LLS} = transformer rated line-to-line secondary kilovolts

kV_P = transformer rated line-to-line primary kilovolts

SCA_S = secondary terminal single-phase short-circuit symmetrical rms current, in amperes

(3) Calculate the short-circuit current at the end of the secondary cable from these equations:

$$Z_{FS} = \frac{kV_{LLS} \cdot 1000}{SCA_S}$$

$$SCA_{LL} = \frac{V_{LLS}}{Z_{FS} + Z_{C1} + Z_{C2}}$$

where

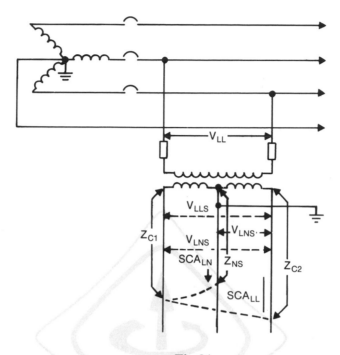

Fig 31
Single-Phase Transformer Connected Across
Single-Phase Tap from a Three-Phase Line

SCA_S and kV_{LLS} are as defined previously
SCA_{LL} = short-circuit current of line-to-line single-phase fault, in amperes
$V_{LLS} = 1000 \times kV_{LLS}$
Z_{FS} = impedance of circuit up to secondary terminals, in ohms
Z_{C1}, Z_{C2} = impedance of single-phase conductors, outgoing and return, from
  transformer secondary terminals to point of the fault, in ohms

For line-to-neutral faults on the circuit supplied by the single-phase transformer (Fig 21), the transformer impedance values ($\%X_T$ and $\%R_T$) should be altered according to the multipliers in Table 14. The formulas developed for line-to-line faults are then used with these altered values except that kVA_T, V_{LLS}, and kV_{LLS} are one-half of the values used before, and Z_{NS} is used instead of either Z_{C1} or Z_{C2}.

2.8 Equipment Evaluation and Coordination Using Calculated Short-Circuit Values. After the calculated values of first-cycle and interrupting rms symmetrical short-circuit current are determined, each piece of equipment should be evaluated to be sure its short-circuit rating is adequate. Active equipment, devices that interrupt fault currents, should have first-cycle and interrupting ratings equal to or higher than the calculated values. Passive equipment, such as busway and cable, should have short-circuit withstand ratings equal to or higher than the calculated values. For devices rated for asymmetrical values of momentary fault current, such as medium-voltage fuses, multiply the calculated symmetrical values by appropriate factors to determine the asymmetrical values.

If some equipment is not adequately rated, the system components should be changed in one of three ways. First, the pieces of equipment can be replaced with others that have higher short-circuit ratings. Second, the impedance in the system can be increased to reduce the fault current. This can be done by using additional reactors, adding extra footage to cables and busways, or increasing transformer impedances. Care should be taken to be sure excessive voltage drops are not introduced by the higher impedances. Third, some equipment can be protected by upstream current-limiting devices or other series devices. The suitability of each application involving these devices should be determined by the manufacturers of the devices in accordance with recognized testing procedures.

The calculated values of short-circuit current are also needed for checking time-current coordination between protective devices. Two devices in series should be sized and set to coordinate up to the calculated maximum short-circuit current. If a large impedance, such as a transformer or a long run of cable, is between the two devices, then even instantaneous devices may possibly be set for selective coordination. Calculated values are also necessary to determine how some devices are sized and set to protect other equipment or themselves. Overcurrent relays have thermal damage characteristics that may require them to be set for fast operation at locations where available short-circuit currents are high. Transformers also have thermal damage characteristics that require their protective devices to clear secondary faults before specified lengths of time. These and other aspects of coordination are covered more fully in Chapter 14.

2.9 References. The following publications shall be used in conjunction with this chapter.

[1] ANSI C57.12.10-1977, American National Standard Requirements for Transformers 230 000 Volts and Below, 833/958 through 8333/10 417 kVA, Single-Phase, and 750/862 through 60 000/80 000/100 000 kVA, Three-Phase.[5]

[2] ANSI C57.12.22-1980, American National Standard Requirements for Pad-Mounted, Compartmental-Type, Self-Cooled, Three-Phase Distribution Transformers with High-Voltage Bushings; High-Voltage, 34 500 GrdY/19 920 Volts and Below; 2500 kVA and Smaller.

[3] ANSI C57.12.40-1982, American National Standard Requirements for Secondary Network Transformers, Subway and Vault Types (Liquid Immersed).

[4] ANSI/IEEE Std 141-1986, IEEE Recommended Practice for Electric Power Distribution for Industrial Plants (Red Book).

---

[5] ANSI publications can be obtained from the Sales Department, American National Standards Institute, 1430 Broadway, New York, NY 10018.

[5] ANSI/IEEE C37.010-1979, IEEE Application Guide for AC High-Voltage Circuit Breakers Rated on a Symmetrical Current Basis.

[6] ANSI/IEEE C37.13-1981, IEEE Standard for Low-Voltage AC Power Circuit Breakers Used in Enclosures.

[7] ANSI/IEEE C37.14-1979, IEEE Standard for Low-Voltage DC Power Circuit Breakers Used in Enclosures.

[8] ANSI/IEEE C57.12.00-1980, IEEE Standard General Requirements for Liquid-Immersed Distribution, Power and Regulating Transformers.

[9] NEMA 201-1976, Primary Unit Substations.[6]

[10] NEMA 210-1976, Secondary Unit Substations.

[11] KAUFMANN, R. H. The Magic of I^2t. *IEEE Transactions on Industry and General Applications,* vol IGA-2, Sept/Oct 1966, pp 384–392.

2.10 Bibliography

[B1] BEEMAN, D. L., Ed. *Industrial Power Systems Handbook.* New York: McGraw-Hill, 1955, chap 2.

[B2] CLARKE, E. *Circuit Analysis of AC Power Systems: Vol 1, Symmetrical and Related Components.* New York: Wiley, 1943.

[B3] *Electrical Transmission and Distribution Reference Book.* East Pittsburgh, PA: Westinghouse Electric Corporation, 1964.

[B4] KERCHNER, R. M. and CORCORAN, G. F. *Alternating-Current Components,* 4th ed. New York: Wiley, 1960, chap 12.

[B5] MILLER, D. F., Application Guide for Shunt Capacitors on Industrial Distribution Systems at Medium-Voltage Levels. *IEEE Transactions on Industry Applications,* vol IA-12, no 5, Sept/Oct 1976, pp 444–459.

[B6] OHLSON, R. O. Procedure for Determining Maximum Short-Circuit Values in Electrical Distribution Systems. *IEEE Transactions on Industry and General Applications,* vol IGA-3, Mar/Apr 1967, pp 97–120.

[B7] REED, M. B., *Alternating Current Circuit Theory,* 2nd ed. New York: Harper and Row, 1956.

[B8] STEVENSON, W. D., JR. *Elements of Power Systems Analysis.* New York: McGraw-Hill, 1962.

[B9] WAGNER, C. F. and EVANS, R. D. *Symmetrical Components.* New York: McGraw-Hill, 1933.

---

[6] NEMA publications can be obtained from the National Electrical Manufacturers Association, 2101 L Street NW, Washington, DC 20037.

3. Instrument Transformers

3.1 Current Transformers. A current transformer transforms line current into values suitable for standard protective relays and isolates the relays from line voltages. A current transformer has two windings, designated as primary and secondary, which are insulated from each other. The various types of primary windings are covered below. The secondary is wound on an iron core. The primary winding is connected in series with the circuit carrying the line current to be measured, and the secondary winding is connected to protective devices, instruments, meters, or control devices. The secondary winding supplies a current in direct proportion and at a fixed relationship to the primary current.

3.1.1 Types. The four common types of construction are as follows.

(1) Wound Type Current Transformer. One that has a primary winding consisting of one or more turns mechanically encircling the core or cores. The primary and secondary windings are insulated from each other and from the core(s) and are assembled as an integral structure (Fig 32).

(2) Bar Type Current Transformer. One that has a fixed, insulated, straight conductor in the form of a bar, rod, or tube that is a single primary turn passing through the magnetic circuit and is assembled to the secondary, core and winding (Fig 33).

(3) Window Type Current Transformer. One that has a secondary winding insulated from and permanently assembled on the core, but has no primary winding as an integral part of the structure. Complete or partial insulation is provided for a primary winding in the window through which one or more turns of the line conductor can be threaded to provide the primary winding (Fig 34).

(4) Bushing Type Current Transformer. One that has an annular core and a secondary winding insulated from and permanently assembled on the core, but has no primary winding or insulation for a primary winding. This type of current transformer is for use with a fully insulated conductor as the primary winding.

A bushing type current transformer usually is used in equipment where the primary conductor is a component part of other apparatus.

3.1.2 Ratios. ANSI/IEEE C57.13-1978 [1][7] designates certain ratios as standard. These ratios are shown in Tables 15 and 16. Note that the standard rated secondary current in all instances is 5 A.

---

[7] The numbers in brackets correspond to those of the references listed at the end of this chapter. The numbers in brackets with the prefix "B" correspond to those of the bibliography listed at the end of this chapter

Fig 32
Wound Type Current Transformer

Fig 33
Bar Type Current Transformer

Fig 34
Window Type Current Transformer

3.1.3 Application. The general considerations for the application of current transformers follow.

(1) Continuous-Current Rating. The maximum continuous-current rating should be equal to or greater than the rating of the circuit in which the current transformer is used. The magnitude of inrush current should also be considered, particularly with respect to its affect upon meters, relays, and other connected devices.

(2) Continuous-Thermal-Current Rating Factor (RF). The specified factor by which the rated primary current of a current transformer can be multiplied to obtain the maximum primary current that can be carried continuously without exceeding the limiting temperature rise from 30 °C ambient air temperature. (When current transformers are incorporated internally as parts of larger transformers or power circuit breakers, they shall meet allowable average winding and hot spot temperatures under the specific conditions and requirements of the larger apparatus.) The rating factor (RF) shall be 1.0, 1.33, 1.5, 2.0, 3.0, or 4.0, for example, a 100:5 current transformer with an RF equal to 1.5 may be operated up to current levels of 150:7.5 (150 A primary current: 7.5 A secondary current).

(3) Thermal Short-Time Rating. This is the symmetrical rms primary current that the current transformer can carry for 1 s with the secondary winding short circuited, without exceeding the limiting temperature in any winding.

(4) Mechanical Short-Time Rating. This is the maximum current the current transformer is capable of withstanding without damage with the secondary short circuited. It is the rms value of the alternating-current component of a completely displaced (asymmetrical) primary current wave. Mechanical limits need only be checked for wound type current transformers.

(5) Nominal System Voltage. Current transformers are capable of operating continuously at 10% above rated nominal system voltage. Standard nominal system voltages for most industrial applications are 600, 2400, 4800, 8320, 13 800, and 14 400 V.

(6) Basic Impulse Insulation Levels Versus Nominal System Voltage. The values are given in Table 17.

3.1.4 Accuracy. Protective-relay performance depends on the accuracy of transformation of the current transformers, not only at load currents, but also at all fault current levels. The accuracy at high overcurrents depends on the cross section of the iron core and the number of turns in the secondary winding. The greater the cross section of the iron core, the more flux can be developed before saturation. Saturation results in a rapid increase of ratio error. The greater the number of secondary turns, the lower the flux required to force the secondary current through the relay.

ANSI/IEEE C57.13-1978 [1] designates the relaying accuracy class by use of one letter, either C or T, and the classification number. C means that the percent ratio correction can be calculated, and T means that it has been determined by test. The classification number indicates the secondary terminal voltage that the transformer will deliver to a standard burden (as listed in Table 18) at 20 times normal secondary current without exceeding a 10% ratio correction. Further-

Table 15
Current Transformer Ratings, Multiratio Bushing Type

Current Ratings (amperes)	Secondary Taps	
600:5	50:5	X2-X3
	100:5	X1-X2
	150:5	X1-X3
	200:5	X4-X5
	250:5	X3-X4
	300:5	X2-X4
	400:5	X1-X4
	450:5	X3-X5
	500:5	X2-X5
	600:5	X1-X5
1200:5	100:5	X2-X3
	200:5	X1-X2
	300:5	X1-X3
	400:5	X4-X5
	500:5	X3-X4
	600:5	X2-X4
	800:5	X1-X4
	900:5	X3-X5
	1000:5	X2-X5
	1200:5	X1-X5
2000:5	300:5	X3-X4
	400:5	X1-X2
	500:5	X4-X5
	800:5	X2-X3
	1100:5	X2-X4
	1200:5	X1-X3
	1500:5	X1-X4
	1600:5	X2-X5
	2000:5	X1-X5
3000:5	300:5	X3-X4
	500:5	X4-X5
	800:5	X3-X5
	1000:5	X1-X2
	1200:5	X2-X3
	1500:5	X2-X4
	2000:5	X2-X5
	2200:5	X1-X3
	2500:5	X1-X4
	3000:5	X1-X5
4000:5	500:5	X1-X2
	1000:5	X3-X4
	1500:5	X2-X3
	2000:5	X1-X3
	2500:5	X2-X4
	3000:5	X1-X4
	3500:5	X2-X5
	4000:5	X1-X5
5000:5	500:5	X2-X3
	1000:5	X4-X5
	1500:5	X1-X2
	2000:5	X3-X4
	2500:5	X2-X4
	3000:5	X3-X5
	3500:5	X2-X5
	4000:5	X1-X4
	5000:5	X1-X5

Table 16
Ratings for Current Transformers with One or Two Ratios

Single Ratio (amperes)	Double Ratio with Series—Parallel Primary Windings (amperes)	Double Ratio with Taps in Secondary Winding (amperes)
10:5	25 × 50:5	25/50:5
15:5	50 × 100:5	50/100:5
25:5	100 × 200:5	100/200:5
40:5	200 × 400:5	200/400:5
50:5	400 × 800:5	300/600:5
75:5	600 × 1200:5	400/800:5
100:5	1000 × 2000:5	600/1200:5
200:5	2000 × 4000:5	1000/2000:5
300:5		1500/3000:5
400:5		2000/4000:5
600:5		
800:5		
1200:5		
1500:5		
2000:5		
3000:5		
4000:5		
5000:5		
6000:5		
8000:5		
12 000:5		

more, the ratio correction should not exceed 10% at any current from 1–20 times rated current at standard burden used as the basis of relay accuracy ratings. The standard designated secondary terminal voltages are 10, 20, 50, 100, 200, 400, and 800 V. For instance, a transformer with a relaying accuracy class of C200 means that the percent ratio correction can be calculated and that it does not exceed 10% at any current from 1–20 times the rated secondary current at a standard burden of 2.0 Ω. (Maximum terminal voltage $= 20 \cdot 5 \text{ A} \cdot 2 \ \Omega = 200 \text{ V}$.)

3.1.5 Burden. Burden, in current transformer terminology, is the load connected to the secondary terminals and is expressed either as volt-amperes (VA) and power factor at a specified value of current, or as total ohms impedance with the effective resistance and reactive components. The term *burden* is used to differentiate the current transformer load from the primary circuit load. The power factor referred to is that of the burden and not of the primary circuit. For the purpose of comparing various transformers, ANSI has designated standard burdens to be used in the evaluation process (refer to Table 18).

3.1.6 Secondary Excitation Characteristics and Overcurrent Ratio Curves. Secondary excitation characteristics, as published by the manufacturers, are in the form of excitation current versus secondary voltage (Fig 35). The values are obtained either by calculation from the transformer design data and core-loss curves or by average test values from a sample of current transformers. The test is an open-circuit excitation current test on the secondary terminals, using a variable rated frequency sine wave and recording rms current versus rms voltage.

For class T transformers, typical overcurrent ratio curves are plotted between primary and secondary current over the range from 1–22 times normal primary current for all the standard burdens[8] up to the standard burden that causes a ratio correction of 50% (Fig 36).

3.1.7 Polarity. Polarity marks designate the relative instantaneous directions of currents. At the same instant of time that the primary current is entering the marked primary terminal, the corresponding secondary current is leaving the similarly marked secondary terminal, having undergone a magnitude change within the transformer (Fig 37). The $H1$ and $X1$ terminals are usually marked with white dots or with "$H1$" and "$X1$." As can be seen in Fig 37, one can consider the marked secondary conductor a continuation of the marked primary line as far as instantaneous current direction is concerned.

3.1.8 Connections. There are three ways that current transformers are usually connected on three-phase circuits: (1) wye, (2) vee or open delta, and (3) delta.

(1) Wye Connection. In the wye connection a current transformer is placed in each phase with phase relays (51) in two or three secondaries to detect phase faults. On grounded systems, a relay (51N) in the current transformer common wire detects any ground or neutral load currents. If neutral load currents are not to be detected by the 51N relay as ground-fault currents, a fourth current transformer is placed in the neutral conductor. Secondary currents are in phase with primary currents (Fig 38).

(2) Vee Connection. This current transformer connection is basically a wye with one leg omitted, using only two current transformers. Applied as shown in Fig 39, this connection detects three-phase and phase-to-phase faults. A zero sequence current transformer (window or bushing type) and relay (51GS) is required to detect ground-fault currents. All three-phase conductors and the neutral (if present) must pass through the current transformer.

(3) Delta Connection. This connection uses three current transformers with the secondaries connected in delta before the connections are made to the relays. The delta connection is used for power transformer differential-relay protection schemes where the power transformer has delta-wye connected windings. The current transformers on the delta side are connected in wye, and the current transformers on the wye side are connected in delta. Note that delta connected current transformers produce a current to the relays equal to $\sqrt{3}$ times the current transformer secondary current. This should be considered when selecting the primary ratings of current transformers that are to be delta connected. Figs 38 and 40 would be combined to complete the differential-relay connections with the ground wire connected to each relay-operating coil.

---

[8]Except B-0.9 and B-1.8.

Table 17
Basic Impulse Insulation Levels for Current Transformers

Nominal System Voltage (kV)	Maximum Line-to-Ground Voltage (kV)	BIL and Full Wave Crest (kV)
0.6	0.38	10
2.4	1.53	45
4.8	3.06	60
8.32	5.29	75
13.8	8.9	110 or 95
25.0	16.0	150 or 125
34.5	22.0	200 or 150

Table 18
Standard Burdens for Current Transformers with 5 A Secondaries*

Burden** Designation	Resistance Ohms	Inductance Millihenrys	Impedance Ohms	Volt-Amperes (at 5 A)	Power Factor
Metering Burdens					
B-0.1	0.09	0.116	0.1	2.5	0.9
B-0.2	0.18	0.232	0.2	5.0	0.9
B-0.5	0.45	0.580	0.5	12.5	0.9
B-0.9	0.81	1.04	0.9	22.5	0.9
B-1.8	1.62	2.08	1.8	45	0.9
Relaying Burdens					
B-1	0.5	2.3	1.0	25	0.5
B-2	1.0	4.6	2.0	50	0.5
B-4	2.0	9.2	4.0	100	0.5
B-8	4.0	18.4	8.0	200	0.5

*If a current transformer is rated at other than 5 A, ohmic burdens for specification and rating may be derived by multiplying the resistance and inductance of the table by $[5/(\text{ampere rating})]^2$, the VA at rated current and the PF remaining the same.

**These standard burden designations have no significance at frequencies other than 60 Hz.

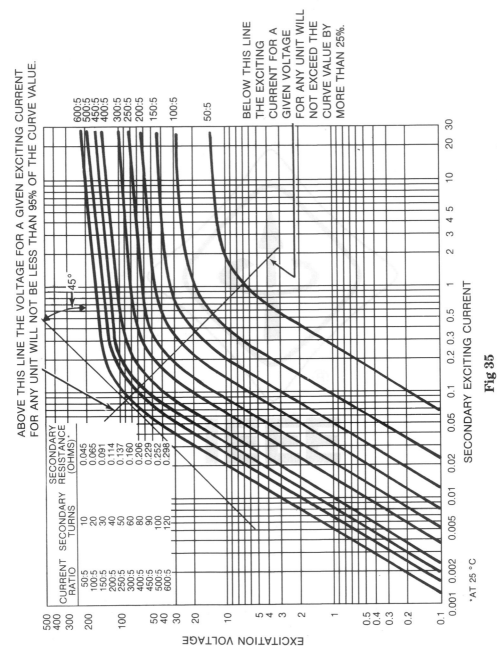

**Fig 35
Secondary Excitation Curves for Various Turn Ratios of a Specific Current Transformer**

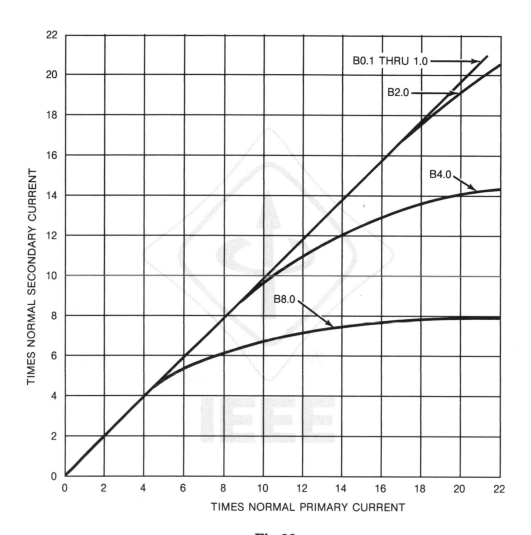

**Fig 36
Typical Overcurrent Ratio Curves for Class T Transformers
for Burdens of 0.1 Thru 8.0 Ohms (Except for B0.9 and B1.8)**

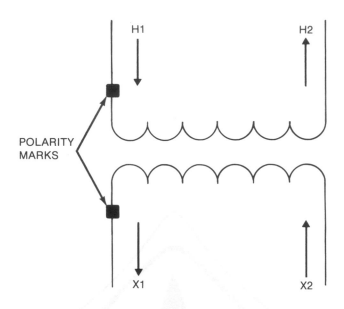

Fig 37
Current Transformer Polarity Diagram

In addition, delta-connected current transformers connected to overcurrent relays are used to provide complete phase protection for zigzag grounding transformers.

3.1.9 Examples of Accuracy Calculations. The following examples assume no residually connected relay (51N in Fig 39) is present. Calculations to account for this relay are shown in [2], Chapter 5.

Example 1: Calculation of a 600:5 Multiratio Bushing Type Current Transformer. Consider a 600:5 multiratio bushing type current transformer with excitation characteristics as shown in Fig 35. It is connected for a 600:5 ratio and to a secondary circuit containing a phase overcurrent relay with instantaneous attachment, a watthour meter, and an ammeter. The circuit contains 50 ft of No 12 wire, and the primary circuit has a capability of 24 000 A of fault current.

From instruction books for the devices and wire resistance tables, the following data are obtained.

(1) Phase relay, time unit, 4–12 A, has a burden of 2.38 VA at 4 A at 0.375 power factor (146 VA at 40 A at 0.61 power factor).

(2) Phase relay, instantaneous unit, 10–40 A, has a burden of 4.5 VA at 10 A setting (40 VA at 40 A setting at 0.20 power factor).

(3) Watthour meter has a burden of 0.77 W at 0.54 power factor at 5 A.

(4) Ammeter has a burden of 1.04 VA at 5 A at 0.95 power factor.

(5) Wire burden equals 0.08 Ω at 1.0 power factor.

(6) Current transformer secondary resistance = 0.298 Ω at 25 °C.

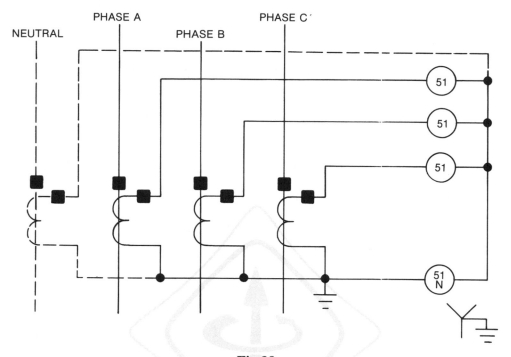

Fig 38
Wye-Connected Current Transformers

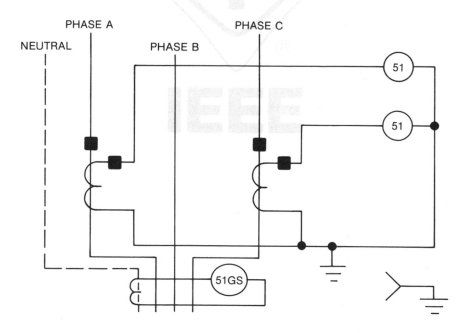

Fig 39
Vee- or Open-Delta-Connected Current Transformers

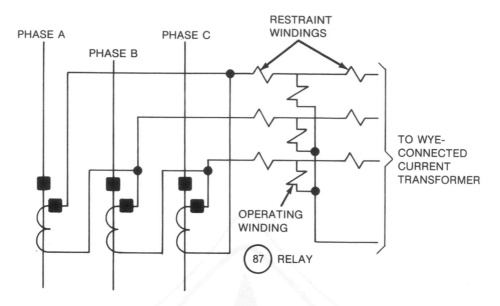

Fig 40
Delta-Connected Current Transformers

The steps for determining the performance of the transformer for this application are as follows:

(1) Determine the current transformer secondary burden.

(2) Determine the voltage necessary to operate the relay at its maximum applicable current.

(3) Determine the current transformer exciting current from Fig 35 and calculate the error.

Step (1). As stated previously, the burden is expressed in volt-amperes (VA) at a given power factor (PF), or as total ohms impedance with the effective resistance and reactance components. Since most of the devices connected to current transformers contain magnetic paths that become saturated, the burden should be calculated for the maximum specific current involved. On circuits where instantaneous elements are involved, the setting of the instantaneous element is the determining factor for establishing the maximum significant current. When instantaneous elements are not involved, the maximum current available is the determining factor.

Since the relay is equipped with an instantaneous element, it is assumed it can be set at its maximum tap setting of 40 A. In this case, primary fault current will amount to $40 \cdot 600/5$ or 4800 A. Thus, the secondary burden should be determined at a secondary current of 40 A.

Device 1, Relay, Time Unit: 146 VA at 40 A at 52.4°.
$Z = 146/(40)^2 = 0.091 \ \Omega$
$0.091 \underline{/52.4°} = 0.0555 + j0.0721$

126

Device 2, Relay, Instantaneous Unit: 40 VA at 40 A at 78.5°.
$Z = 40/(40)^2 = 0.025 \ \Omega$
$0.025\underline{/78.5°} = 0.005 + j0.0245$
Device 3, Watthour Meter: 0.77 W at 5 A at 57.3°.
$VA = W/PF = 0.77/0.54 = 1.43 \ VA$
$Z = 1.43/(5)^2 = 0.057 \ \Omega$
$0.057\underline{/57.3°} = 0.031 + j0.048$

Since a watthour meter also has an iron-core magnetic circuit, the power factor at 8 times current is approximately 0.94. Thus, at 40 A,

Z = resistance/power factor
   $= 0.031/0.94 = 0.033 \ \Omega$
$R + jX = 0.031 + j0.011$
$VA = I^2Z = (40)^2 \cdot 0.033 = 52.8 \ VA$

Device 4, Ammeter: 1.04 VA at 5 A at 18°.

$Z = 1.04/5^2 = 0.042 \ \Omega$
$0.042\underline{/18°} = 0.040 + j0.013$

Since an ammeter applies basically an air-core magnetic circuit, no saturation is present at 8 times current. Thus, at 40 A,

$VA = I^2Z = (40)^2 \cdot 0.042 = 67.2 \ VA$

Device 5, Wire: 0.08 Ω at 1.00 PF. Thus, at 40 A,

$VA = I^2R = (40)^2 \cdot 0.08 = 128 \ VA$

Device 6, Current Transformer Secondary Resistance: 0.298 Ω at 1.00 PF. Thus, at 40 A,

$VA = I^2R = (40)^2 \cdot 0.298 = 476.8 \ VA$

Totalizing for Devices 1–6 at 40 A:

Device	VA	Impedance
1	146	0.0555 + j0.0721
2	40	0.005 + j0.0245
3	52.8	0.031 + j0.011
4	67.2	0.040 + j0.013
5	128	0.08
6	476.8	0.298
	910.8	0.5095 + j0.1206

$Z_1 = 910.8/(40)^2 = 0.569 \ \Omega$
$Z_2 = 0.5095 + j0.1206 = 0.524 \ \Omega$
Note that Z_1 compares favorably with the more accurate Z_2.

Step (2). The current transformer voltage necessary to produce a secondary current of 40 A through the above burden is IZ.

$IZ_1 = 40 \cdot 0.569 = 22.8$ V

$IZ_2 = 40 \cdot 0.524 = 21.0$ V

Step (3). From Fig 35, find the secondary exciting current I_e. At 22.8 V $I_{e1} = 0.032$ A, and at 21.0 V $I_{e2} = 0.032$ A. The percent ratio error is given by

$(I_e / I_s) \cdot 100 = (0.032/40) \cdot 100$
$$= 0.08\%$$

Thus, for this application the current transformer is more than adequate.

Example 2: Using the 100:5 A Tap on the 600:5 Multiratio Current Transformer. The total requirement of apparent power in volt-amperes would change as the current transformer secondary burden is reduced. The current transformer secondary resistance is 0.065 Ω at 25 °C. The apparent power of the current transformer is

VA $= I^2R = (40)^2 \cdot 0.065 = 104$ VA

Thus, the total value is

VA $= 910.8 - (476.8 - 104) = 538$ VA

The voltage required for the 100:5 ratio is

$538/40 = 13.5$ V

and

$Z = 538/(40)^2 = 0.336$ Ω

From Fig 36 for 13.5 V $I_e = 0.5$ A and the percent error is

$(I_e / I_s) \cdot 100 = (0.5/40) \cdot 100 = 1.25\%$

Thus, again, for the assumed application, the 100:5 ratio is adequate for the relaying involved.

Example 3: Using the 100:5 Tap with 100 A Instantaneous Setting. If the 100:5 ratio were applied with a relay that would require operation at 100 A instead of 40 A, the current transformer burden is assumed to be approximately as much as that calculated for the 40 A current (0.336 Ω). The current transformer voltage necessary to produce 100 A would be approximately 33 V. As evident from Fig 35, the 100:5 ratio would not develop 33 V, except at much higher exciting currents than shown. Thus, the 100:5 current transformer would not be applicable on a circuit where operation at 100 A secondary current is necessary for correct system operation.

From these examples it can be seen that a major criterion for the selection of the current transformer ratio is the magnitude of the maximum secondary current. However, the selected transformer ratio should also be examined for short-time thermal and short-time mechanical current ratings of the current transformer against the maximum available short-circuit current.

3.1.10 Saturation. Abnormally high primary currents, residual flux, high secondary burden, or a combination of these factors will result in the creation of a high flux density in the current transformer iron core. When this density reaches or exceeds the design limits of the core, saturation results. At this point, the accuracy of the current transformer becomes very poor, and the output waveform may be distorted by harmonics. This results in the production of a secondary current lower in magnitude than would be indicated by the current transformer ratio. Saturation effects in themselves are not usually dangerous to properly designed equipment. The greatest danger is loss of protective device coordination. In other words, if current transformers saturate on a branch circuit, the branch circuit breakers may not be tripped and this would result in the operation of the upstream main circuit breaker leading to outage of an entire plant system on a fault that should have been cleared by the branch circuit breaker. To avoid or minimize saturation effects, secondary burden should be kept as low as possible. Where fault currents of more than 20 times the current transformer nameplate rating are anticipated, a different current transformer, or different current transformer ratio, or less burden may be required.

A comprehensive review of saturation and its effect on transient response of current transformers is presented in IEEE Publication 76 CH 1130-4 PWR [3].

3.1.11 Safety Precautions. An important precaution with respect to current transformers is that they should not be operated with the secondary circuit open because hazardous crest voltages may result. Any current transformer that has been subjected to open secondary circuit operation should be examined for possible damage before being placed back in service.

3.2 Voltage (Potential) Transformers.

A voltage (potential) transformer is basically a conventional transformer with primary and secondary windings on a common core. Standard voltage transformers are single-phase units designed and constructed so that the secondary voltage maintains a fixed relationship with primary voltage. The required rated primary voltage of a voltage transformer is determined by the voltage of the system to which it is to be connected and by the way in which it is to be connected. Most voltage transformers are designed to provide 120 V at the secondary terminals when nameplate-rated voltage is applied to the primary. Standard ratings are shown in Tables 19 and 20. Special ratings are available for applications involving unusual connections.

Voltage transformers are capable of continuous and accurate operation when the voltage applied across the primary is within 10% (plus or minus) of rated primary voltage.

Standard accuracy classifications of voltage transformers range from 0.3–1.2, representing percent ratio corrections to obtain a true ratio. These accuracies are high enough so that any standard transformer will be adequate for protective relaying purposes as long as it is applied within its open-air thermal and voltage limits. Standard burdens for voltage transformers with a secondary voltage of 120 V are shown in Table 21.

Thermal burden limits, as given by transformer manufacturers, should not be exceeded in normal practice since transformer accuracy and life will be adversely affected. Thermal burdens are given in volt-amperes and may be calculated by simple arithmetic addition of the volt-ampere burdens of the devices connected to the transformer secondary. If the sum is within the rated thermal burden, the transformer should perform satisfactorily over the range of voltages from 0–110% of nameplate voltage.

Table 19
Ratings and Characteristics of Voltage Transformers*

Rated Primary Voltage for Rated Voltage Line-to-Line (V)	Marked Ratio	Basic Impulse Insulation Level (kV Crest)
120 for 208Y	1:1	10
240 for 416Y	2:1	10
300 for 520Y	2.5:1	10
120 for 208Y	1:1	30
240 for 416Y	2:1	30
300 for 520Y	2.5:1	30
480 for 832Y	4:1	30
600 for 1040Y	5:1	30
2400 for 4160Y	20:1	60
4200 for 7280Y	35:1	75
4800 for 8320Y	40:1	75
7200 for 12 470Y	60:1	110 or 95
8400 for 14 560Y	70:1	110 or 95

*Voltage transformers for application with 100% of rated primary voltage across the primary winding when connected line-to-line or line-to-ground.

Table 20
Ratings and Characteristics of Voltage Transformers**

Rated Primary Voltage for Rated Voltage Line-to-Line (V)	Marked Ratio	Basic Impulse Insulation Level (kV Crest)
120 for 120Y	1:1	10
240 for 240Y	2:1	10
300 for 300Y	2.5:1	10
480 for 480Y	4:1	10
600 for 600Y	5:1	10
2400 for 2400Y	20:1	45
4800 for 4800Y	40:1	60
7200 for 7200Y	60:1	75
12 000 for 12 000Y	100:1	110 or 95
14 000 for 14 000Y	120:1	110 or 95
24 000 for 24 000Y	200:1	150 or 125
34 500 for 34 500Y	300:1	200 or 150

**Voltage transformers primarily for line-to-line service; may be applied line-to-ground or line-to-neutral at a winding voltage equal to the primary voltage rating divided by $\sqrt{3}$.

Table 21
Standard Burdens for Voltage Transformers

Designation	Characteristics on Standard Burdens*		Characteristics on 120 V Basis		
	Volt-Amperes	Power Factor	Resistance (ohms)	Inductance (henries)	Impedance (ohms)
W	12.5	0.10	115.2	3.04	1152
X	25	0.70	403.2	1.09	576
Y	75	0.85	163.2	0.268	192
Z	200	0.85	61.2	0.101	72
ZZ	400	0.85	30.6	0.0503	36
M	35	0.20	82.3	1.07	411

*These burden designations have no significance except at 60 Hz.

Voltage transformers are normally identified for polarity by marking a primary terminal $H1$ and a secondary terminal $X1$. Alternatively, these points may be identified by distinctive color markings. The standard voltage relationship provides that the instantaneous polarities of $H1$ and $X1$ are the same.

Where balanced system load and, therefore, balanced voltages are anticipated, voltage transformers are usually connected in open delta. Where line-to-neutral loading is expected, voltage transformers are more often connected wye–wye, particularly where metering is required. Many protective devices require specific delta or wye voltages; therefore, it is desirable to make a study of requirements before choosing the connection scheme. Wye–delta or delta–wye connections are occasionally used with certain special relays, but these connections are infrequent in industrial use. Where ungrounded power systems are in use, voltage transformers connected wye-broken delta are sometimes used for ground detection. When so connected, the transformers can seldom be used for any other purpose. Broken delta connections used on ungrounded systems should normally include a loading resistor in the secondary to mitigate possible ferroresonance between the system capacitance and the VT [B2], [2].

The application of fuses to voltage transformer circuits has been a subject of discussion for many years. The main purpose of a voltage transformer primary fuse is to protect the power system by de-energizing failed voltage transformers. General practice now calls for a current-limiting fuse or equivalent in the primary connection where this connection is made to an ungrounded conductor of the system. Figure 41 shows a typical voltage transformer with fuses.

Voltage transformer secondary fusing practices cannot be so clearly defined. It is usually impossible to select primary fuses that will protect the transformer from most overloads or faults in the external secondary circuit. Secondary fuses selected to interrupt at loadings below the thermal burden rating can provide such protection. Where branch circuits are tapped from voltage transformer secondaries to supply devices located at a distance from the voltage transformer, it may be desirable to fuse the branch at a reduced rating.

131

**Fig 41
Voltage Transformer**

3.3 References. The following publications shall be used in conjunction with this chapter.

[1] ANSI/IEEE C57.13-1978, IEEE Standard Requirements for Instrument Transformers.

[2] *Applied Protective Relaying*. Newark, NJ: Westinghouse Electric Corporation, 1976.

[3] IEEE Publication 76 CH 1130-4 PWR, Transient Response of Current Transformers.[9]

3.4 Bibliography

[B1] *Electric Utility Engineering Reference Book: Vol 3, Distribution Systems*. Trafford, PA: Westinghouse Electric Corporation, 1965.

[B2] FINK, D. G. and BEATY, H.W. *Standard Handbook for Electrical Engineers*. New York: McGraw-Hill, 1978.

[B3] *Instrument Transformer Burden Data*. Schenectady, NY: General Electric Company, GET-1725, 1961.

[B4] *Manual of Instrument Transformers*. Schenectady, NY: General Electric Company, GET-97, 1975.

[B5] MASON, C. R. *The Art and Science of Protective Relaying*. New York: Wiley, 1956.

---

[9] IEEE publications can be obtained from the Institute of Electrical and Electronics Engineers, Service Center, Piscataway, NJ 08854.

4. Selection and Application of Protective Relays

4.1 General Discussion of Protective System. Power systems should be designed so that protective relays operate to sense and isolate faults quickly to limit the extent and duration of service interruptions. Protective relays are important in industrial power systems because they can prevent large losses of production due to unnecessary equipment outages or unnecessary equipment damage occurring as a result of a fault or overload. Other considerations are safety, property losses, and replacements. Protective relays have been called the *watchdogs* or *silent sentinels* of a power system.

Many people are familiar with the simple, clapper type relays, but there are many other types of relays. Protective relays are classified by the variable they monitor, or by the function they perform. For instance, an overcurrent relay senses current and it operates when the current exceeds a predetermined value. Another example would be a thermal overload relay that senses the temperature of a system component, either directly and indirectly (as a function of current), or both, and operates when the temperature is above a rated value.

The application of relays is often called an "art" rather than a science because there is judgement involved in making selections. The selection of protective relays requires compromises between conflicting objectives, such as maximum protection, minimum equipment cost, reliable protection, high-speed operation, simple designs, high sensitivity to faults, insensitivity to normal load currents, selectivity in isolating only a small, faulty part in the system, and capable of operating properly for several system operating conditions. The cost of the protective relays should be balanced against the risks involved if the protection is not applied. Planning for the protection system should be considered in the power system design stages to insure that a good system can be implemented.

A separate zone of protection is normally established around each system element so that any failure occurring within that zone will cause tripping of circuit breakers in that zone to isolate the faulted zone from the rest of the system. This logically divides the system into protective zones for generators, transformers, busses, transmission lines, distribution lines or cable circuits, and motors. The primary objective is to have the faulted zone's protective devices operate first, but, if there are protective equipment failures in that zone, some form of backup protection is provided to trip out the adjacent zones surrounding the fault. The protection zones generally overlap to insure that each system element is protected. In addition to backup relay coordination, there should be selective coordination by current magnitude, time, fault type, direction, temperature, etc.

Within a zone, two forms of protection may be provided; one is considered primary, and the other one is considered backup. The primary relay(s) may be high-speed (1–3 cycles exclusive of breaker operating time), and the backup relay(s) could be the slower and less expensive time overcurrent. Generally, separate primary and backup relaying becomes more of a consideration in a zone as the system voltage or equipment power capacity increases, that is, when reliability becomes more important.

The various protective relay functions have been given identifying device function numbers, with appropriate suffix letters when necessary. These numbers are listed in ANSI/IEEE C37.2-1979 [2][10] and are used in diagrams, instruction books, and specifications. There are many possible numbers that can be used, but only the most commonly used ones are listed in Table 22A, along with the function that each number represents; for a complete list and more detailed definition of the functions, refer to the ANSI standard mentioned above.

Table 22B lists the commonly used suffix letter applied to each number denoting the circuit element being protected or the application. Each of the relay types listed in Table 22A will be discussed in some detail in this chapter, along with their operating principles and applications. In the relay identification numbering system, a circuit breaker is given number 52.

4.2 Fundamental Relay—Operating Principles and Characteristics.

All relays will operate in response to one or more electrical quantities to open or close contacts or, in the case of some solid-state relays, to trigger thyristors. Electromechanical relays have been used for years and have established a reputation for simplicity, reliability, security, low maintenance, and long life. However, in recent years, solid-state relays are being used advantageously in some applications. Some of the advantages are lower burden, improved dynamic performance characteristics, high seismic-withstand capability, and reduced panel space. Many of the protection functions can be accomplished equally well by either electromechanical or solid-state relays. The specific application should dictate which type of relay is used.

Electromechanical relays have only two different operating principles: (1) electromagnetic attraction and (2) electromagnetic induction. Electromagnetic-attraction relays operate by having either a plunger drawn by a solenoid or an armature drawn to a pole of an electromagnet. This type of relay will operate from either an ac or a dc current or voltage source and is used for instantaneous or high-speed trips.

Electromagnetic-induction relays use the principle of the induction motor, whereby torque is developed by induction into a rotor. This principle is used in a

---

[10] The numbers in brackets correspond to those of the references listed at the end of this chapter.

Table 22A
Abbreviated List of Commonly Used Relay Device Function Numbers [1]

Relay Device Function No	Protection Function
21	Distance
25	Synchronizing
27	Undervoltage
32	Directional Power
40	Loss of Excitation (Field)
46	Phase Balance (Current Balance, Negative Sequence Current)
47	Phase-Sequence Voltage (Reverse Phase Voltage)
49	Thermal (Generally Thermal Overload)
50	Instantaneous Overcurrent
51	Time-Overcurrent
59	Overvoltage
60	Voltage Balance (Between Two Circuits)
67	Directional Overcurrent
81	Frequency (Generally Underfrequency)
86	Lockout
87	Differential

Table 22B
Commonly Used Suffix Letters Applied to Relay Function Numbers

Suffix Letter	Relay Application*
A	Alarm Only
B	Bus Protection
G	Ground-Fault Protection (Relay CT in a System Neutral Circuit) or Generator Protection
GS	Ground-Fault Protection (Relay CT Is Toroidal or Ground Sensor Type)
L	Line Protection
M	Motor Protection
N	Ground Fault Protection (Relay Coil Connected in Residual CT Circuit)
T	Transformer Protection
V	Voltage

*Examples:
(1) 87T - Transformer Differential Relay
(2) 51G - Time-Overcurrent Relay Used for Ground Fault Protection
(3) 49M - Motor Winding Overload (or Over Temperature) Relay

watthour meter, where the rotor is a disk. The actuating force developed on the rotor is a result of the interaction of the electromagnetic fluxes applied and the flux produced by eddy currents that are induced in the rotor.

Induction type relays can only be used in ac applications, and the rotor is normally a disk or a cylinder. Time overcurrent and time under/overvoltage relays commonly are of the disk design, while cup (cylinder) structures are often found in high-speed overcurrent, directional relays, differential relays, and distance relays.

4.3 Distance Relay—Device No 21

4.3.1 Application. Distance relays are widely used for primary and backup protection on subtransmission and transmission lines where high-speed relaying is desired, normally on circuits having voltages above 34.5 kV.

4.3.2 Principles of Operation. *Distance relay* is a generic term applied to ohmic relays that use voltage and current inputs to provide an output signal if there is a fault within a predetermined distance from the relay location. Distance may be obtained indirectly from a signal that is proportional to the voltage-to-current ratio as a measure of impedance, from a signal that is proportional to the imaginary component of the voltage-to-current ratio as a measure of reactance, or from a signal that is proportional to the current-to-voltage ratio as a measure of the admittance to the fault. The major advantage of a distance relay is that it responds mainly to system impedance instead of the magnitude of current. Thus, the distance relay has a fixed distance reach as contrasted with overcurrent units for which the reach varies as short-circuit levels and system configurations change.

Electromechanical-distance relays utilize an induction cup type construction to achieve operating times of 1–1.5 cycles. Static distance relays have an inherent operating time of 0.25–0.5 cycles. A fixed time delay is added when required for selectivity by using an external timer relay.

Distance relay characteristics can be shown graphically in terms of two variables, R and X (or Z and Θ), where R is the resistance, X is the reactance, Z is the impedance, and Θ is the angle by which current lags voltage. The relay characteristics and the line impedance can be plotted on the same R–X diagram for analysis purposes. In examining R–X diagrams, it should be recalled that regions of positive R and X represent impedances in a defined tripping direction, while the third quadrant (negative R and X) contain impedance *behind the relay,* or in the nontripping direction. The origin of the R–X diagram is placed at the relay location.

4.3.3 Reactance Type Distance Relay. Reactance type relays measure the reactive component of system complex impedance. The generic reactance relay characteristic appears on the R–X diagram as a straight line parallel to the R-axis, as shown in Fig 42.

Operation of the generic reactance relay occurs when the reactance from the relay to the point of fault, X_2 in Fig 42, is less than or equal to the reactance X_1 Ω; reactance X_1 is the reactance setting of the relay. The relay also responds to any reactance in the negative direction; reactance relays are inherently nondirection-

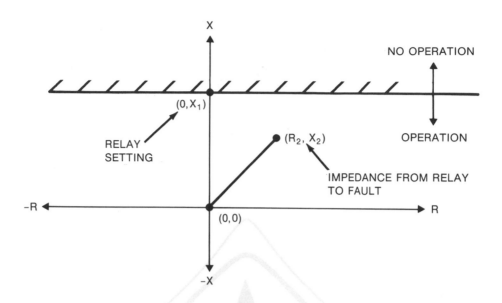

Fig 42
R-X Diagram for the Generic Reactance Relay

al. Operation is practically unaffected by arc resistance, but reactance relays may operate on load current and hence should be used in conjunction with other relays to restrict their reach along the R-axis and in the reverse, *negative reactance* direction.

4.3.4 Impedance Type Distance Relay. The *impedance type relay* measures the magnitude of the complex impedance. The generic impedance relay characteristic is a circle on the R–X diagram as shown in Fig 43.

Operation of the generic impedance relay occurs when the reactance and resistance (impedance) from the relay to the point of fault, $Z_2 = R_2 + JX_2$ in Fig 43, lies within the circle having radius $\sqrt{R_1^2 + X_1^2}$; impedance $Z_1 = R_1 + JX_1$ is the setting on the relay.

To make the impedance relay directional, the generic impedance relay must be used in conjunction with other relays to restrict their reach in the reverse direction (third quadrant of the R–X diagram).

4.3.5 Mho Type Distance Relay. An mho type relay measures complex admittance, but unlike *impedance relays,* mho relays are directional. The mho type distance relay has a circular characteristic, as shown in Fig 44.

Relay operation occurs when the impedance from the relay to the fault $(Z_2 = R_2 + JX_2)$ lies inside the mho characteristic. Since the circular characteristic falls mainly in the first quadrant of the R–X diagram, the mho relay is directional.

For special applications, a mho characteristic may be shifted in either forward or reverse directions. For loss of field applications (see 4.7), the mho relay characteristic is centered on the X-axis, with its center offset in the negative direction, as shown in Fig 45(b).

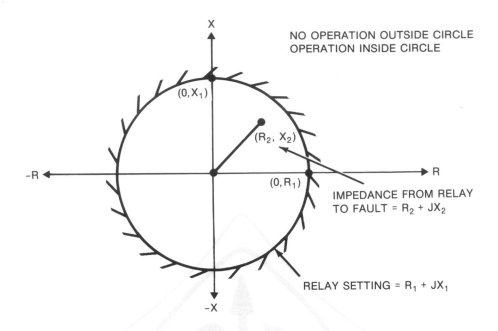

Fig 43
R-X **Diagram for the Generic Impedance Relay**

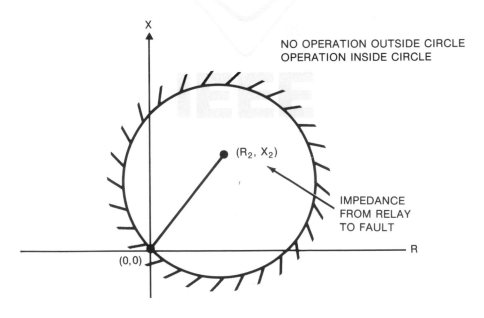

Fig 44
R-X **Diagram for the Mho Relay**

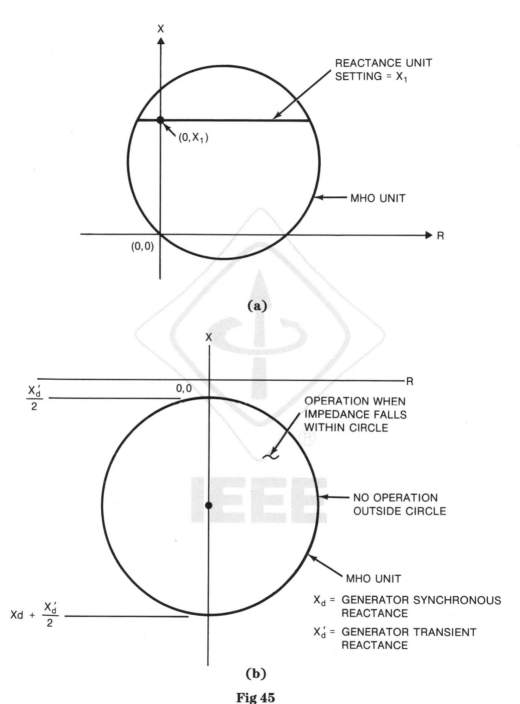

Fig 45
(a) *R-X* Diagram for the Mho-Supervised Reactance Relay
(b) *R-X* Diagram for Mho Relay Being Applied As a
Loss-of-Field Relay for a Synchronous Generator

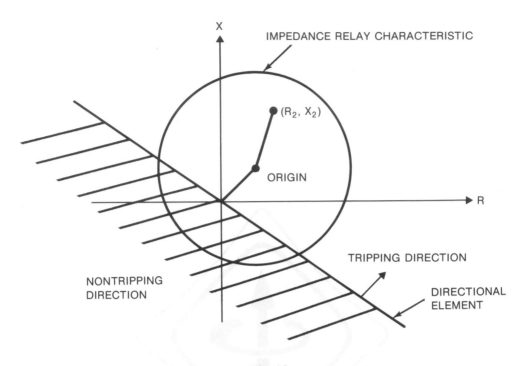

Fig 46
Directional Impedance Relay R-X Diagram

4.3.6   Mho-Supervised Reactance Relay Characteristic. For applications on short lines, a composite consisting of an ohm (reactance) unit and a mho unit in one case is often used. The mho unit, called the *starting unit,* provides a directional function. The tripping contacts of the mho unit and ohm unit are in series so that relay tripping is confined to those areas where both characteristics overlap, reference to Fig 45(a).

Breaker tripping occurs if the reactance from the relay to the point of fault is less than or equal to X_1 *and* the impedance from the relay to the fault is within the mho characteristic.

4.3.7   Directional Impedance Relay Characteristic. One type of directional impedance relay that is sometimes used is shown in Fig 46. The origin of the relay's circular characteristic is shifted into the first quadrant, and a directional element is added. Breaker tripping occurs when the impedance between the relay and the fault is within the relay's unshaded circular characteristic.

4.4   Synchronism Check and Synchronizing Relays—Device No 25

4.4.1   Application. These relays are applied when two or more sources of power are to be connected to a common bus; see Figs 47(a) and (b). The success of connecting two sources together depends very largely upon securing small and preferably diminishing differences in the voltage magnitudes, phase angles, and frequencies of the two sources at the time they are to be connected together.

140

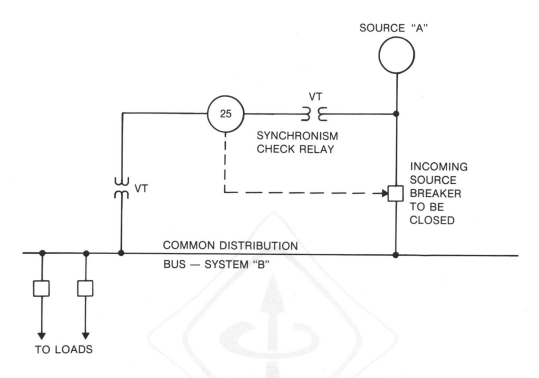

Fig 47
(a) Application of Synchronism Check Relay to Connect Two Systems Together

Synchronism check (also called syncro-verifier) relays *permit* automatic or manual closing of a circuit breaker only when the systems on each side of the circuit breaker are very nearly in synchronism; see Fig 47(a). Because the relay is relatively slow, even at low time dial settings, it is normally *not* used for synchronizing generators. Synchronism check relays are typically used at locations where the probability is small that the systems on each side of the open breaker are not in synchronism.

Synchronizing relays may be used to automatically close or supervise the closing of a circuit breaker whose function is to connect a generator to a system or to connect two separate systems; see Fig 47(b). The same synchronizing relay may be used to control more than one breaker at a station by switching the relay wiring to the potential and control circuits of the unit being synchronized. Unlike the synchronism check relay, the synchronizing relay can energize the breaker's closing coil at the precise point of synchronism.

Manual synchronization requires training, use of good judgement, experience, and the careful attention of the operator. Switchgear and generating equipment have been damaged as a result of misjudgement by an operator. Bent shafts of industrial size turbine generators occur all too often when operators close circuit breakers when the systems are too far out of phase. Therefore, manual synchro-

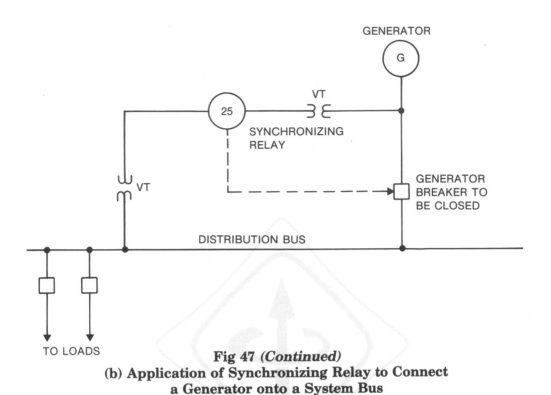

Fig 47 *(Continued)*
**(b) Application of Synchronizing Relay to Connect
a Generator onto a System Bus**

nizing is not recommended unless it is supervised with a relay that performs a synchronizing verification.

4.4.2 Synchronism Check Relay. It is recommended practice to use a synchronizing check relay as a permissive device to supervise manual closing of a breaker between two systems when they are in synchronism elsewhere. In this scheme, the operator performs the normal manual synchronizing functions, but operations are supervised by the relay so that the incoming breaker cannot be closed at the improper moment. A normally open contact of a synchronism check relay is put in series with the breaker's closing circuit.

Relay supervision of manual synchronism is accomplished in the following way. When the operator is satisfied that the systems are in synchronism, the breaker closing switch is operated to connect the two sources together. The relay monitors the voltages on each side of the breaker, and the relay's normally open contact closes; hence, the breaker is closed, after a preset time-delay period when (1) the phase angle difference between the two systems remains less than a preset value and (2) the voltage difference between the two systems remains within preset values.

The time delay is adjustable by using a time dial; the higher the time dial, the longer the delay. Normally, a low time dial is used, so the slip frequency between the two systems must be small in order for the contacts to close within the delay period. Synchronism check relays are available with fixed and adjustable closing angles. Typically, the closing angle is 20 degrees. The relay may not permit closure because the magnitude and phase angle of the voltages on each side of the open breaker are different due to the load flow in the system; however, the relay can be preset to permit closure under these conditions.

4.4.3 Synchronizing Relays. A synchronizing relay can be used for synchronizing an incoming generator to a power system in two ways: relay supervision of manual synchronization or automatic synchronization. The selection of either scheme should be based on the advantages and disadvantages of each for the system under consideration.

4.4.3.1 Relay Supervision of Manual Synchronization. It is recommended practice to include, as a minimum, a normally open contact of a synchronizing relay that performs synchronizing verification in series with the closing circuit of a generator breaker, so the relay acts like a permissive device. In this scheme, shown in Fig 47(b), the operator performs all the normal manual synchronizing functions, but cannot hold the breaker control switch closed until the relay senses that the systems are in synchronism; the generator must be in synchronism with the system *before* the breaker control switch is closed. Synchronizing relays have an adjustable closing angle that is typically set between 10 and 30 degrees. The relay closes its contacts when the phase angle between the two sources is less than this set closing angle and remains closed until this angle is exceeded. A late closing signal could permit the sources to be connected at an angle greater than the set closing angle because of the delay time introduced by energization of the breaker. A time-delay relay may be added, which will limit the time during which the operator can close the circuit breaker.

4.4.3.2 Automatic Synchronizing. Automatic synchronizing is applied to generating equipment where the station is unattended, where the element of human error must be ruled out in the start-up procedures of a generating unit, or where consistent, accurate, and rapid synchronization is preferred. The relays used are multifunction devices that sense the differences in phase angle, voltage magnitude, and frequency of the sources on both sides of an incoming generator breaker, and initiate corrective signals to adjust the generator frequency and voltage until the systems are in synchronism. The relay sends a signal to close the incoming source breaker in advance of the generator coming into synchronism with the running system, so that when the breaker is closed the systems will be exactly in synchronism.

When the generator is to be connected to the system, the appropriate synchronizing switch is selected and closed. The synchronizing equipment performs the following functions automatically:

(1) A speed matching relay element senses the frequency difference between the sources and adjusts the governor with raised or lowered signals to control the speed of the incoming generator and thereby matches the frequency of the generator with the frequency of the running system bus.

(2) A voltage matching relay element compares the running system and incoming generator voltages and provides raising or lowering signals to the excitation system of the incoming generator so that its voltage matches the running system voltage.

(3) As the phase angle approaches zero, that is, synchronism, the relay energizes the breaker's closing circuit at an advance angle determined by the relay so that when the breaker contacts close, the two systems are in exact synchronism. The synchronizing relay itself has at least two adjustable settings that should be made for correct performance. There is an adjustment to permit the relay to accommodate breaker closing time, for example, 0.05 to 0.4 s and an adjustment for setting the maximum phase angle advance, from 0 to 30–40 degrees. The advance closing angle is calculated by the following expression:

$$\Theta = 360\, s \cdot t$$

where
   Θ = advance angle, degrees
   s = slip frequency, cycles/s
   t = breaker closing time, s

For example, for systems coming into synchronism rapidly, that is, $s = 0.5$ cycles/s, the closing circuit must be energized well in advance of synchronism. If the breaker has a 0.15 s closing time, the advance angle required would be 27 degrees. If the slip frequency is much lower, then the advance angle is much smaller when the contacts close to energize the breaker. For a 0.1 cycle/s slip frequency, the closing angle is now 5.4 degrees. Thus, very precise control of the point of synchronism can be obtained.

Several different schemes for automatic synchronizing can be developed depending on the economics, reliability, and operating system requirements. By using electromagnetic relays, several relays will be required to perform all functions. However, recent solid-state relays have become available for automatic synchronizing, which provide all the functions in one unit.

4.5 Undervoltage Relays—Device No 27

4.5.1 Application. An undervoltage relay is one that is calibrated on decreasing voltage to close a set of contacts at a specified voltage. The typical uses for this relay function include:

(1) Bus Undervoltage Protection. The relay may either alarm or trip voltage sensitive loads, such as induction motors, whenever the line voltage drops below the calibrated setting. A time-delay relay is normally used to enable it to ride through momentary dips, thus preventing nuisance operation. An instantaneous undervoltage relay with its contacts connected in series with the time-delay undervoltage relay contacts may also be used to provide a fast reset time, thus preventing the inertia (overtravel) of the time-delay relay from tripping the circuit.

(2) Source Transfer Scheme. The relay is used to initiate transfer and, when desired, retransfer of a load from its normal source to a standby or emergency power source. Due to the possibility of a motor load, this relay has a time delay in order to preclude out-of-synchronism closures.

(3) Permissive Functions. An instantaneous undervoltage relay is used as a permissive device to initiate or block certain action when the voltage falls below the dropout setting.

(4) Backup Functions. A time-undervoltage relay may be used as a backup device following the failure of other devices to operate properly. For example, a long time-delay relay may be used to trip an isolated generator and its auxiliaries if the primary protective devices fail to do so.

(5) Timing Applications. A time-undervoltage relay can be used to insert a precise amount of time delay in an operating sequence. Certain protective functions such as a negative sequence voltage relay may require a time delay to prevent nuisance tripping.

4.5.2 Principles of Operation. Undervoltage relays may be either the electromechanical or solid-state design.

(1) Time-undervoltage relays of the electromechanical design are generally of the induction disk type. When the applied voltage is above the pickup, the normally closed contacts will open and will be maintained open as long as the voltage remains above the dropout voltage. When the voltage is reduced to the dropout value and below, the relay contacts begin to close. The operating time is inversely related to the applied voltage, and several ranges of time delay are available. Typical operating characteristics are shown in Fig 48. There are frequency compensated models available that maintain constant operating characteristics over a specified range of frequency variation.

(2) Time-undervoltage relays of the solid-state design provide similar inverse time operating characteristics as the electromechanical design. The relays are frequency compensated and are capable of withstanding high levels of seismic stress without malfunction.

(3) Instantaneous undervoltage relays of the electromechanical design are built in two basic types. The first is a high-speed cylinder design that has a dropout time of less than 1.5 cycles, and a dropout voltage that can be accurately set over a wide calibration range. In a three-phase design the relay may also respond to a reverse phase sequence condition. The second type consists of a dc hinged-armature telephone type relay rectified by a full-wave bridge for ac operation. A Zener diode provides for an accurate operating point, the value of which is determined by a rheostat. The dropout voltage is adjustable over a specified range and the operating time is approximately one cycle at 0 V.

(4) AC electromechanical hinged armature relays cannot generally be used as undervoltage relays, since to do so, they would have to remain continuously picked up when voltage is nominal. AC relays operated in this fashion would attempt to drop out every half-cycle (at voltage 0) and the resulting vibration could cause early relay fatigue failure.

4.5.3 Adjustments

(1) Time-Delay Relays. The dropout (tap) voltage is adjustable by means of discrete taps over a specified range. Various tap ranges are available depending on the application. The operating time is adjustable by means of a time dial setting. In the induction disk design it is continuously adjustable and in the solid-state design it is established in discrete steps. Several ranges of time delay are available. The operating time is specified at zero applied voltage when set on the maximum time dial setting.

(2) Instantaneous Relays. These relays have dropout settings that are adjust-

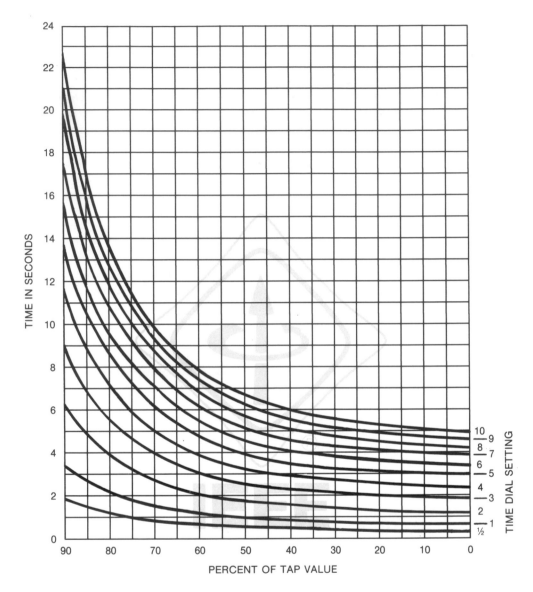

Fig 48
Typical Time-Voltage Characteristic of an Undervoltage Relay

able over a specified voltage range depending on the application. The method of adjustment will vary depending on the construction of the relay as discussed in 4.5.2. Some instantaneous solid-state relays have a fixed time delay adjustable over a short range.

4.6 Directional Power Relay—Device No 32

4.6.1 Application. As the name implies, a directional power relay functions when the real power component (watts) flow in a circuit exceeds a preset level in a specified direction. Typical uses are:

(1) Source Power Flow Control. On systems having in-plant generation operating in parallel with the utility supply, a reverse power relay sensing the incoming power from the utility can be set to detect (and alarm or trip) when the generator begins to supply power to the utility company. Plants designed to sell surplus power to the utility would not use this relay.

(2) Antimotoring of Generators. This relay is used to detect the motoring power into a generator that has not been disconnected from the system following a shutdown. See Chapter 11 for further information.

(3) Reverse Power Flow. A sensitive high-speed relay can be used to detect line-to-ground faults on the delta side of a transformer bank, shown in Figs 49 and 50, by detecting the in-phase component of the transformer magnetizing current. This would occur when another relay in the system had tripped the transformer's primary breaker and the transformer was being energized through its secondary circuit.

4.6.2 Principles of Operation. Electromechanical units are three types:

(1) A single-phase induction cup power directional unit with or without an auxiliary timing element

(2) A single-phase induction disk power directional element that provides an inherent time delay

(3) A polyphase directional unit consisting of three induction disk elements on a common vertical shaft

Maximum torque on the relay element occurs when the relay current is at a designated angle relative to relay voltage; the maximum torque angle depends on the relay design. The relay is connected to the VT's and CT's so the maximum relay torque occurs at unity power factor of the load in the designated tripping direction. Figure 51 shows the proper connection for a reverse power relay having a maximum torque angle of 90 degrees. A solid-state unit is available with adjustable angles of maximum torque.

In applying this relay, there is a need to delay relay operation to prevent undesired operations resulting from generator swings relative to the utility. The relay only need be fast enough to permit a successful reclosure from the remote end of the line.

4.7 Loss of Excitation—Device No 40

4.7.1 Application. This device is used to protect a synchronous motor or generator against loss of excitation. Common protection used for smaller motors are two types: (1) an instantaneous direct-undercurrent relay that monitors field current, or (2) a relay that monitors the relative angle between voltage and current, thereby responding to a power factor. On large synchronous motors (normally above 3000 hp) and most generators, an impedance measuring relay operating from current and voltage at the machine stator terminals is used. The

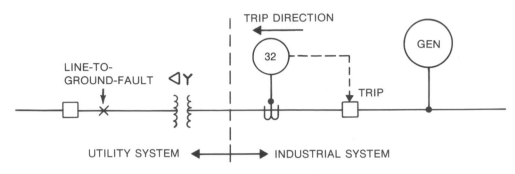

Fig 49
Application of Directional Power Relay Used to Detect Line-to-Ground Faults
(Example 1)

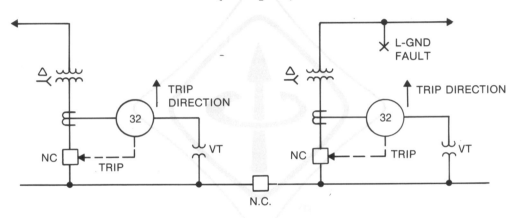

Fig 50
Application of Directional Power Relay Used to Detect Line-to-Ground Faults
(Example 2)

distance unit has either an impedance or mho characteristic; when excitation is lost, the apparent impedance seen by the relay traces a path into the relay's tripping zone. See Fig 45(b) for the type of mho unit characteristic used for protection of a generator. Additional discussion is given in Chapter 11.

4.7.2 Construction. The undercurrent relay is a dc polarized relay or a highly sensitive D'Arsonval type contact-making dc millivoltmeter. The power factor relay is solid-state. The impedance measuring relay is an induction cylinder unit having directional characteristics.

4.8 Phase Balance Current Relay—Device No 46

4.8.1 Application. Phase balance relays provide motor or generator protection against unbalanced phase currents that are caused by an open fuse or conductor in a motor branch circuit, an open fuse or conductor in the primary of a delta–wye connected transformer serving a group of motors, and, for generators, unbalanced load conditions or single-phase switching in the distribution/ transmission systems. There are two types of phase balance relays that are normally applied, current balance and negative sequence overcurrent. The current balance relay operates when the difference in the magnitude of rms currents in two phases exceeds a given percentage value. The negative sequence current relay operates on magnitude of negative sequence current, but is calibrated in terms of $I_2^2 t$, the thermal energy produced by this current. In order to set the negative sequence relay, the $I_2^2 t$ characteristic of the machine must be specified.

4.8.2 Current Balance Relay Principles of Operation

4.8.2.1 Electromagnetic Type. The electromechanical type of relay consists of two or three induction disk elements each having two current coils, as shown in Fig 52(a) and (b). These coils are connected to different phases so that a closing torque is produced on the disk that is proportional to the difference or unbalance between the currents in the two phases. The amount of unbalance current required to close the contacts may be a fixed percentage, typically 25%, or it may be a variable percentage, as shown by the operating characteristic in Fig 53.

4.8.2.2 Solid-State Type. The relay may be a part of an ac motor protective device that has several protective functions within the same unit, or as a separate protection module. The relay determines the difference between line currents, and trips when the difference exceeds a preset percentage of full-load current, or when the difference exceeds a preset ampere value (depending on the relay manufacturer). Trip time is either inversely proportional to the phase unbalance current, or a definite time.

4.8.3 Negative Sequence Relay Principles of Operation

4.8.3.1 Electromechanical Type. The relay consists of an induction disk overcurrent relay and a negative sequence filter so that the relay responds only to negative sequence currents. The relay characteristics are extremely inverse, which provide essentially a constant $I_2^2 t$ line. The typical operating characteristic is shown in Fig 54.

4.8.3.2 Solid-State Type. This relay is similar in construction to the electromechanical type except that it has a solid-state overcurrent unit. A typical connection diagram is shown in Fig 52(c). This relay provides an alarm signal at a sensitive, pretrip value of $I_2^2 t$ in addition to the trip setting. Figure 54 shows typical characteristics of this relay.

4.9 Phase Sequence Voltage Relay—Device No 47

4.9.1 Application. The phase sequence relays are used to protect ac machines from undervoltage and to prevent starting on open or reverse phase sequence. Phase sequence relays may also provide overvoltage protection. Some phase sequence relays will not give single-phase protection once the motor is running since the dynamic action of the motor will support the open phase voltage at or near its rated value. Often, Device No 47 will monitor the bus voltage, thus protecting a group of motors.

4.9.2 Principles of Operation. The electromechanical version is an induction disk polyphase voltage relay. All units normally have an undervoltage

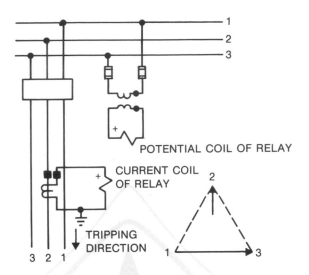

Fig 51
Directional Power Relay Connection Diagram; *I* Leads *E* by 90°

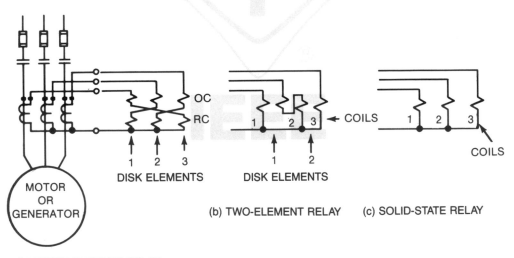

(b) TWO-ELEMENT RELAY   (c) SOLID-STATE RELAY

(a) THREE-ELEMENT RELAY

OC = OVERCURRENT COIL
RC = RESTRAINT COIL

Fig 52
Current Balance Relay Connection Diagrams

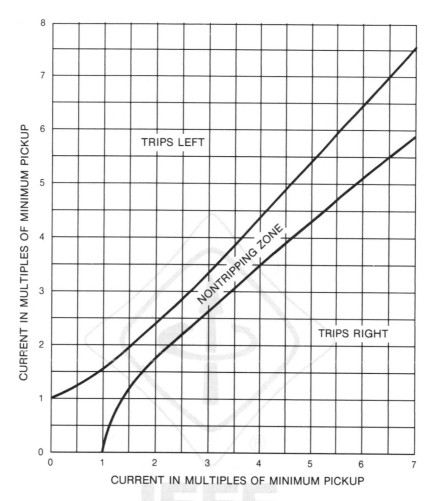

Fig 53
Typical Current Balance Relay Operating Characteristic

pickup tap setting; some units are also available with time-dial and overvoltage tap settings. Operating time is inversely related to applied voltage.

4.10 Machine or Transformer Thermal Relay—Device No 49

4.10.1 Application. Thermal relays are used to protect motors, generators, and transformers from damage due to excessive long-term overloads.

4.10.2 Principles of Operation. Three types of thermal relays are available:

(1) Replica type temperature relays operating from current transformers

(2) Bridge type relays operating from resistance temperature detectors (RTD's) located in the protected equipment

(3) A combination relay operating from a current signal biased by an RTD signal

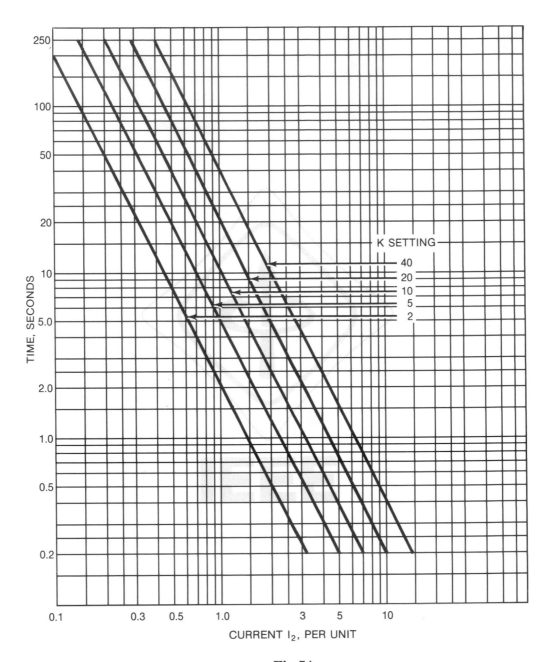

Fig 54
Typical Negative Sequence Current Relay Operating Characteristic

Replica type temperature relays usually consist of a coiled thermostatic metal spring mounted on a shaft and a heater element that monitors the output of a current transformer in a power circuit. The characteristics of the heater element and metal spring approximate the heating curve of the machine or transformer. These relays may or may not be ambient compensated. This type of thermal relay is normally applied to small (less than 1500 hp) motors where motor RTD's are not generally included in the protected motor. However, if the motor is important, a more elaborate protection scheme, discussed below, should be used. Some replica type relays have limited adjustments. Instantaneous attachments are available for some relays.

Bridge type temperature relays operate on the Wheatstone bridge principle using an RTD as a sensor to precisely measure the temperature in a certain part of a motor or generator stator. This relay may be applied to larger, more important motors, generators and transformers where it is desirable to monitor the actual temperature in the windings. Normally, six RTD's are installed and the relay is set to monitor the hottest one as determined by test. Some temperature relays have a 10 °C differential feature to prevent reenergizing equipment until the winding temperature has dropped.

Solid-state thermal relays, some using microprocessor technology, also generate the motor heating curves. In some cases, the heating curves are modified by inputs from winding RTD's; this provides very precise protection for motors and combines the best features of both relay types. The relay is available either as an individual module or combined with other functions to provide complete motor protection in a multifunction module. Refer to Chapter 9 on motor protection for more details.

4.11 Time-Overcurrent and Instantaneous Overcurrent Relays—Device Nos 50, 51, 50/51, and 51 V

4.11.1 Application. By far, the most commonly used protective relays are the time- and instantaneous overcurrent relays. They are used as both primary and backup protective devices and are applied in every protective zone in the system. Specific application information can be found in Chapters 7–14, which describe the protection of major system components.

The time-overcurrent relay is selected to give a desired time-delay tripping characteristic versus applied current, whereas instantaneous overcurrent relays are selected to provide high-speed tripping (0.5–2 cycles). The instantaneous unit may be applied by itself or mounted in the same enclosure as the time-overcurrent relay; in the latter case, it is referred to as an instantaneous trip attachment.

4.11.2 Instantaneous Overcurrent Relay. The instantaneous overcurrent relays used in industry historically have been of the electromagnetic attraction and induction cylinder type. Solid-state overcurrent relays have become available in recent years, and their characteristics are generally comparable to electromagnetic relays, except that solid-state relays can provide faster reset times.

There are two types of instantaneous relays using the principle of electromag-

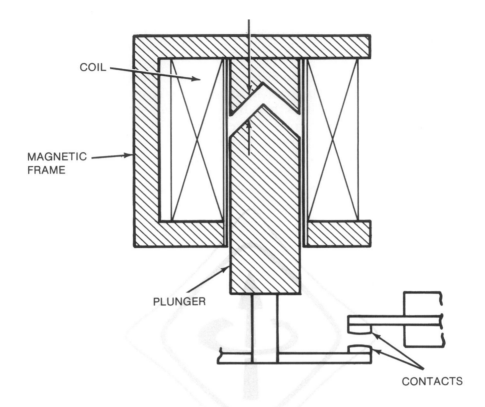

COIL

MAGNETIC
FRAME

PLUNGER

CONTACTS

Fig 55
Solenoid or Plunger Type Instantaneous Relay

netic attraction. One is called the solenoid or plunger-type, and the other is called the clapper or hinged armature type; see Figs 55 and 56. The basic elements of the solenoid type are a solenoid and a movable plunger of soft iron. The pickup current is determined by the position of the plunger in the solenoid. A calibration screw may be provided to adjust the position of the plunger. Up to three relays may be mounted in a common enclosure.

In a clapper type, a hinged armature that is held open by a restraining spring is attracted to the pole face of an electromagnet. The magnetic pull of the electromagnet is proportional to the coil current. The pickup current is that coil current required to overcome the tension of the spring, and it may be calibrated over a specified range. In some applications, the pickup current can be varied by adjusting the position of a slug in the pole. The clapper type relay is normally the one found in an induction relay case when a "50/51" (time-overcurrent with instantaneous) function is specified.

4.11.3 Induction Type Time-Delay Overcurrent Relay. The most commonly used time-delay relays for system protection use the induction disk principle. The same principle is used on alternating-current watthour meters, and, when applied to relay construction, it provides many varieties of time-current characteristics, depending on differences in electrical and mechanical

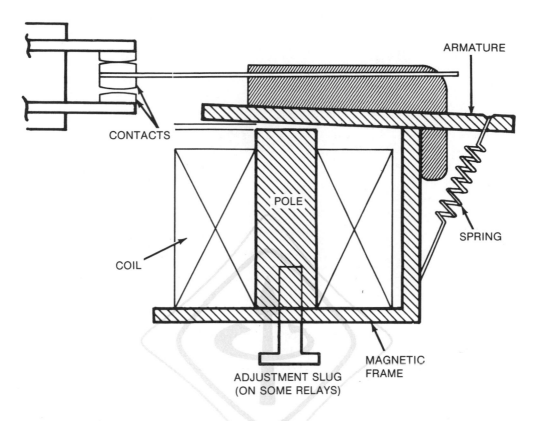

Fig 56
Clapper (or Hinged Armature) Type Instantaneous Relay

design. The principal component parts of an induction-type overcurrent relay are shown in Fig 57.

The elements of an induction disk type relay are shown in Fig 58. The disk is mounted on a rotating shaft, restrained by a spring. The moving contact is fastened to the shaft. The operating torque on the disk is produced by an electromagnet having a main (operating current) coil and a lag coil, which produce the out-of-phase magnetic flux. A damping magnet provides restraint after the disk starts to move and contributes to the desired time characteristic. There are two adjustments, the pickup current tap and the time dial. The pickup current is determined by a series of discrete taps that are furnished in several current ranges. The time dial setting determines the initial position of the moving contact when the coil current is less than the tap setting. Its setting controls the time necessary for the relay to close its contact. A relay constructed on these principles has an inverse time characteristic. This means that the relay operates slowly on small values of current above the tap setting; but, as the current increases, the time of operation decreases. If the current continues to increase, the time delay will become a constant value due to saturation of the

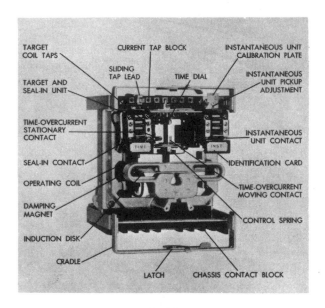

Fig 57
Induction Disk Overcurrent Relay
with Instantaneous Attachment
(Relay Removed from Drawout Case)

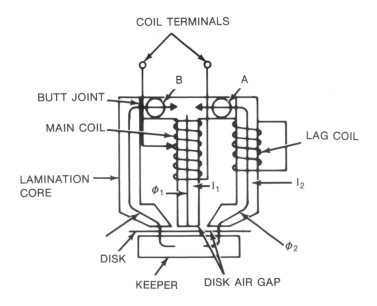

Fig 58
Elementary Induction Type Relay

OVERCURRENT RELAYS

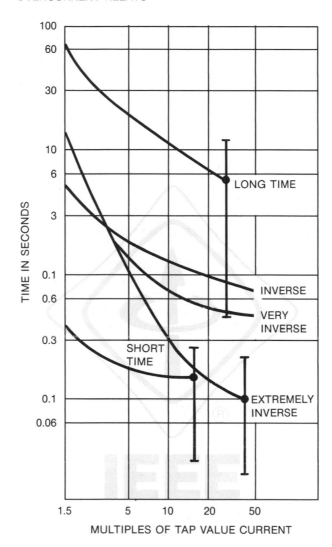

**Fig 59
Comparison of Typical Curve Shapes
for Overcurrent Relays**

electromagnet. Different time-current curves can be obtained by modifications of electromagnetic design; some of these typical curves are shown in Fig 59.

An auxiliary seal-in relay is incorporated into the relay case to lighten the current-carrying duty of the moving contact. It also operates the target indicator.

4.11.4 Solid-State Overcurrent Relay. Some new overcurrent relay design utilizes solid-state technology. Time-current curves are obtained through the use of RC or digital timing circuits. Time-current characteristic curves and tap ranges are similar to those provided in induction relays. Solid-state overcurrent relays have the same application as induction relays and are particularly useful where severe vibration specifications or seismic shock are imposed or where fast reset is required. In addition, solid-state overcurrent relays can provide faster reset times and have no significant overtravel. *Overtravel* is the term used to describe the continuation of operation, due to the inertia of the mechanical disk, after the fault current has been removed.

4.11.5 Overcurrent-Relay Types and Their Characteristic Curves. Time-overcurrent relays are available with many different current ranges and tap settings. The range of tap settings that are typically available are shown in Table 23.

The relays can be specified to have either single or double circuit closing contacts.

Figure 59 compares the basic shapes of five typical induction disk relay curves at the number 5 time dial. Manufacturer's printed time-current curves show the relay operating times for a full range of time dial settings and multiples of tap current applied to the relay.

4.11.6 Special Types of Overcurrent Relays. By adding different elements to the basic overcurrent relay, special types of overcurrent relays are derived.

4.11.6.1 Voltage-Controlled and Voltage-Restrained Overcurrent Relays. These relays are used in generator circuits for backup for external faults. When an external fault occurs, the system voltage collapses to a relatively low value; but when an overload occurs, the voltage drop is relatively small. These relays utilize the voltage to modify the time-current characteristics so that the relay will ride out permissible power swings but will give fast response when tripping off faults.

There are two variations of this relay—a voltage-controlled and a voltage-restrained overcurrent relay. In the voltage-controlled overcurrent relay, an

Table 23
Typical Tap Ranges and Settings of
Time-Overcurrent Relays

Tap Range	Tap Settings
0.5–2.5 (or 0.5–2)	0.5, 0.6, 0.8, 1.0, 1.2, 1.5, 2.0, 2.5
0.5–4	0.5, 0.6, 0.7, 0.8, 1.0, 1.2, 1.5, 2.0, 2.5, 3.0, 4.0
1.5–6 (or 2–6)	1.5, 2, 2.5, 3, 3.5, 4, 5, 6
4–16 (or 4–12)	4, 5, 6, 7, 8, 10, 12, 16
1–12	1.0, 1.2, 1.5, 2.0, 2.5, 3.0, 3.5, 4, 5, 6, 7, 8, 10, 12

auxiliary undervoltage element controls the operation of the induction disk element. When the applied voltage drops below a predetermined level, an undervoltage contact is closed in a shaded pole circuit, permitting the relay to develop torque and operate as a conventional overcurrent relay. Thus, the undervoltage unit supervises the operating coil, permitting it to operate only when the voltage is below a preset value.

The voltage restrained relay has a voltage element that provides restraining torque proportional to voltage and thus actually shifts the relay pickup current. Hence, the relay becomes more sensitive the larger the voltage drop (during faults), but is relatively insensitive at nominal voltage. The relay is set so it rides through permissible power swings at nominal voltage.

There is additional discussion in Chapter 11 on the application of this relay for generator protection.

4.12 Overvoltage Relay—Device No 59

4.12.1 Application. An overvoltage relay is one that is calibrated on increasing voltage to close a set of normally open contacts at a specified voltage. The typical uses for this relay function are as follows:

(1) Simple Overvoltage Bus Protection. The relay may either alarm or trip voltage sensitive loads or circuits in order to protect them against sustained overvoltage conditions.

(2) Ground-Fault Detection. There are two methods that are currently used. One method measures the zero-sequence voltage across the corner of a broken delta secondary of three voltage transformers (VT's) that are connected delta-grounded wye. A low pickup relay is used since there is normally no voltage across the relay. During a ground fault on a high resistance grounded neutral or ungrounded system, the applied voltage will cause the relay to operate in a predetermined time period. A resistor may be required across the relay to prevent damage to the VT due to ferroresonance.

The second method measures the actual voltage across a high ohmic value resistance that is connected between the system neutral and ground. The voltage appearing across the relay (and the resistor) during a ground fault may be several times the pickup voltage such that the relay can be set to operate in a specific time. The maximum continuous operating voltage limit of the relay should not be exceeded. In medium voltage systems, a stepdown transformer is used with windings rated 120 V or 240 V. Using a 240 V secondary of a 4160 V/240 V transformer will produce a maximum of 139 V on the secondary during a ground fault. Often, a 150 V meter is used with this relay.

4.12.2 Principles of Operation. Overvoltage relays may be either the electromechanical or solid-state design.

(1) Time-overvoltage relays of the electromechanical design are generally the induction disk type. When the applied voltage is above the pickup voltage, the normally open contacts begin to close at a rate dependent on the percentage of voltage above the pickup value. This results in a typical inverse operating characteristic. There are frequency compensated models available that maintain constant operating characteristics over a specified range of frequency variation.

The low pickup relays used in ground-fault applications have filters in the coil circuits tuned to filter out third harmonic voltages when applied to generator neutrals. This will make them less sensitive to third harmonic voltages that may be present under normal conditions.

(2) Time-overvoltage relays of the solid-state design provide similar inverse operating characteristics as the electromechanical design. The relays are frequency compensated and are capable of withstanding high levels of seismic stress without malfunction.

(3) Instantaneous overvoltage relays are typically plunger type relays where the armature is adjustable on the plunger rod to vary the pickup over the adjustment range. The operating time is approximately one cycle for voltage greater than 1.5 times the pickup setting. This type of relay may be used either as an overvoltage or an undervoltage relay simply by calibrating the relay at the desired pickup or dropout voltage, although operation in the undervoltage mode is not normally recommended. Many of these relays have a dropout-to-pickup ratio of 90–98%.

4.12.3 Adjustments

(1) Time Delay Relays. The pickup (tap) voltage is adjustable by means of discrete taps over a specified range. Various tap ranges are available depending on the application. The operating time is adjustable by means of a time dial setting. In the induction disk design, it is continuously adjustable and in the solid-state design it is determined by discrete taps. Several ranges of time delay are available and these are specified at a given percentage of tap setting, that is, 200%, and at a given time dial setting.

(2) Instantaneous Relays. These have pickup settings that are adjustable over a specified voltage range, depending on the application. The method of adjustment will depend on the construction of the relay.

4.13 Voltage Balance Relay—Device No 60. A voltage balance relay may be used to block relays or other devices that will operate incorrectly when a voltage transformer fuse blows. Two sets of voltage transformers are required that are normally connected to the same primary source during the time when blown fuse protection is required. The relay is connected as shown in Fig 60. Normally, open contacts are typically used to ring an alarm, and normally closed contacts are used to open trip circuits of relays subject to misoperation, such as voltage restrained relays and synchronizing relays. Typical examples would be to prevent incorrect operation of the loss of excitation relay (Device 40), or the backup overcurrent relay (Device 51V) in a generator circuit because of loss of voltage on a voltage transformer. The operating time is adjusted at the factory (200 ms is typical), and it is sufficiently fast to disable these relays before they have a chance to trip the circuit breaker.

A voltage balance relay may also be used to detect a small voltage unbalance in a three-phase system. The principal application of this relay is to protect three-phase motors from the damage that may be caused by single-phase operation caused by a blown fuse or open conductor. The relay can detect single-phase conditions for light loads as well as heavy loads by detecting the negative sequence component of voltage. Typically, an external timer relay is also required.

4.14 Directional Overcurrent Relay—Device Nos 67 and 87G

4.14.1 Application. Directional overcurrent relays are used to provide sensitive tripping for fault currents in one direction and nontripping for load or fault currents in the other (normal) direction. Typical applications of this relay include:

(1) Detection of uncleared faults on the utility line where fault current can be back-fed through the industrial system from in-plant generation or a second utility line, as illustrated in Fig 60(b). The fault current magnitude fed from in-plant generators and motors *to* the utility line will normally be smaller than when it is fed *from* the utility line to the plant; therefore, a sensitive relay setting is required to respond to faults on the utility system.

(2) Protection of a network of distribution lines (not radial feeders) where tripping in a given direction to provide selective operation is required. In Fig 60(a), the tripping direction is for faults within the line section that are above the pickup setting of the relay. For faults on other lines from the bus at substation A, the operating current in the relay at substation A will reverse, and the relay will not operate. Both phase and ground relays are normally used.

(3) Sensitive high-speed ground-fault protection of transformers and generators, as shown in Fig 60(c) and (d). The directional control gives the relay the characteristics of the differential protective scheme described in 4.17 and makes it particularly useful. Product type directional relays may be used for this application.

(4) Other applications where the desired objectives can be achieved by distinguishing between the direction of current flow.

In all applications, a reference or polarizing input is required to provide the directional control. The polarizing input may be either a current or voltage or both. Current polarizing input is obtained from a current transformer in the neutral grounding conductor of a generator or transformer, as shown in Fig 60(c) and (d). An auxiliary CT may be required to match the CT ratio of the operating current when the relay is connected for differential protection. The auxiliary CT is used to provide sufficient operating current during faults within the protection zone and sufficient restraint for faults outside the protection zone. Potential polarizing input for phase relays is obtained from voltage transformers, either two units connected line-to-line in open delta or three units connected line-to-ground in wye–wye, as shown in Fig 60(a). The zero-sequence potential required for polarizing ground relays is obtained using three VT's connected wye–delta, with the potential coil connected in series with the secondary windings. This is referred to as the *broken delta* or *corner of the delta* connection. Three auxiliary VT's may be used, connected as shown in Fig 60(b), or fully rated VT's may be used.

4.14.2 Principles of Operation. An electromagnetic relay consists of a conventional induction disk time-overcurrent element and an instantaneous power directional element arranged as shown in Fig 61. The various time-delay

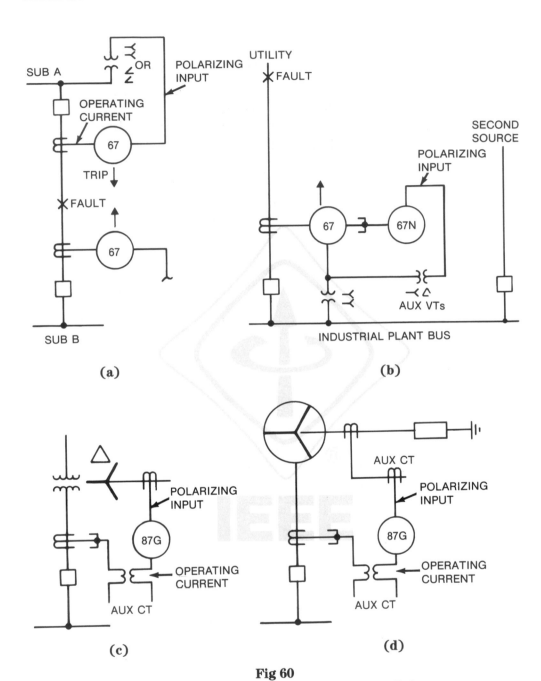

Fig 60
Typical Applications for Directional Current Relays
(a) Line Protection Using Directional Phase Relays
(b) Protection of Industrial Plant Bus from Uncleared Utility Line Faults
(c) Directional Ground-Fault Protection of Transformer
Using Product Type Relays
(d) Directional Ground-Fault Protection of Generator
Using Product Type Relays

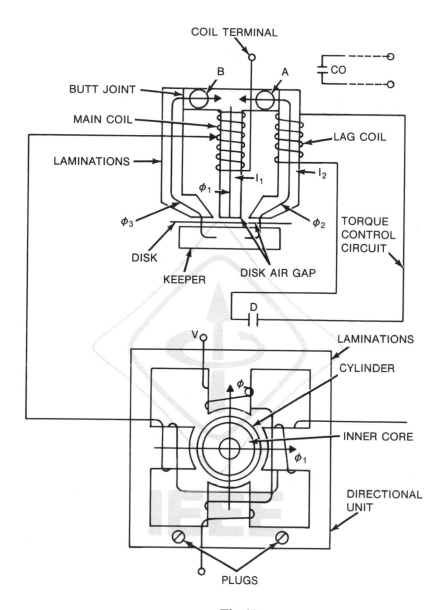

Fig 61
Directionally Controlled Overcurrent Relay

characteristics may be selected as described in 4.11.5. Operation of this element is controlled by the directional element. When the current is flowing in the tripping direction, the directional contacts close, which are in the lag coil circuit, thus enabling the overcurrent element to operate when the current exceeds its tap setting. The relay does not start to operate until the directional element operates, even when the operating current is above the pickup setting.

163

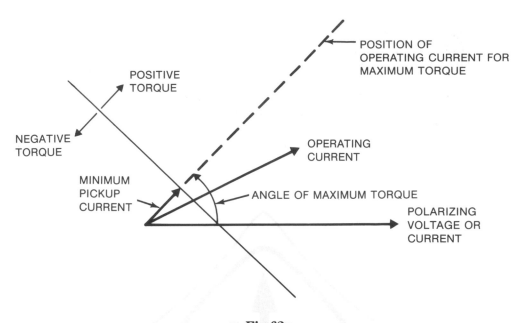

Fig 62
Characteristics of a Directional Element

The directional element has an operating current coil and a polarizing coil. The latter is energized by either voltage or current in order to determine the direction of current flow. Some units are dual polarized, having both a voltage and current coil. Maximum positive torque is produced (in tripping direction) when the angle between the operating coil current and the polarizing coil quantity is equal to the maximum torque angles of the relay. This characteristic of the directional element is shown in Fig 62.

For example, maximum torque may be produced when the operating current leads the voltage by 45 degrees. In a current-polarized relay, maximum torque may occur when the two currents are in-phase (zero phase angle). These angles of maximum torque will vary, so that manufacturer's data should be obtained. The relay is then connected to the CT and VT circuits, so that during the fault conditions being protected, the relay will produce maximum torque for tripping.

A directional instantaneous overcurrent element is optionally available for mounting within the enclosure. Its operation is supervised by the same directional element used for the time-overcurrent element.

4.14.3 Instantaneous Directional Overcurrent Relay. An electromagnetic relay has an instantaneous induction cup element that is controlled by an instantaneous power directional element, as described in 4.14.2. The operating current is adjustable over a selected range, and the directional characteristics must be identified and applied in the same manner as described in 4.14.2.

4.14.4 Product Type Directional Ground Relay. This relay operates on the product of the current in the operating coil and the voltage/current in the polarizing coil. It provides very sensitive protection in the desired direction of current flow. The operating element of an electromagnetic relay may be either an induction disk element having an adjustable time delay for selectivity, or it may be an induction cup element for high-speed operation. The directional characteristics of the relay should be determined in order to assure correct application. The application of product type relays in directional applications on network system is generally complex; this kind of relay is normally reserved for use in ground-fault protection of generators and transformers.

4.15 Frequency Relays—Device No 81

4.15.1 Application. A frequency relay is a device that functions on a predetermined value of frequency—either under or over normal system frequency or rate of change of frequency. When it is used to function on a predetermined value below nominal frequency, it is generally called an *underfrequency relay,* and when it functions on a predetermined value above nominal, it is called an *overfrequency relay.* Both functions are often included in the same case, but are utilized for different purposes.

It is highly desirable to apply underfrequency relays whenever the loads are supplied either exclusively by local generators or by a combination of local generation and utility tie; see Fig 63(a), (b), and (c). When a major generator drops off line unexpectedly in a system, supplied only by local generation [Fig 63(a)], the underfrequency relay(s) automatically open(s) plant load breaker(s) so the load matches, or is less than, the remaining generation. Otherwise, moderate-to-severe overloads on the remaining generator(s) could plunge the plant into a blackout before the operator can react. When the utility disconnects a plant system that has local generation [Fig 63(b)], the underfrequency relay(s) automatically open(s) plant load breaker(s) so the load matches, or is less than, the local generation. In the case that the utility does not disconnect the plant during an underfrequency condition, the plant's generators will begin to supply the utility system loads, causing overloading of the local generators. To prevent this from happening, an underfrequency relay may be used to supervise an extremely sensitive directional power relay; see Fig 63(c). The underfrequency and reverse power relay trip contacts are connected in series in the trip circuit of the incoming breaker so that both relays would have to operate together in order to trip the incoming breaker.

When there is an overload (the load exceeds the available generation), the generator(s) begin(s) to slow down and the frequency drops. The underfrequency relay operates at a specific (preset) frequency below nominal to trip off a predetermined amount of load so the most critical load will remain running with the available generation. More than one underfrequency relay may be used to permit a number of steps of load shedding, depending on the severity of the overload. For instance, $X\%$ of the load may be removed at 59.5 Hz, $Y\%$ of the load removed at 59 Hz, and $Z\%$ of the load removed at 58.5 Hz for a three-step load shedding scheme. The number of load shedding steps, the amount of load shed at each step, and the frequency settings for each step should be determined by a systems study. Also, it is necessary to assign a priority to each load, so the load

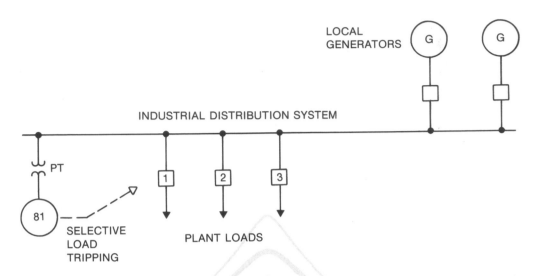

LOCAL GENERATORS

INDUSTRIAL DISTRIBUTION SYSTEM

PT

81

SELECTIVE
LOAD
TRIPPING

PLANT LOADS

Fig 63
(a) Load Shedding Scheme for System with Only Local Generation

will be removed on a priority basis, with the lowest priority loads being removed first.

Overfrequency relays are often applied to generators. These relays protect against overspeed during start-up or when the unit is suddenly separated from the system with little or no load. Relay contacts either sound an alarm or remove power input to the turbine.

4.15.2 Construction. Three types of frequency relays are available: induction disk, induction cup (cylinder), and solid-state or microprocessor relays.

The induction disk relay is subjected to two ac fluxes whose phase relationship changes with frequency to produce contact opening torque above the frequency setting and closing torque below it. A time-dial is used to adjust the initial contact separation that determines the operating time for a given applied frequency. The greater the rate at that the frequency drops, the faster the relay operates for a given time dial setting. The induction disk underfrequency relay is accurate to within 0.1–0.2 Hz and is designed for applications where high tripping speed is not essential.

The induction cup underfrequency relay is more accurate and faster than the induction disk model. The operating principle is the same as the induction disk relay. Two ac fluxes, whose phase relationship changes with frequency, produce contact closing torque in the cup unit when the frequency drops below the setting. The contacts have a fixed initial separation; the greater the rate of frequency decline, the faster the contacts close. The contacts may close in as little as 5–6 cycles after application of the underfrequency potential. Because phase shifts in the ac potential supply due to faults or fault clearing may cause incorrect operation, at least 6 cycles of intentional delay should be added before tripping. The frequency accuracy of this relay is about ±0.1 Hz.

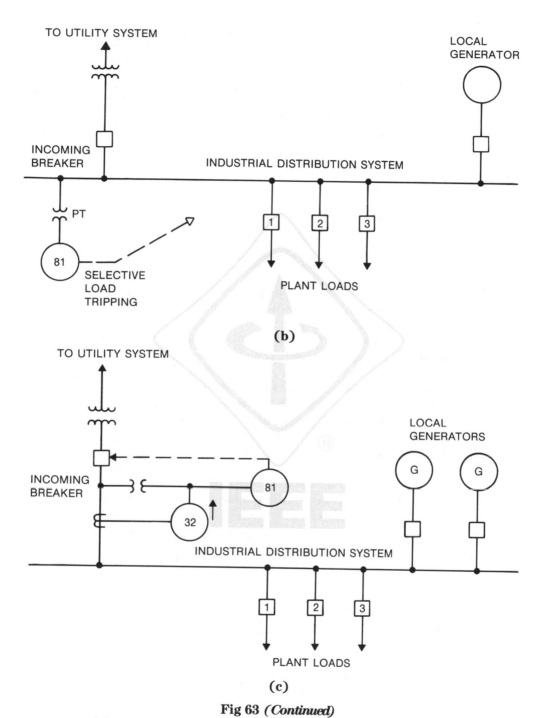

Fig 63 *(Continued)*
(b) Load Shedding Scheme for Loss of an External System
(c) Tripping Incoming Breaker for Power Outflow to External Loads

Solid-state or microprocessor relays operate on a specific frequency(s) with definite operating times. There are also solid-state relays that operate on the rate of change of frequency. To prevent misoperation, the relay waits until a predetermined number of consecutive underfrequency cycles before it operates. In addition, the relay will not operate when the voltage drops below a predetermined level.

4.15.3 Adjustment. Induction disk and induction cup underfrequency relays usually may be adjusted at 90–100% of rated frequency; overfrequency relays from 100 to 110%. Solid-state or microprocessor relays may have more than one-step operation with a much wider range of adjustment.

4.16 Lockout Relay—Device No 86. Although this relay is not a protective relay, it is included in this section because it is used widely in conjunction with relaying schemes. This relay is a high-speed, multicontact, manually or electrically reset auxiliary relay for multiplying contacts, increasing contact rating, isolating circuits, and tripping and locking out breakers. The relay is operated by differential relays, such as a transformer or bus differential, and other protective relays. The lockout relay in turn trips all the source and feeder breakers as required to isolate the fault. The relay must be reset before any of the breakers can be reenergized. The manual reset prevents reclosing the breakers before the fault is cleared.

4.17 Differential and Pilot Wire Relays—Device No 87

4.17.1 Application of Differential Relays. A differential relay operates by summing the current flowing into and out of a protected circuit zone. Normally, the current flowing into a circuit zone equals the current flowing out, in which case no differential current flows in the relay. If a fault occurs in the circuit zone, part of the current flowing in will be deflected into the fault, and the current flowing out of the circuit element will be less than the current flowing in; thus, there is a differential current that will flow in the relay. If the differential current is above a preset value, the relay will trip. Differential protection may be applied to any section of a circuit and is used extensively to detect and initiate the isolation of internal faults in large motors, generators, lines or cables, transformers, and busses. It detects these faults immediately and is not affected by overloads or faults outside the differentially protected section.

Differential relays generally will not detect turn-to-turn coil failures on motors, generators, or transformers.

Differential relays provide high-speed, sensitive, and inherently selective protection. The types of relays used are:

(1) Overcurrent differential
(2) Percentage differential
  (a) Fixed percentage (restraint) differential
  (b) Variable percentage (restraint) differential
  (c) Harmonic restraint percentage differential
(3) High-impedance differential relay
(4) Pilot wire differential.

The correct selection and application of current transformers used in differential protection schemes is critical to their proper operation. The proper matching of relay and current transformer characteristics is a prime design requirement; see Chapter 3.

4.17.2 Overcurrent Differential Relays—Device No 87. An overcurrent differential relay operates on a fixed current differential and can be easily affected by current transformer errors. It is the least expensive form of differential relaying, but it has the least sensitive settings compared to other forms, especially for detecting low-level ground faults.

Figure 64(a) shows differential protection applied on one phase (three relays —one per phase—are required). Both ends of the protection zone must be available for the installation of the current transformers (CT's). Under normal conditions, the current flowing in each current transformer secondary winding will be the same, and the differential current flowing through the relay operating winding will be zero. In case of an internal fault in the zone, the CT currents will no longer be the same and the differential current will flow through the relay operating coil, causing it to trip.

Current transformers do not always perform exactly in accordance with their ratios. This difference is caused by minor variations in manufacture, differences in secondary loadings, and differences in magnetic history. Where there is a prolonged dc component in the primary fault current, such as invariably occurs close to generators, the current transformers will not saturate equally and a substantial relay operating current can be expected to flow. Hence, if overcurrent relays are used, they have to be set so that they do not operate on the maximum error current, which can flow in the relay during an external fault. To meet this problem without sacrificing sensitivity, the percentage differential type relay is usually used.

A high-speed, economical differential overcurrent relay can be applied to motor protection for phase-to-phase and phase-to-ground faults. Figure 64(b) shows how one toroidal CT per phase will measure the phase current, producing a differential current to the relay for a fault. Fault currents as low as 2 A may be detected, and this application should follow the manufacturer's recommendations concerning CT and relay.

4.17.3 Percentage Differential Relays—Device Nos 87T, 87B, 87M, and 87G

4.17.3.1 Application. Percentage differential relays are generally used in transformer, bus, motor, or generator applications. The advantage of this relay is that it is insensitive to high currents flowing into faults outside its protection zone when CT errors are more likely to produce erroneous differential currents. However, the relay is highly sensitive to faults within its zone of protection.

There are three types of relays: fixed percentage, variable percentage, and harmonic-restraint percentage differential relays. The fixed percentage and variable percentage are used for all the applications mentioned above, but the harmonic-restraint percentage differential relay is used only for transformer applications.

The variable percentage relay is more sensitive to detect low-level faults

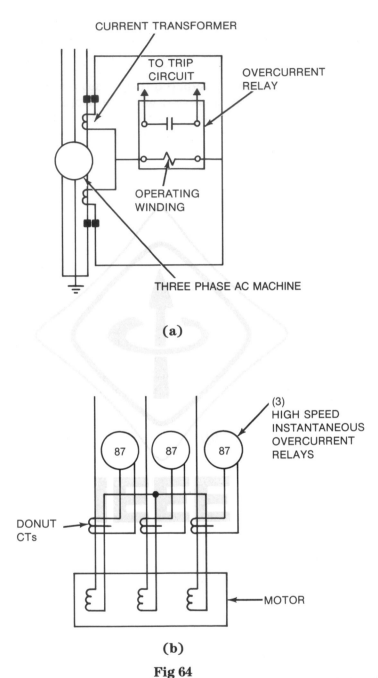

Fig 64
(a) Overcurrent Relay Used for Differential Protection, One Phase Shown
(b) Differential Protection of a Motor Using Donut CTs and Overcurrent Relays

within its protection zone and is less likely to have nuisance tripping for severe faults outside its protection zone than the fixed percentage relay. Generator protection type relays are usually the variable percentage type. Less sensitive percentage differential relays should be selected for transformer protection compared with the percentage differential relays used for bus, motor, or generator applications. This is done to prevent nuisance tripping due to magnetizing inrush current through only the power transformer's primary circuit CT's during energization. For transformers, the standard sensitivities of these relays approach current values, which may be as high as 50% of the transformer's full load current.

The harmonic restraint percentage differential relay has the feature of offering more restraint to tripping when transformer inrush currents are present compared with the standard percentage differential relay. Hence, the relay can achieve fault current sensitivities of between 20 and 30% of the transformer's rated current. Since the inrush currents are rich in harmonics, with second harmonic predominant, the second and higher order harmonic currents are used to restrain the relay on inrush.

4.17.3.2 Operating Principles of an Electromechanical Percentage Differential Relay. The relay uses the induction principle. It is connected, as shown in Fig 65. Under normal conditions, current circulates through the current transformers and relay restraining coils "R1" and "R2"; no current flows through the operating coil "O." The current in the relay restraining coils produces a restraining or contact-opening torque. An internal fault in the protected machine will unbalance the secondary currents, forcing a differential current I_o through the relay operating coil.

For a fixed percentage differential relay, the amount of differential or operating current required to overcome the restraining torque and close the relay is a fixed (constant) percentage of the restraining current. The operating characteristic for this relay is shown in Fig 66(a). As an example, for a fixed 10% percentage differential relay, the relay would trip if the operating current was greater or equal to 10% of the restraint current. In a variable percentage relay, the operating current required to operate the relay is a variable percentage of the restraining current, having a higher percentage at high fault current levels. The operating characteristic for this relay is shown in Fig 66(b).

The number of restraint elements in the relay is a function of application for which the relay is designed. A generator or motor differential relay will contain two restraint elements, where a relay intended for bus or transformer protection may have multiple elements. All relays are single-phase units, thus requiring three relays for a complete installation.

4.17.3.3 Operating Principles of a Solid-State Percentage Differential Relay. The solid-state type percentage differential relay consists of various solid-state functional circuit elements, connected together to provide a three-phase relay. The elements consist of a restraint circuit, operating circuit, sensing circuit, amplifier circuit, trip circuit, and indicating circuit. The restraint circuit senses each phase current and produces an output voltage proportional to the phase current with the largest magnitude. The operating circuit senses the

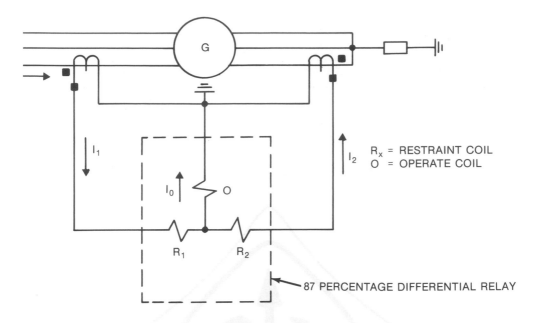

R_x = RESTRAINT COIL
O = OPERATE COIL

87 PERCENTAGE DIFFERENTIAL RELAY

Fig 65
Basic Relay Connections (One Phase) for Fixed Percentage
Restraint Differential Relay

differential current for each phase and produces an output voltage proportional to the differential current with the largest magnitude. The relay receives outputs of the restraint circuit and the operating circuits, and combines them into an output that reflects the differences of the two. This output is applied to the amplifier circuit. If the magnitude is sufficient to trigger the amplifier, its signal causes the trip circuit thyristor to gate (fire), thus tripping the circuit breaker. A signal from the amplifier also triggers the indicator circuit.

4.17.3.4 Percentage Differential Relay Taps and CT's. One type of percentage differential relay that is used for transformer protection has taps on the restraint windings that allow for different current transformer ratios to be used in the power transformer primary and secondary circuits. If the taps are not available on the relay used for transformer protection, tapped auxiliary CT's are required. Figure 67 shows a connection diagram for a fixed percentage differential relay applied to protect a transformer.

Percentage relays used for bus protection applications typically do not have taps. Normally, all CT's must have the same ratio and characteristics; however, solid-state relays are available for use with CT's that have different ratios.

4.17.3.5 Harmonic Restraint Percentage Differential Relays—Device No 87T. The connection diagram for delta–wye transformer protection is shown in Fig 68.

The electromechanical relay consists of transformer and rectifier units connected in restraint and operating configurations. The output of these units is

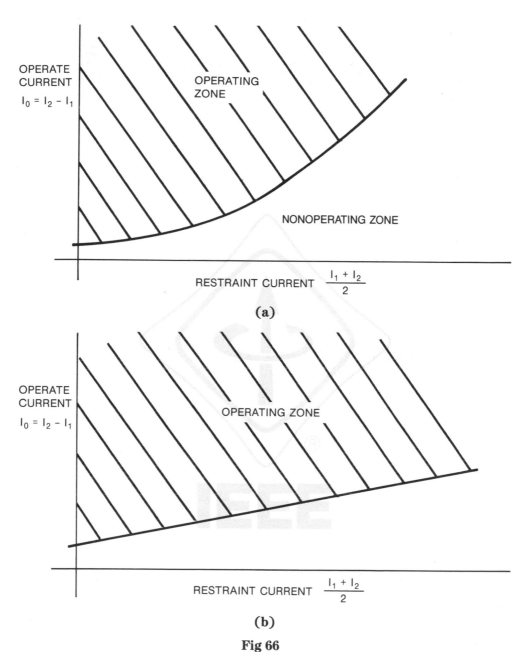

Fig 66
(a) Typical Operating Characteristic of a Fixed Percentage Differential Relay
(b) Typical Operating Characteristic of a Variable Percentage Differential Relay

applied to the differential unit, causing it to close its trip contact when the operating current exceeds the restraint current by an amount greater than the relay characteristic.

The harmonic restraint element is constructed similarly, except that filters block the fundamental frequency current to the restraint unit while directing harmonic current to the restraint unit. The operating unit of this element receives only fundamental frequency current, while harmonics are blocked. This causes the relay to be insensitive to the harmonic current that flows during

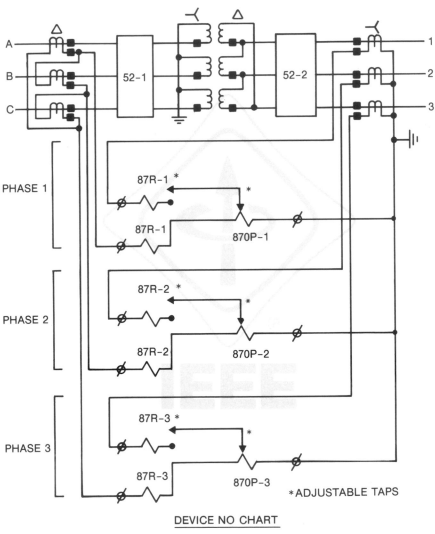

*ADJUSTABLE TAPS

DEVICE NO CHART

87	—	TRANSFORMER PERCENTAGE DIFFERENTIAL RELAY
87R	—	RESTRAINT COIL
870P	—	OPERATING COIL
52	—	POWER CIRCUIT BREAKERS

Fig 67
Fixed Percentage, Transformer Differential Relay Diagram

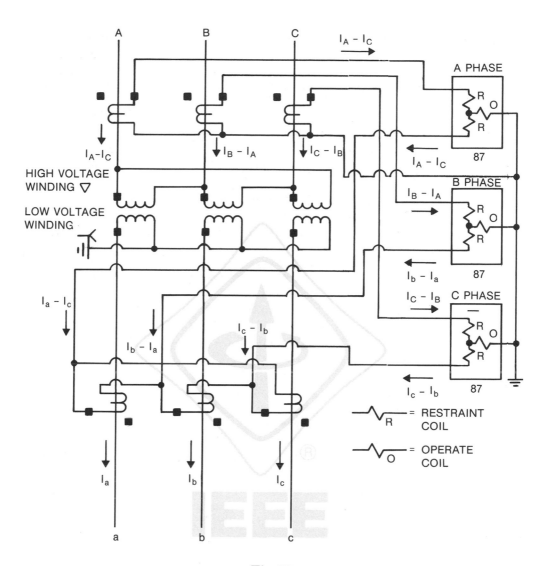

Fig 68
87 Harmonic Restraint Transformer Differential Relay Connection
Diagram for a Delta–Wye Connected Transformer

transformer energization. An instantaneous trip unit is included in the operating circuit to provide fast operating times on very high internal faults. These relays have current taps that are used to correct for mismatch between the currents from the CT's in the power transformer's primary and secondary circuits. Relay sensitivity can be adjusted by selecting an appropriate *slope* tap unless the relay has a variable percentage characteristic.

The overall relay operating time is between 1.0 and 2.0 cycles.

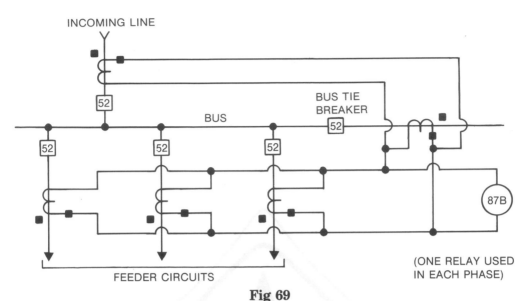

Fig 69
High-Impedance Differential Relay Used In Bus Protection, One Phase
Shown

4.17.4 High-Impedance Differential Relays—Device No 87B. The high-impedance differential relay, used primarily for bus protection, avoids the problem of unequal current transformer performance by loading transformers with a high-impedance relay unit. For faults outside the protected zone (external faults), there is a high degree of error in the current transformers in the faulted circuit. A higher than normal voltage is developed across the relay (typically having an impedance of 1700–2600 Ω), and hence, a higher voltage is impressed across the CT increasing the CT excitation current. As a result, CT error currents are forced through the equivalent magnetizing impedance of the current transformers rather than through the high impedance of the relay. However, for faults within the protected zone, the CT error currents are small, the CT magnetizing impedances appear to be almost infinite, and the current flows through the relay coil.

The connection diagram for the high-impedance type differential relay, designed to operate on bus circuits, is shown in Fig 69. The electromechanical relay consists of an overvoltage unit and an instantaneous overcurrent unit of either the plunger or clapper type. The overvoltage unit is connected across the paralleled secondaries of the current transformers. The magnitude of voltage across the relay is a function of the fault location, that is, internal or external to the protected zone, the resistance of the CT secondary leads and CT, the current transformer performance, the CT ratio, and the magnitude of fault current. The overvoltage unit will operate when the voltage exceeds the pickup setting. When there is a fault in the relay's protection zone, the CT current is directed to the high-impedance relay. A nonlinear resistor in the relay limits the voltage developed across the relay to a value that will not overstress the relay's

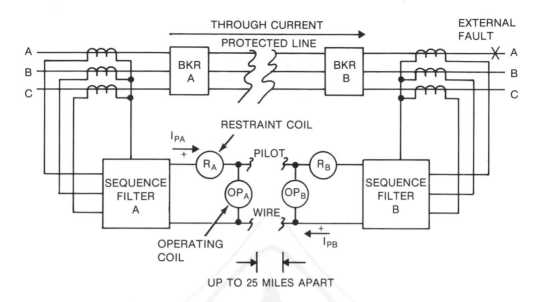

Fig 70
Simplified Connections for a Pilot Wire Differential Relay

insulation. This nonlinear element limits the voltage by permitting a large current to flow through it. The instantaneous overcurrent unit in the relay is connected in series with the nonlinear resistor and operates when the current exceeds its pickup setting. This relay provides fast tripping times of 0.5–1.5 cycles on very severe faults within its protection zone.

4.17.5 Pilot Wire Differential Relays—Device No 87L

4.17.5.1 Applications. The differential relays that have been discussed earlier cannot be used to protect long lines or cables because of the distances required to bring current transformer leads and breaker tripping wires to the relay from both ends of the line. Therefore, a special type of relay called a *pilot wire* differential relay is used to protect lines.

The pilot wire differential relay is a high-speed relay designed for phase and ground protection for two and three terminal transmission and distribution lines. They are generally applied on short lines, normally less than 25 miles long. Their operating speed is approximately 20 ms. One of the typical pilot wire relays is discussed below.

4.17.5.2 Operating Principles (Current Pilot Wire Relay). Pilot wire type differential relaying is a relay system consisting of two identical relays located at each end of a line; refer to Fig 70. The relays are connected together with a two-conductor pilot wire. The output from three individual phase current transformers is applied to a sequence network that produces a composite current that is proportional to the line current and has a polarity related to line current flow direction. Each relay contains a restraint element and an operate

177

element. The restraint elements are in series with the pilot wire, while the operate elements of each relay are in parallel with the pilot wire. Note that the circuit is basically that of the percentage restraint differential relay with the operating circuit broken into parallel circuits separated by pilot wires. This relay is available in both electromechanical and solid-state designs.

When the fault is external to the relay's protective zone, reference to Fig 70, current will flow in the pilot wire through each relay's restraint coils, but not through the relay's operating coil. If the fault is within the relay's protective zone and current is flowing into the fault from both directions, the direction of pilot wire current I_{PA} will remain the same but the direction of current I_{PB} will reverse, forcing current to flow into each relay's operating coil. If the fault current flows through breaker A only, the relay at A will still pass sufficient current through the pilot wire to operate the relay at breaker B.

The relay is designed to give complete phase and ground protection. The ground protection is derived from the residual connection of the line current transformers, and its sensitivity depends on the CT ratio. A static pilot wire type relay is available that accepts an input from a low-ratio ground sensor CT. This feature allows for a sensing low-level low ground current that is useful when applied on a resistance grounded system.

Compliance with the manufacturer's application instructions is necessary to provide a total system of protection.

4.17.5.3 Pilot Wire Specifications. To insure the pilot relay system is reliable, there should be detailed specifications on the construction and installation of the pilot wires. Construction requirements should include wire size, insulation, twist length, shielding, and jacketing. Installation instructions should include: overhead or underground (if overhead include spacing below any power lines), splicing, where to ground the shield, protection of pilot wire shield against excessive currents, protection of cable and relays from overvoltages, and how to treat spare wires that are in the same conductor bundle as the pilot wires.

There is much controversy on what the exact specifications should be, since there are many variations that have worked successfully.

The following is an example specification that has proved satisfactory at a number of industrial sites, for both overhead and underground applications.

4.17.5.3.1 Pilot Wire Construction Requirements

(1) Wire size—6 pairs of AWG No 19 solid annealed copper conductors.[11]

(2) Insulation—Each conductor shall be insulated with 0.015" polyethylene. The conductor shall be bound with a nonhydroscopic binder tape, over which shall be extruded a high dielectric polyethylene inner jacket. The jacket shall have a nominal thickness of 0.045".

(3) Twist length—The pairs shall be twisted with a minimum twist length of 7". The twist length of each pair shall be different.

(4) Shield—Over the inner jacket shall be applied a spiral wound shield tape with a minimum overlap of 20% of the tape width. The tape shall be 0.005" thick copper or 0.008" thick aluminum.

---

[11] Maximum pilot wire loop resistance should be less than 2000 Ω and maximum capacitance is 1.5 μF for a two-terminal system.

(5) Overall jacket—An overall jacket shall be applied over the shield tape. The thickness of the jacket over the shield tape shall be 0.0060″. The overall jacket shall encompass the cable and messenger in a "figure 8" configuration. The jacket compound shall be applied so that it completely floods the interstices of the messenger. The outer jacket material shall be pigmented for protection from radiation and may be either polyvinyl chloride or polyethylene.

(6) Messenger—The messenger shall be 1/4″, 7-strand, extra strength steel (minimum breaking strength—6650 lbs) with Class A zinc coating. Messenger shall comply with ASTM Specification A-475-69 [3].

(7) The cable shall comply with the requirements of ICEA Standard S-61-402 [4] for Type D control cable for pilot wire duty.

(8) DC high potential test—In addition to the testing required by the reference specifications, the completed cable shall be subjected to a 20 000 V dc high potential test between conductors and shield and between shield and the messenger.

4.17.5.3.2 Installation Instructions

(1) Ground the messenger wire at each pole.

(2) Ground the shield at each terminal.

   (a) Connect shield to station ground

   (b) Shield should be continuous end-to-end

   (c) Protect shield from possible transient overcurrent during system faults with parallel power conductor. Since there may be potential differences in grounds during fault conditions, a conductor should be connected in parallel with the shield to carry the current that may flow between grounds rather than permit the current to flow through the shield.

(3) Splices (if required) shall be in-line using Scotchmold Epoxy splice kits or equal.

(4) Insulating transformers shall not have a midpoint grounded on a high voltage side.

(5) Neutralizing transformers shall not be used.

(6) Carbon gaps shall not be used.

(7) Mutual drainage reactors or gas discharge, or both, tubes shall not be used.

(8) If any pair in cable is used for purposes other than pilot wire, it must be:

   (a) Twisted pair with twist length different from pilot wire pair

   (b) Ungrounded

   (c) Terminated at each end in 10 kV (minimum) insulating transformers or equivalent.

(9) Unused pairs in cable shall be short-circuited and grounded at only one end.

(10) Terminal blocks are acceptable *only* if they are mounted on suitable standoff insulators and *only* if suitable clearances to ground and to shields are maintained.

4.17.5.4 Application Guidelines.

Other devices are required and applied with each relay terminal to provide a complete system. These include a milliammeter, switch and auxiliary transformer for a testing; insulating transformer for pilot wire isolation; and pilot wire supervision relays for detection and alarm of pilot wire problems. An optional neutralizing reactor is applied where the difference between station ground and remote ground can exceed 600 V rms during power system faults. This rise in ground potential appears across the neutralizing transformer inserted in the pilot wire. In addition, an optional drainage reactor is applied to drain off longitudinally induced voltages that may occur by lightning surges (not a direct stroke) or the parallel association of the pilot wire with faulted power circuits. By forcing equal current flow from the two wires into ground, it minimizes wire-to-wire voltages.

4.17.5.5 Taps. Current taps are provided to give adjustable minimum trip selection and sequence filter circuit taps that permit phase and ground sensitivity selection.

14.18 References. The following publications shall be used in conjunction with this chapter.

[1] ANSI/IEEE Std 141-1986, IEEE Recommended Practice for Electric Power Distribution for Industrial Plants.

[2] ANSI/IEEE C37.2-1979, IEEE Standard Electrical Power System Device Function Numbers.

[3] ASTM A-475-69, Specification for Zinc-Coded Steel Wire Strand.[12]

[4] ICEA S-61-402, Thermal Plastic-Insulated Wiring Cable for the Transmission and Distribution of Electrical Energy.[13]

[5] *Applied Protective Relaying,* Westinghouse Electric Corporation, Relay Instrument Division, 1976.

[6] *The Art of Protective Relaying,* General Electric Company, Philadelphia, PA, 1957, Bulletin GET-1768.

[7] MASON, C. R. *The Art and Science of Protective Relaying.* New York: Wiley, 1956.

[8] *Power System Protection—Use of R-X Diagram,* General Electric Company, Philadelphia, 1966, Bulletin GET-2230B.

14.19 Bibliography

[B1] ANSI/IEEE Std 241-1983, IEEE Recommended Practice for Electric Power Systems in Commercial Buildings.

[B2] ANSI/IEEE C37.90-1978, IEEE Relays and Relay Systems Associated with Electric Power Apparatus.

---

[12] ASTM publications are available from the Sales Department, American Society of Testing and Materials, 1916 Race St., Philadelphia, PA 19103.

[13] ICEA publications are available from the Insulated Cable Engineers Association, P.O. Box 411, South Yarmouth, MA 02664.

[B3] ANSI/IEEE C37.91-1985, IEEE Guide for Protective Relay Applications to Power Transformers.

[B4] ANSI/IEEE C37.95-1973, IEEE Guide for Protective Relaying of Utility-Consumer Interconnections.

[B5] ANSI/IEEE C37.96-1976, IEEE Guide for AC Motor Protection.

[B6] ANSI/IEEE C37.97-1979, IEEE Guide for Protective Relay Applications to Power System Buses.

[B7] ANSI/IEEE C37.99-1980, IEEE Guide for Protection of Shunt Capacitor Banks.

[B8] ANSI/IEEE C37.103 (in preparation: P815), IEEE Guide for Differential and Polarizing Relay Circuit Testing.

[B9] ANSI/IEEE C57.13-1978, IEEE Standard Requirements for Instrument Transformers.

[B10] ANSI/IEEE C57.105-1978, IEEE Guide for Application of Transformer Connections in Three-Phase Distribution Systems.

[B11] ANSI/NFPA 70-1984, National Electrical Code.

[B12] ANSI/UL 1053-1976, Safety Standard for Ground-Fault Sensing and Relaying Equipment.

[B13] ANDERSON, P. M. *Analysis of Faulted Power Systems.* Iowa State University Press: Ames, IA, 1973.

[B14] ANDRICHAK, J. G. Polarizing Sources for Directional Ground Relays. *Presented to the Electric Council of New England,* South Portland, ME, Oct 11, 1973.

[B15] *AIEE Committee Report.* Bibliography of Industrial System Coordination and Protection Literature. *IEEE Transactions on Industry Applications,* vol IA-82, Mar 1963, pp 1–2.

[B16] BARKLE, J. E. and GLASSBURN, W. E. Protection of Generators Against Unbalanced Currents. *AIEE Transactions,* vol 72, pt III, 1953, pp 282–286.

[B17] BERDY, J. Loss of Excitation Protection for Modern Synchronous Generators. *IEEE Transactions on Power Apparatus and Systems,* vol PAS-94, 1975, pp 1457–1463, 1481–1483.

[B18] BLACKBURN, J. L. Ground Relay Polarization. *AIEE Transactions on Power Apparatus and Systems,* vol PAS-71, pt III, Dec 1952, pp 1088–1095.

[B19] BLACKBURN, J. L. Voltage Induction in Parallel Transmission Circuits. *IEEE Transactions on Power Apparatus and Systems,* vol PAS-81, pt III, 1962, pp 921-929.

[B20] BOOTHMAN, D. R., ELGAR, E. C., REHDER, R. H. and WOODDALL, R.

J. Thermal Tracking—A Rational Approach to Motor Protection. *IEEE Transactions on Power Apparatus and Systems,* vol PAS-93, no 5, Sept/Oct 1974, pp 1335-1344.

[B21] BOTTRELL, G. W. and YU, L.Y. Motor Behavior Through Power System Disturbances. *IEEE Transactions on Industry Applications,* vol IA-16, no 5, Sept/Oct 1980, pp 600–605.

[B22] BRIGHTMAN, F. P. More About Setting Industrial Relays. *AIEE Transactions on Power Apparatus and Systems,* vol PAS–73, pt III–A, 1954, pp 397–406.

[B23] BRIGHTMAN, F. P. Selecting AC Overcurrent Protective Device Settings for Industrial Plants. *AIEE Transactions on Industry Applications,* vol IA-71, pt II, Sept 1952, pp 203–211.

[B24] BURKE, J. J., KOCH, R. F., and POWELL, L. J. A Comparison of Static and Electromechanical Time-Overcurrent Relay Characteristics, Application and Testing. *Paper presented at the Pennsylvania Electric Association Relay Committee Meeting,* Philadelphia, PA, Feb 14, 1975.

[B25] CALHOUN, H. J. and HOINOWSKI, E. R. Solid-State CO Past and Present. *Presented at the Pennsylvania Electric Association Relay Committee Meeting,* Hershey, PA, May 27, 1976.

[B26] CHUMAKOV, W. V. and DOWNS, C. L. Solid-State Relaying Applied to High-Voltage Distribution in Mining. *Conference Record of 1979 IEEE IAS Mining Industry Technical Conference,* Pittsburgh, PA, June 7–8, 1979, pp 136–153.

[B27] CLARK, H. K. Load Shedding for Industrial Plants. *Conference Record of the 1973 IEEE IAS Annual Meeting,* Oct 8–11, 1973, pp 725–733.

[B28] COFFMAN, A. W. Transient Voltage Phenomena in Relay and Control Circuits. *Presented at the Pennsylvania Electric Association Relay Committee Meeting,* Feb 25, 1977, Reading, PA, pp 72–79.

[B29] COOKE, J. L. Fundamentals of Symmetrical Components. *Paper presented to 19th Annual Conference for Protective Relay Engineers,* Texas A&M University, Apr 25–27, 1966.

[B30] DALZIEL, C. F. and STEINBACK, E. W. Underfrequency Protection of Power Systems for System Relief—Load Shedding—System Splitting. *AIEE Transactions on Power Apparatus and Systems,* vol PAS–78, pt III, Dec 1959, pp 1227–1238.

[B31] DODDS, G. B. and MARTER, W. E. Reactance Relays Discriminate Between Load-Transfer Currents and Fault Currents on 2300 V Station Service Generator Bus. *AIEE Transactions,* vol 71, pt III, Dec 1952, pp 1124–1128.

[B32] EINVALL, C. H. and LINDERS, J. R. A Static Directional Overcurrent Relay with Unique Characteristics and Its Applications. *IEEE PES Winter Meeting,* New York, Jan 25–30, 1976, Conference Paper A76-073-7.

[B33] *Electrical Transmission and Distribution Reference Book.* Westinghouse Electric Corporation, East Pittsburgh, PA, 1964.

[B34] ESTWICK, C. F. Real Power and Imaginary Power in AC Circuits. *AIEE Transactions,* vol 72, pt III, Feb 1953, pp 27–35.

[B35] FAWCETT, D. V. How to Select Overcurrent Relay Characteristics. *IEEE Transactions on Industry Applications,* vol IA-82, May 1963, pp 94–104.

[B36] FAWCETT, D. V. The Tie Between a Utility and an Industrial When the Industrial Has Generation. *AIEE Transactions on Industry Applications,* vol IA-77, pt II, July 1958, pp 136–143.

[B37] GILBERT, M. M. and BELL, R. N. Directional Relays Provide Differential Type Protection on Large Industrial Plant Power System. *AIEE Transactions on Industry Applications,* vol IA-74, pt II, Sept 1955, pp 220–227.

[B38] GLASSBURN, W. E. and SHARP, R. L. A Transformer Differential Relay with Second-Harmonic Restraint. *AIEE Transactions,* vol 77, pt III, 1958, pp 913–918.

[B39] GLASSBURN, W. E. and SONNEMAN, W. K. Principles of Induction Disk Relay Design. *AIEE Transactions,* vol 72, pt III, 1953, pp 23–27.

[B40] GLEASON, L. L. and ELMORE, W. A. Protection of 3-Phase Motors Against Single-Phase Operation. *AIEE Transactions,* vol 77, pt III, 1958, pp 1112–1120.

[B41] GOFF, L. E., ROOK, M. J., and POWELL, L. J. Pilot-Wire Relay Applications in Industrial Plants Utilizing a New Static Pilot Wire Relay. *Conference Record 1979 IEEE IAS Annual Meeting,* pp 1195–1203.

[B42] GRAHAM, D. J., BROWN, P. G., and WINCHESTER, R. L. Generator Protection with a New Static Negative Sequence Relay. *IEEE Transactions on Power Apparatus and Systems,* vol PAS-94, no 4, July/Aug 1975, pp 1208–1213.

[B43] GRIFFIN, C. H. Development of Fault-Withstand Standards for Power Transformers. *Presented to the Pennsylvania Electric Association Relay Committee Meeting,* Erie, PA, May 23, 1980.

[B44] GRIFFIN, C. H. and POPE, J. W. Generator Ground Fault Protection Using Overcurrent, Overvoltage and Undervoltage Relays. *Presented at IEEE PES 1982 Winter Meeting,* New York, Jan 31–Feb 5, 1982, Conference Paper 82 WM 171-7.

[B45] HARDER, E. L., KLEMMER, E. H., SONNEMANN, W. K., and WENTZ, E. C. Linear Couplers for Bus Protection. *AIEE Transactions,* vol 61, 1942, pp 241–248.

[B46] HARDER, E. L. and MARTER, W. E. Principles and Practices of Relaying in the United States. *AIEE Transactions,* vol 67, pt II, 1948, pp 1005–1023.

[B47] HEIDBREDER, J. F. Induction Motor Temperature Characteristics. *AIEE*

Transactions, vol 77, pt III, 1958, pp 800–804.

[B48] HOGREBE, L. F. Switching Motor Loads in Emergency Power Systems. *Specifying Engineer,* May 1979.

[B49] IEEE Committee Report 76 CH-1130-4 PWR. Transient Response of Current Transformers, Jan 1976.

[B50] IEEE Committee Report 81 WM-122-1. Bibliography of Relay Literature, 1978–1979. *Presented at IEEE PES Winter Meeting,* Atlanta, GA, Feb 1–6, 1981.

[B51] IEEE Committee Report 81 WM 121-3. Review of Recent Practices and Trends in Protective Relaying. *IEEE PES Winter Meeting,* Atlanta, GA, Feb 1–6, 1981.

[B52] IEEE Project P490. Generator Ground Protection Guide (in preparation: PES Power System Relay Committee).

[B53] *IEEE Special Publication S-117. Application and Protection of Pilot-Wire Circuits for Protective Relaying,* 1960.

[B54] KELLY, A. R. Relay Response to Motor Residual Voltage During Automatic Transfers. *AIEE Transactions on Industry Applications,* vol IA-74, pt II, Sept 1955, pp 245–252.

[B55] KOTHEIMER, W. C. and MANKOFF, L. L. Electromagnetic Interference and Solid-State Protective Relays. *IEEE Transactions on Power Apparatus and Systems,* vol PAS-96, no 4, July/Aug 1977, pp 1311–1317.

[B56] LATHROP, C. M. and SCHLECKSER, C. E. Protective Relaying on Industrial Power Systems. *AIEE Transactions,* vol 70, pt II, 1951, pp 1341–1345.

[B57] LAZAR, I. Protective Relaying for AC Generators. *Power Engineering,* vol 83, no 2, Feb 1979, pp 70–73.

[B58] LAZAR, I. Protective Relaying for Motors. *Power Engineering,* vol 82, no 9, Sept 1978, pp 66–69.

[B59] LAZAR, I. Transformer Protection and Relaying in Industrial Power Plants. *Power Engineering,* vol 83, no 5, May 1979, pp 56–61.

[B60] LENSNER, H. W. Protective Relaying Systems Using Pilot-Wire Channels. *AIEE Transactions on Power Apparatus and Systems,* vol PAS-52, Feb 1961, pp 1107–1120.

[B61] LINDERS, J. R. Characteristics of Industrial Power Systems as They Affect Protective Relaying. *Conference Record 1973 IEEE IAS Annual Meeting,* Oct 8–11, 1973, pp 825–831.

[B62] LINDERS, J. R. Electric Wave Distortions: Their Hidden Costs and Containment. *IEEE Transactions on Industry Applications,* vol IA-15, no 5, Sept/Oct 1979, pp 458–71.

[B63] LITTLE, D. W., POTOCHNEY, G. J., and PINKLEY, R. A. A Time-

Overcurrent Relay with Solid-State Circuitry. *IEEE PES Winter Meeting,* New York, Jan 26–31, 1975, Conference Paper C75-061-7.

[B64] LOVE, D. J. Ground Fault Protection for Electric Utility Generating Station Medium Voltage Auxiliary Power System. *Presented at IEEE PES Summer Meeting,* Mexico City, July 1977, Paper F77 693-5; and also in *IEEE Transactions on Power Apparatus and Systems,* Mar/Apr 1978.

[B65] MARTINY, W. T., McCOY, R. M., and MARGOLIS, H. B. Thermal Relationships in an Induction Motor Under Normal and Abnormal Operation. *AIEE Transactions,* vol 80, pt III, 1961, pp 66–76.

[B66] MATHUR, BAL K. A Closer Look at the Application and Setting of Protective Devices. *Conference Record paper 1979 IEEE I&CPS Technical Conference,* 79 CH 1460–51A, pp 107–118.

[B67] MENTLER, S. A Half-Cycle Static Bus Protection Relay Using Instantaneous Voltage Measurement. *IEEE Transactions on Power Apparatus and Systems,* vol PAS-94, no 3, May/June 1975, pp 939–944.

[B68] NOCHUMSON, C. J. and SCHWARTZBURG, W. E. Transfer Considerations in Standby Generator Application. *Conference Record IEEE IAS Annual Meeting,* Oct 1–5, 1978, pp 1190–1198.

[B69] OAKES, P. A. Special Circuits for Ground Relay Current Polarization from Autotransformers Having Delta Tertiary. *AIEE Transactions on Power Apparatus and Systems,* vol 78, pt III, Dec 1959, pp 1191–1196.

[B70] POPE, J. W. PURPA and the Customer-Utility Interconnections. *Presented at the Conference for Protective Relay Engineers,* Texas A&M University, Apr 13–15.

[B71] POWELL, L. J. Current Transformer Burden and Saturation. *Conference Record IEEE IAS 1977 Annual Meeting,* pp 127–137.

[B72] *Protective Relays Application Guide.* QEC Measurements Ltd, Stafford, England, 1975.

[B73] ROOK, M. J., GOFF, L. E., POTOCHNEY, G. J., and POWELL, L. J. Application of Protective Relays on a Large Industrial-Utility Tie with Industrial Cogeneration. *Presented at IEEE PES Winter Meeting,* Atlanta, GA, Feb 1–6, 1981, Paper 81 WM 158-5.

[B74] ROTHE, F. S. and CONCORDIA, C. Transient Characteristics of Current Transformers During Faults—II. *AIEE Transactions,* vol 66, 1947, pp 731–734.

[B75] SEELEY, H. T., VON ROESCHLAUB, F. Instantaneous Bus-Differential Protection Using Bushing Current Transformers. *AIEE Transactions,* vol 67, 1948, pp 1709–1718.

[B76] SHULMAN, J. M., ELMORE, W. A., and BAILEY, K. D. Motor Starting Protection by Impedance Sensing. *IEEE Transactions on Power Apparatus and*

Systems, vol PAS-97, no 5, Sept/Oct 1978, pp 1689–95.

[B77] SMITH, D. H. Problems Involving Industrial Plant-Utility Power System Interties. *IEEE Transactions on Industry Applications,* vol IA-11, no 6, Nov/Dec 1975.

[B78] SONNEMANN, W. K. Phasors for Directional Relays. *Presented at 21st Annual Conference for Protective Relay Engineers,* Texas A&M University, Apr 22–24, 1968.

[B79] SONNEMANN, W. K. A Study of Directional Element Connections for Phase Relays. *AIEE Transactions,* vol 69, 1950, pp 1438–1451.

[B80] STOLLER, S. In-Plant Generator—Design Considerations for Industrial Facilities. *IEEE Transactions on Industry Applications,* vol IA-12, no 3, May/ June 1976, pp 226–231.

[B81] TREMAINE, R. L. and BLACKBURN, J. L. Loss-of-Field Protection for Synchronous Machines. *AIEE Transactions,* vol 73, pt III-A, 1954, pp 765–777.

[B82] WAGNER, C. F. and EVANS, R. D. *Symmetrical Components.* New York: McGraw-Hill, 1933.

[B83] WALDRON, J. E. Motor Characteristics and Protection. *Presented at 29th Conference for Protective Relay Engineers,* Texas A&M University, Apr 12–14, 1976.

[B84] WALDRON, J. E. Innovations in Solid-State Protective Relays. *IEEE Transactions on Industry Applications,* vol IA-14, no 1, Jan/Feb 1978, pp 39–47.

[B85] WALDRON, J. E. and ZOCHOLL, S. E. Design Considerations for a New Solid-State Transformer Differential Relay with Harmonic Restraint. *Presented at 5th Annual Western Protective Relay Conference,* Sacramento, CA, Oct 15–18, 1978.

[B86] WALSH, G. F. The Effects of Reclosing on Industrial Plants. *Proceedings of the American Power Conference,* vol 23, 1961, pp 768–778.

[B87] WARRINGTON, A. C. Application of the OHM and MHO Principle to Protective Relaying. *AIEE Transactions,* vol 65, 1946, pp 378–86, 490.

[B88] WARRINGTON, A. C. *Protective Relays—Their Theory and Practice.* New York: Wiley, 1968.

[B89] WENTZ, E. C. and SONNEMANN, W. K. Current Transformers and Relays for High-Speed Differential Protection, with Particular Reference to Offset Transient Currents. *AIEE Transactions,* vol 59, Aug 1940, pp 481–488, discussions pp 1144–1148.

[B90] WEST, D. J. Current Transformer Application Guidelines. *Conference Record IEEE IAS 1977 Annual Meeting,* pp 110–126.

[B91] WILSON, J. R. Power-Reactive Versions of R–X Diagrams. *Presented at the*

21st Annual Conference for Protective Relay Engineers, Texas A&M University, Apr 22–24, 1968.

[B92] ZOCHOLL, S. E. Solid-State Overcurrent Relay with Conventional Time Current Curves. *IEEE PES Winter Meeting,* New York, Jan 30–Feb 4, 1972, Paper C72-042-5.

[B93] ZOCHOLL, S. E. and PENCINGER, C. J. Analysis of Overcurrent Relay Testing. *Presented at Pennsylvania Electric Association Relay Committee Meeting,* Wilkes-Barre, PA, Oct 6, 1977.

[B94] ZOCHOLL, S. E. and WALDRON, J. E. Solid-State Relays for Generator Protection. *Presented at 6th Annual Western Protective Relay Conference,* Spokane, WA, Oct 16–18, 1979.

5. Fuses

5.1 General Discussion. A fuse may be defined as *a device that protects a circuit by fusing open its current-responsive element when an overcurrent or short-circuit current passes through it.* A fuse has these functional characteristics:

(1) It combines both the sensing and interrupting elements in one self-contained device.

(2) It is direct acting in that it responds to a combination of magnitude and duration of circuit current flowing through it.

(3) It normally does not include any provision for manually making and breaking the connection to an energized circuit, but requires separate devices, such as an interrupter switch, to perform this function.

(4) It is a single-phase device. Only the fuse in the phase or phases subjected to overcurrent will respond to de-energize the affected phase or phases of the circuit or equipment that is faulty.

(5) After having interrupted an overcurrent, it is renewed by the replacement of its current-responsive element before restoration of service.

5.2 Fuse Terms and Definitions. The following alphabetically arranged terms may be found in other industry publications on fuses. These definitions are for the purpose of ready reference.

ampere rating. That current that the fuse will carry continuously without deterioration and without exceeding temperature rise limits specified for that fuse.

arcing time. The arcing time of a fuse is the time elapsing from the melting of the current-responsive element (such as the link) to the final interruption of the circuit. This time will be dependent upon such factors as voltage and reactance of the circuit. Refer to Fig 71.

ferrule. The cylindrical-shaped fuse terminal that also encloses the end of the fuse. In low-voltage fuses, the design is only used in fuses rated up to and including 60 A. The ferrule may be made of brass or copper, and may be plated with various materials.

fuse-link. In British terminology only, a complete enclosed cartridge fuse; in such cases the addition of the *carrier,* or holder, completes the fuse. (See *link.*) In the US, a renewable fusible element for fuse cutouts.

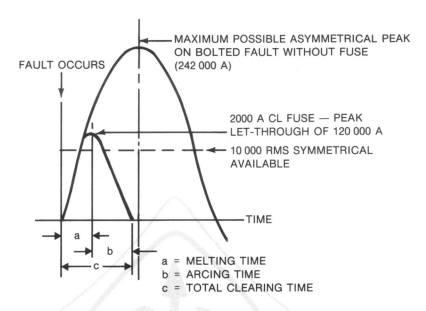

FAULT OCCURS

MAXIMUM POSSIBLE ASYMMETRICAL PEAK
ON BOLTED FAULT WITHOUT FUSE
(242 000 A)

2000 A CL FUSE — PEAK
LET-THROUGH OF 120 000 A

10 000 RMS SYMMETRICAL
AVAILABLE

TIME

a

b

c

a = MELTING TIME
b = ARCING TIME
c = TOTAL CLEARING TIME

Fig 71
Typical Current Limitation Showing Peak Let-Through
and Total Clearing Time

HRC. In British and Canadian terminology, *high rupturing capacity,* equivalent to US *high interrupting capacity* and generally indicating capability of interruption of at least 100 000 rms A for low-voltage fuses.

I^2t. The measure of heat energy developed within a circuit during the fuse's melting or clearing. Generally stated as *melting I^2t* or *clearing I^2t.*

interrupting rating. A rating based upon the highest rms alternating current that the fuse is required to interrupt under the conditions specified. The interrupting rating, in itself, has no direct bearing on any current-limiting effect of the fuse.

link. The current-responsive element in a fuse that is designed to melt under overcurrent conditions and so interrupt the circuit. A *renewal link* is one intended for use in Class H low-voltage renewable fuses.

melting time. The time required to melt the current-responsive element on a specified overcurrent. Where the fuse is current limiting in less than half-cycle, the melting time may be approximately half or less of the clearing time. (Sometimes referred to as *pre-arcing.*) Refer to Fig 71.

peak let-through current. The maximum instantaneous current through a current-limiting fuse during the total clearing time. Since this is an instantaneous value, it may well exceed the rms available current, but will be less than the peak current available without a fuse in the circuit if the fault level is high enough for it to operate in its current-limiting mode. Refer to Fig 71.

pre-arcing time. See *melting time.*

threshold current. The magnitude of current at which a fuse becomes current limiting, specifically, the symmetrical rms available current at the threshold of the current-limiting range, where the fuse total clearing time is less than half-cycle at rated voltage and rated frequency, for symmetrical closing, and a power factor of less than 20%. Refer to various peak let-through current curves for each type of fuse. The threshold ratio is the relationship of the threshold current to the fuse's continuous-current rating.

total clearing time. The total time between the beginning of the specified overcurrent and the final interruption of the circuit, at rated voltage. It is the sum of the minimum melting time plus tolerance and the arcing time. For clearing times in excess of half-cycle, the clearing time is substantially the maximum melting time for low-voltage fuses. Refer to Fig 71.

voltage rating. The rms alternating current (or the direct current) voltage at which the fuse is designed to operate. All low-voltage fuses will function on any lower voltage, but use on higher voltages than rated is hazardous. For high short-circuit currents, the magnitude of applied voltage will affect the arcing and clearing times and increase the clearing I^2t values.

Low-Voltage Fuses—600 V or Less

5.3 Fuse Standards. The various published fuse standards of the electrical industry are listed below. Each is available from its source and should be studied for detailed requirements.

5.3.1 Fuse standards for low-voltage fuses (600 V or less) by Underwriters Laboratories, Inc, are presently found in ANSI/UL 198B-1982 [10], ANSI/UL 198C-1981 [11], ANSI/UL 198D-1982 [12], ANSI/UL 198E-1982 [13], ANSI/UL 198F-1982 [14], ANSI/UL 198G-1981 [15], ANSI/UL 198H-1982 [16], and UL 198L-1984 [22].[14]

5.3.1.1 ANSI/UL 198B - 1982 [10]. This standard covers standard cartridge fuses, frequently referred to as Class H fuses rated 250 or 600 V, 600 A or less, all to be employed in accordance with ANSI/NFPA 70-1984 [9] (National Electrical Code [NEC]).

This standard states, "these requirements do not cover fuses recognized as being current limiting or fuses having interrupting capacity ratings." It also states (12.9) "neither a fuse nor its carton shall bear a marking which states or implies that it is current limiting," and (12.2) "neither a fuse nor its carton shall be marked with the words direct current or abbreviations thereof" unless found suitable for use on direct current by investigation. (See 5.3.1.8.) The prohibitions concerning current limiting and interrupting rating do NOT apply to fuses listed and labeled as meeting the standards for UL Classes J, R, L, CC, and T. Fuses marked with the words "time delay," "D," "dual element," or any phrase of similar significance must not open in less than 10 s at five times their rating.

---

[14] The numbers in brackets correspond to those of the references listed at the end of this chapter.

The principal requirements of ANSI/UL 198B-1982 [10] are dimension, design, construction, performance, and markings. Under performance, the following are tested for:

(1) Continuous-current carrying ability and temperature rise.

(2) Overload operation within prescribed maximum times at 135 and 200% of the fuse's continuous-current rating.

(3) Time-delay test (optional) for a minimum opening time of 10 s at five times the continuous-current rating. A fuse may be labeled "time delay," "dual element," etc, only if it passes this test.

(4) Short-circuit interrupting capability at 10 000 A rms (ac). No dc testing is presently required.

5.3.1.2 ANSI/UL 198C-1981 [11]. UL Class G, J, L, and CC fuses for this standard's scope state that *"these requirements cover nonrenewable cartridge fuses that limit the peak let-through current and the total ampere-squared seconds and that exhibit current-limiting characteristics above specified values of current.* These fuses are not made in the same dimensions as Class H and Class K fuses. (See the NEC [9], Section 240-60b. See, also, Table 24.)

5.3.1.3 ANSI/UL 198D-1982 [12]. This standard covers fuses made in the same dimensions as Class H fuses, but which have an interrupting rating of 50 000, 100 000, or 200 000 A rms ac. Class K fuses have prescribed values of peak let-through current and I^2t for each case size. Current limitation does occur; however, since they have no required threshold ratio and are interchangeable with Class H fuses, they are not labeled as *current limiting*. See the NEC [9], Section 240-60(b). No *rejection feature* for noninterchangeability is recognized by UL. They are rated up to 600 A in both 250 and 600 V sizes.

Class K fuses are tested for continuous-current-carrying ability, temperature rise, overload opening, and an optional time-delay test of at least 10 s at five times the current rating in order to be labeled as "time delay," "dual element," etc. They are also tested at various short-circuit levels up to their maximum interrupting rating and for compliance with prescribed maximum values of peak let-through current and I^2t values. No dc testing is presently a part of the UL standards. Class K fuses are subdivided into Classes K-1, K-5, and K-9, with increasing values for current limitation (in that order). They are labeled for the class subdivision, interrupting rating, amperes, and maximum voltage. It is important to note that any one manufacturer may produce two or three Class K-5 fuses with different interrupting ratings, or with the same interrupting rating and differing peak let-through and I^2t values distinguished by catalog numbers, but all within the requirements of that class.

5.3.1.4 ANSI/UL 198E-1982 [13]. This standard covers fuses made in much the same dimensions as Class H and K fuses, except for the added rejection feature in Class R fuses. Fuseholders designed to accept Class R fuses will not accept Class H or K fuses or any other class. However, Class R fuses will fit into Class H or K fuseholders, thus allowing upgrading of older systems to the maximum allowed by all other devices in the circuit. The Class R rejection features provide noninterchangeability in accordance with the NEC [9], Section

Table 24
UL Classifications

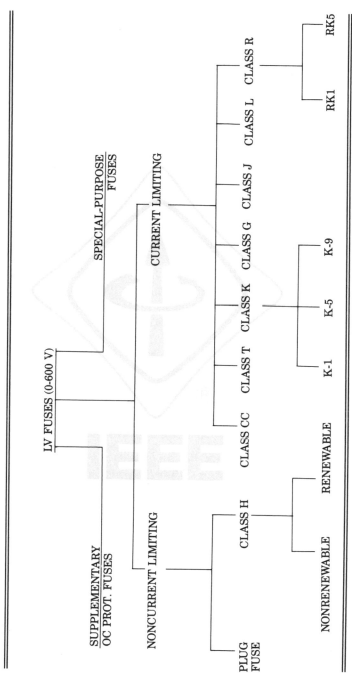

240-60b. All Class R fuses have an interrupting rating of 200 000 A rms ac and are current limiting. Class R fuses have similar characteristics to Class K fuses and have two subclasses, RK1 and RK5, which define the maximum peak current, I^2t let-through, and the threshold ratio of fuses in the subclass. See Table 24. Class R fuses may be provided with time delay.

5.3.1.5 ANSI/UL 198F-1982 [14]. This standard covers edison base and type S base plug fuses, which may or may not be provided with time-delay characteristics. Note that type S base plug fuses have rejection features that limit the ampere rating of fuses that may be installed in a particular fuseholder.

5.3.1.6 ANSI/UL 198G-1981 [15]. Encompassing glass, miniature, micro, and other miscellaneous fuses for supplementary protection, this standard does not pertain to branch circuit fuses.

5.3.1.7 ANSI/UL 198H-1982 [16]. The requirements of this standard cover the newest class of current-limiting fuses—Class T. These fuses have characteristics similar to Class J fuses, but are dimensionally smaller. They are available in ratings up to 1200 A at both 300 and 600 V ac. As current-limiting fuses, they are not dimensionally interchangeable with any other class of fuse.

5.3.1.8 UL-198L-1984 [22]. These requirements cover dc rated Class H fuses, including renewable Class H and Class CC, G, K, RK, J and T fuses. These fuses have a maximum rating of 600 A and can have a voltage rating of 125, 250, 300, or 600 V dc. A Class H fuse has a dc interrupting rating of 10 000 A. Other classes listed in this paragraph may have a maximum interrupting rating of 10 000, 20 000, 50 000, or 100 000 A.

5.3.2 NEMA FU1-1978. This standard was adopted by ANSI C97.1-1972 [5], with the possible addition of other material at a later date.

5.3.3 ANSI C97.1-1972 [5]. ANSI standard designations are referenced after the UL.

5.3.4 ANSI/NFPA 70-1984 [9] (National Electrical Code [NEC]). Some sections of the National Electrical Code that can apply to fuses are:

110-9   Interrupting Capacity Requirements
110-10   Circuit Impedance and Other Characteristics
230-98   Available Short-Circuit Current
240-02   List of Articles Covering Overcurrent Protection for Specific Equipment
240-6   Standard Ampere Ratings
240-11   Definition of Current-Limiting Protective Device
240-50   Plug Fuses, Fuseholders, and Adapters
240-51   Edison Base Fuses
240-53   Type S Fuses
240-54   Types S Fuses, Adapters, and Fuseholders
240-60b Noninterchangeability
240-60c Marking on Fuses
240-61   Fuse Classification

5.4 Fuse Terms and Definitions. The following alphabetically arranged terms may be found in other industry publications on fuses. These definitions are for the purpose of ready reference.

bridge. That narrowed portion of a fuse link that is expected to melt first. One link may have two or more bridges in parallel and in series as well. The shape and size of the bridge is a factor in determining the fuse characteristics under overload and fault current conditions.

Class H fuses. Cartridge fuses were formerly known as "NEC-dimensioned fuses." Class H fuses are tested and listed by Underwriters Laboratories, Inc, under their standard ANSI/UL 198B-1982 [10] in 250 and 600 V ratings with interrupting capabilities of 10 000 A. Class H fuses are not marked as current limiting. UL standards for Class H fuses have a time-delay requirement of at least 10 s opening time at five times rating in order to have the words "time delay" on the label.

Class J fuses. These fuses are rated to interrupt 200 000 A ac and meet the standards of Underwriters Laboratories, Inc, for Class J fuses. They are UL-labeled as "current limiting," are rated only for 600 V (or less) ac, and are of dimensions not interchangeable with other classes. Class J fuses that have a time delay of at least 10 s opening time at five times rated current may have the words "time delay" on the label.

Class K fuses. These fuses meet ANSI/UL 198D-1982 [12] of Underwriters Laboratories, Inc, for Class K as either K-1, K-5, or K-9. These standards have prescribed values for maximum peak let-through currents and I^2t for each subclass, with K-1 having the lowest (most restrictive) values and Class K-9 having the highest (least restrictive) values. Dimensionally the same as Class H fuses, these fuses have no UL-recognized "rejection feature." Their ac interrupting rating appears on their labels as 50 000, 100 000, or 200 000 A. They are not labeled as "current limiting." The words "time delay," "dual element," letter "D," or phrase of similar significance on the label will indicate the manufacturer has met UL's optional testing for this feature. The use of Class K fuses permits equipment and circuits to be applied on systems having potential fault currents in excess of 10 000 A. Some hazards may exist in that they can be replaced with Class H fuses by uninformed personnel under present standards.

Class L fuses. These fuses meet ANSI/UL 198C-1981 [11] of Underwriters Laboratories, Inc, for Class L fuses, have ratings in the range of 601–6000 A, are rated to interrupt 200 000 A ac, are rated only for 600 V or less ac, and are of specified dimensions larger than those of other fuses rated 600 V (or less). They are intended to be bolted to bus bars and are not used in clips. UL has no definition of time delay for Class L fuses; however, many Class L fuses have substantial overload time-current carrying capability. Class L fuse standards do not include 250 V ratings, dc testing, or dc ratings.

Class R fuses. These fuses meet ANSI/UL 198E-1982 [13] of Underwriters Laboratories, Inc, for Class R as either RK1 or RK5. Their interrupting ratings are 200 000 A ac. The standard has prescribed values for maximum peak let-through currents, I^2t and threshold current, with subclass RK1 having the lowest (most restrictive) values as compared to subclass RK5. These fuses have dimensions that provide a one-way physical rejection feature, that is, no other class of fuse will fit into equipment designed to employ Class R fuses; however, Class R fuses can be installed in older Class H or Class K equipment as replacement to upgrade these systems to the maximum allowed by other devices in the system. Class R fuses are available with or without time delay. If marked "time delay" or similar phrase, they are required to have a minimum opening time of 10 s when subjected to a load of five times rated current.

current limiter. A device intended to function only on fault currents of high magnitude and that may not successfully open on lesser overcurrents regardless of time. Such a device should always be used in series with a fuse, contactor, or circuit breaker to protect against overloads and low-level short circuits. Current limiters are typically added to molded-case circuit breakers, power circuit breakers, or instantaneous circuit protectors.

current-limiting fuse. A fuse that will interrupt all available currents above its threshold current and below its maximum interrupting rating, limit the clearing time at rated voltage to an interval equal to or less than the first major or symmetrical loop duration, and limit peak let-through current to a value less than the peak current that would be possible with the fuse replaced by a solid conductor of the same impedance. Note that current-limiting action only becomes effective at a specific value of current. (See **threshold current**.) UL only recognizes and permits labeling of Classes G, J, L, R, CC, and T as *current limiting,* although Class K fuses are, in fact, current limiting. Refer to the NEC [9], Section 240-60b, which prohibits fuse clips for current-limiting fuses accepting noncurrent-limiting fuses.

delay. This term is usually applied to the opening time of a fuse when in excess of one cycle, where the time may vary considerably between types and makes and still be within established standards. This word, in itself, has no specific meaning other than in manufacturers' claims (see **time delay**) unless published standards specify delay characteristics.

dual-element fuse. A cartridge fuse having two or more current-responsive elements of different fusing characteristics in series in a single cartridge. This is a construction/design technique frequently used to obtain a time-delay response characteristic. Labeling a fuse as *dual element* means this fuse meets UL *time-delay* requirements (can carry five times rated current for a minimum of 10 s for Class H, K, J, and R fuses) and in this case defines a time–current response characteristic and not necessarily a dual-element construction technique.

NEC dimensions. Dimensions once stated in the National Electrical Code, but now found in ANSI/UL 198B-1982 [10] and in ANSI/UL 198D-1982 [12]. These dimensions are common to Class H and K fuses and provide interchangeability between manufacturers for fuses and fusible equipment of a given ampere and voltage range.

one-time fuse. Strictly speaking, any nonrenewable fuse, but generally accepted and used to describe any Class H nonrenewable cartridge fuse, with a single (as opposed to dual) fusing element and intended to interrupt not over 10 000 A.

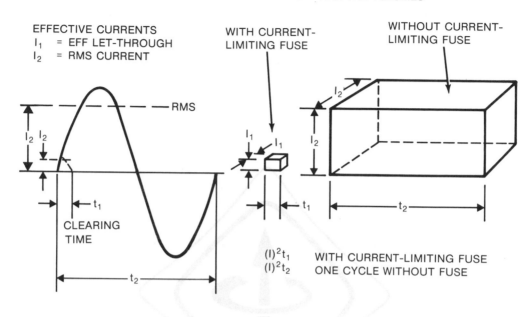

Fig 72
Graphic Representation of I^2t

overload. Generally used in reference to an overcurrent that is not of sufficient magnitude to be termed a short circuit. An overload is normally that overcurrent value from 100% of fuse rating up to ten times fuse rating. (See **short circuit.**)

plug fuses. Plug fuses are rated 125 V and are available with current ratings up to 30 A. Their use is limited to circuits rated 125 V or less. However, they may also be used in circuits supplied from a system having a grounded neutral, and in which no conductor operates at more than 150 V to ground. The NEC [9] requires type S fuses in all new installations of plug fuses because they are tamper resistant and size limiting, thus making it difficult to overfuse.

renewable fuse. A fuse in which the element, usually a zink link, may be replaced after the fuse has opened. Once a very popular item, this fuse is gradually losing popularity due to the possibility of using higher ampere-rated links or multiple links in the field, which can present a hazard.

short-circuit current. An overcurrent usually defined as being in excess of ten times normal continuous rating. (See **overload.**)

time delay. Meaningless unless defined. This term is now used by NEMA, ANSI, and UL to mean, in Classes H, K, J, and R cartridge fuses, a minimum opening time of 10 s on an overload current five times the ampere rating of the fuse. Such a delay is particularly useful in allowing the fuse to pass the momentary starting overcurrent of a motor, yet not hindering the opening of the fuse should the overload persist. In Class G, CC, and plug fuses, the phrase "time delay" is required by UL to be a minimum opening time of 12 s on an overload of twice the fuse's ampere rating. The time-delay characteristic does not affect the fuse's short-circuit current clearing ability.

tube. The cylindrical enclosure of a fuse. Such a tube may be made of laminated paper, special fiber, melamine impregnated glass cloth, bakelite, ceramic, glass, plastic, or other materials.

5.5 UL Classifications. Table 24 shows the various low-voltage fuses covered by UL.

5.6 Standard Dimensions. Figures 73–79 and Tables 25 and 26 show typical dimensions (see fuse manufacturer's data for actual dimensions) for Class H and K, L, G, J, T, CC, and R fuses.

5.7 Typical Interrupting Ratings. Interrupting ratings of low-voltage fuses are:

Class H renewable or nonrenewable—not over 10 kA
Class K fuses—50, 100, or 200 kA
Class RK1 and RK5—200 kA
Class RK1 and RK5 time delay—200 kA
Class J, CC, T, and L—200 kA

5.8 Time-Current Characteristic (TCC) Curves. Time-current characteristic curves are intended to show the relationship between various overcurrent values and some opening function of the fuse. The currents are normally represented across the bottom of the graph. The time values are normally shown along the left (vertical) side and may represent *virtual time* (which must be defined), minimum melting time, average melting time, or total clearing time. Two curves showing different *time* concepts cannot, obviously, be compared on the same basis. Average melting time is assumed to be represented unless otherwise stated. Average current is defined as being subject to a maximum tolerance of plus or minus 10% in current for any given time. Thus, minimum melting is approximately 10% less (in current) than average. In times of approximately 0.1 s or more, maximum melting time is virtually the same as total clearing time. For times between 0.1 s and 0.01 s, total clearing time, relatively, is considerably greater than melting time due to the greater percentage of arcing time. For periods of less than 0.01 s, the melting time may be as little as half the clearing time. In these short time periods, I^2t becomes of increasing importance.

   5.8.1 The fuse curves in Figs 80 and 81 show Class RK-5 time-delay fuses (30–600 A) and Class L current-limiting fuses (800–6000 A).

5.9 Selectivity of Fuses. Since the electrical distribution system is the heart of most industrial, commercial, and institutional type installations, it is imperative that any unnecessary shutdowns of electrical power be prevented. Such incidents

FERRULE-TYPE CARTRIDGE FUSE —
0–60 A

KNIFE-BLADE TYPE CARTRIDGE FUSE —
61–600 A

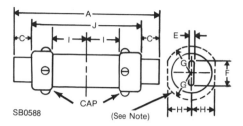

SB0585

FERRULE

SB0588

CAP

(See Note)

Note — The dash line represents the limit of the maximum projection of a screw, rivet head, or the like. It becomes a circle for a fuse rated at more than 200 A.

DIMENSIONS OF KNIFE-BLADE TYPE FUSES IN INCHES (mm)

Rating Volts	Rating Amperes	Overall Length of Fuse[a]	Maximum Outside Diameter of Tube	Minimum Length of Ferrule or Blade	Outside Diameter of Ferrule[b]	Thickness of Blade[c]	Width of Blade[d]	Maximum Dimensions Over Projections[e] Measured Parallel to Blade	Measured at Right Angles to Blade	Minimum Distance From Midpoint of Fuse to Nearest Live Part	Minimum Overall Length of Cylindrical Body[f]
		A	B	C	D	E	F	G	H	I	J
250	0–30	2.00 (50.8)	0.53 (13.5)	0.50 (12.7)	0.562 (14.27)						
	31–60	3.00 (76.2)	0.78 (19.8)	0.625 15.9	0.812 (20.62)						
	61–100	5.87 (149.2)		1.00 (25.4)		0.125 (3.18)	0.750 (19.05)	0.66 (16.7)	0.59 (15.1)	1.03 (26.2)	
	101–200	7.12 (181.0)		1.37 (34.9)		0.188 (4.78)	1.125 (28.58)	0.94 (23.8)	0.84 (21.4)	1.19 (30.2)	4.12 (104.8)
	201–400	8.62 (291.1)		1.87 (47.6)		0.250 (6.35)	1.625 (41.28)	1.20 (30.6)	1.20 (30.6)	1.19 (30.2)	4.62 (117.5)
	401–600	10.37 (263.5)		2.25 (57.1)		0.250 (6.35)	2.000 (50.80)	1.45 (36.9)	1.45 (36.9)	1.53 (38.9)	5.19 (131.8)
600	0–30	5.00 (127.0)	0.78 (19.8)	0.50 (12.7)	0.812 (20.62)						
	31–60	5.50 (139.7)	1.03 (26.2)	0.62 (15.9)	1.062 (26.97)						
	61–100	7.87 (200.0)		1.00 (25.4)		0.125 (3.18)	0.750 (19.05)	0.78 (19.8)	0.72 (18.3)	1.75 (44.4)	
	101–200	9.62 (244.5)		1.37 (34.9)		0.188 (4.78)	1.125 (28.58)	1.06 (27.0)	0.98 (25.0)	2.25 (57.1)	6.12 (155.6)
	201–400	11.62 (295.3)		1.87 (47.6)		0.250 (6.35)	1.625 (41.28)	1.45 (36.9)	1.45 (36.9)	2.50 (63.5)	7.12 (181.0)
	401–600	13.37 (339.7)		2.25 (57.1)		0.250 (6.35)	2.000 (5.08)	1.72 (43.7)	1.72 (43.7)	2.69 (68.3)	8.19 (208.0)

[a]Tolerances: 0–60 A, ±0.03 in (±0.8 mm); 61–200 A, ±0.06 in (±1.6 mm); 201–600 A, ±0.09 in (±2.4 mm).

[b]Column D tolerance: ±0.008 in (±0.20 mm).

[c]Column E tolerance: ±0.003 in (±0.08 mm).

[d]Column F tolerance: ±0.035 in (±0.89 mm).

[e]The maximum overall dimension of a screw ring for a renewable fuse, the position of which with respect to the position of the knife blade cannot be predetermined, shall be no more than the value specified for dimension H.

[f]The length of the cylindrical body may be less than the indicated value if other acceptable interference means, pins through the blades, collars, and the like, are provided to prevent mounting the fuse in a fuseholder that will accommodate a fuse rated in the next lower bracket of current ratings.

Fig 73
Class H and K Dimensions

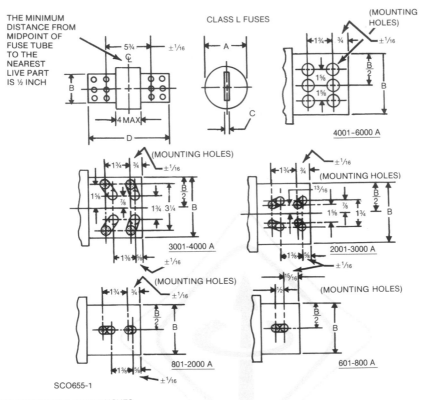

THE MINIMUM DISTANCE FROM MIDPOINT OF FUSE TUBE TO THE NEAREST LIVE PART IS ½ INCH

CLASS L FUSES

(MOUNTING HOLES)

4001–6000 A

(MOUNTING HOLES)

3001–4000 A

(MOUNTING HOLES)

2001–3000 A

(MOUNTING HOLES)

801–2000 A

(MOUNTING HOLES)

601–800 A

SCO655-1

ALL DIMENSIONS ARE IN INCHES
ALL TOLERANCES ARE ±1/64 INCH UNLESS OTHERWISE NOTED
⅝±1/32 INCH SLOTS OR HOLES MAY BE USED. SHADED HOLES INDICATE DRILLING REQUIRED FOR CLASS "L" FUSE MOUNTING.

Inches	1/64	1/32	1/16	½	⅝	¾	13/16	⅞	15/16	1⅜	1⅝	1¾	3¼	4	5¾
mm	0.4	0.8	1.6	12.7	15.9	19.0	20.6	22.2	23.8	34.9	41.3	44.4	82.6	101.6	146.0

DIMENSIONS OF CLASS L FUSES IN INCHES (mm)[a]

Cartridge Size in Amperes	Maximum Diameter A	Width of Blades B	Thickness of Blades C	Overall Length D
601–800	2.53 (64.3)	2.0 (50.8)	0.375 (9.5)	8.63 (219.1)
801–1200	2.78 (70.6)	2.0 (50.8)	0.375 (9.5)	10.75 (273.0)
1201–1600	3.03 (77.0)	2.38 (60.3)	0.44 (11.1)	10.75 (273.0)
1601–2000	3.53 (89.7)	2.75 (69.8)	0.50 (12.7)	10.75 (273.0)
2001–2500	5.03 (127.8)	3.50 (88.9)	0.75 (19.0)	10.75 (273.0)
2501–3000	5.03 (127.8)	4.0 (101.6)	0.75 (19.0)	10.75 (273.0)
3001–4000	5.78 (146.8)	4.75 (120.6)	0.75 (19.0)	10.75 (273.0)
4001–5000	7.16 (181.8)	5.25 (133.4)	1.0 (25.4)	10.75 (273.0)
5001–6000	7.16 (181.8)	5.75 (146.0)	1.0 (25.4)	10.75 (273.0)

[a] Tolerances: B, ±0.06 inch (±1.6 mm); C, ±0.03 inch (±0.8 mm); D, ±0.09 inch (±2.4 mm).

Fig 74
Class L Fuse Dimensions

CLASS G FUSES

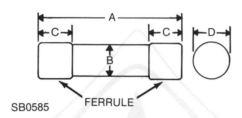

SB0585   FERRULE

DIMENSIONS OF CLASS G FUSES IN INCHES (mm)[a]

Rating		Overall Length of Fuse	Maximum Outside Diameter of Tube	Minimum Length of Ferrule	Outside Diameter of Ferrule
V	A	A	B	C	D
300	0–15	1.31 (33.3)	0.38 (9.5)	0.28 (7.1)	0.406 (10.31)
300	16–20	1.41 (35.7)	0.38 (9.5)	0.28 (7.1)	0.406 (10.31)
300	21–30	1.62 (41.3)	0.38 (9.5)	0.28 (7.1)	0.406 (10.31)
300	31–60	2.25 (57.1)	0.38 (9.5)	0.50 (12.7)	0.406 (10.31)

[a]Tolerances: A, ±0.03 in (±0.8 mm); D, ±0.006 in (±0.15 mm).

**Fig 75
Class G Fuse Dimensions**

FERRULE-TYPE CLASS J FUSES —
0-60 AMPERES

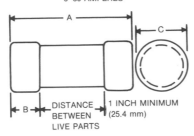

DIMENSIONS OF FERRULE-TYPE CLASS J FUSES
IN INCHES (mm)[2]

Cartridge Size in Amperes	Overall Length A	Minimum Length of Ferrule B	Outside Diameter of Ferrule C
0-30	2.25 (57.1)	0.50 (12.7)	0.812 (20.62)
31-60	2.37 (60.3)	0.63 (15.9)	1.062 (26.97)

[a]Tolerances: A, ±0.03 inch (±0.8 mm); C, ±0.008 inch (±0.20 mm).

KNIFE-BLADE TYPE CLASS J FUSES —
61-600 AMPERES

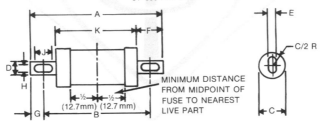

DIMENSIONS OF KNIFE-BLADE-TYPE CLASS J FUSES IN INCHES (mm)[a]

Cartridge Size in Amperes	Overall Length A	Distance Between Centers of Slot B	Maximum Diameter[b] C	Width of Blades D	Thickness of Blades E	Length of Blades F	Distance From End of Blade to Center of Slot G	Width of Slot H	Length of Slot J	Length of Tube K
61-100	4.62 (117.5)	3.62 (92.1)	1.13 (28.6)	0.750 (19.05)	0.125 (3.18)	1.00 (25.4)	0.50 (12.7)	0.281 (7.14)	0.375 (9.52)	2.62 (66.7)
101-200	5.75 (146.0)	4.38 (111.1)	1.63 (41.3)	1.125 (28.58)	0.188 (4.78)	1.37 (34.9)	0.69 (17.5)	0.281 (7.14)	0.375 (9.52)	3.00 (76.2)
201-400	7.12 (181.0)	5.25 (133.4)	2.13 (54.0)	1.625 (41.28)	0.250 (6.35)	1.87 (47.6)	0.94 (23.8)	0.406 (10.32)	0.531 (13.49)	3.37 (85.7)
401-600	8.00 (203.2)	6.00 (152.4)	2.63 (66.7)	2.000 (50.80)	0.375 (9.52)	2.12 (54.0)	1.00 (25.4)	0.531 (13.49)	0.688 (17.48)	3.75 (95.2)

[a] Tolerances: A, ±0.09 in (±2.4 mm); B, ±0.06 in (±1.6 mm); D, ±0.035 in (±0.89 mm); E, ±0.003 in (±0.08 mm); F, ±0.03 in (±0.8 mm); G, ±0.03 in (±0.8 mm); H, ±0.005 in (±0.13 mm); J, plus 0.062, minus 0.000 in (plus 1.57 mm, minus 0.00 mm); K, ±0.03 in (±0.8 mm).

[b] C/2 includes maximum dimension over projection.

**Fig 76
Class J Fuse Dimensions**

FERRULE-TYPE FUSES

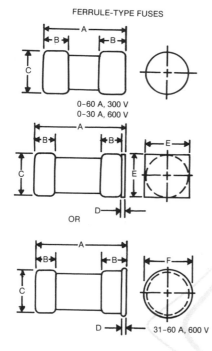

0–60 A, 300 V
0–30 A, 600 V

OR

31–60 A, 600 V

SB 1284

Rating		Overall Length of Fuse[a]	Length of Ferrule[b]	Outside Diameter of Ferrule[c]	Thickness of Rejection Feature[d]	Width of Rejection Feature[d,e]	Diameter of Rejection Feature[e,f]
Volts	Amperes	A	B	C	D	E	F
300	0–30	0.880 (22.35)	0.280 (7.11)	0.406 (10.31)	—	—	—
	31–60	0.880 (22.35)	0.280 (7.11)	0.563 (14.30)	—	—	—
600	0–30	1.500 (38.10)	0.280 (7.11)	0.563 (14.30)	—	—	—
	31–60	1.560 (39.62)	0.410 (10.41)	0.812 (20.62)	0.062 (1.57)	0.812 (20.62)	0.994 (25.25)

DIMENSIONS OF FERRULE-TYPE FUSES IN INCHES (mm)

[a]Tolerances: 0–60 A, 300 V, ±0.020 in (±0.51 mm); 0–60 A, 600 V, ±0.040 in (±1.02 mm).
[b]Column B tolerance: ±0.020 in (±0.51 mm).
[c]Column C tolerance: ±0.006 in (±0.15 mm). Diameter of tube is less than ferrules.
[d]Columns D and E tolerance: ±0.006 in (±0.15 mm).
[e]Rejection feature may be either square or round.
[f]Column F tolerance: minus 0.006 in plus 0.016 (minus 0.15 mm plus 0.41 mm).

KNIFE-BLADE TYPE FUSES

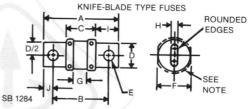

ROUNDED EDGES

SEE NOTE

Note — The dashed line represents limit of the maximum 0.063 in (1.58 mm) projection of screw, rivet head or the like.

DIMENSIONS OF KNIFE-BLADE TYPE FUSES IN INCHES (mm)

Rating		Overall Length of Fuse[a]	Distance Between Mounting Centers[b]	Maximum Length of Body	Width of Blade[c]	Diameter of Mounting Holes[d]	Maximum Diameter of Fuse	Minimum Length of Insulated Body	Thickness of Blade[e]	Minimum Length of Blade	Distance of Mounting Holes From End[c]
Volts	Amperes	A	B	C	D	E	F	G	H	I	J
300	61–100	2.156 (54.76)	1.556 (39.52)	0.850 (21.59)	0.750 (19.05)	0.281 (7.14)	0.828 (21.03)	0.250 (6.35)	0.125 (3.18)	0.646 (16.41)	0.300 (7.62)
	101–200	2.438 (61.93)	1.895 (43.05)	0.850 (21.59)	0.875 (22.22)	0.344 (8.74)	1.078 (27.38)	0.250 (6.35)	0.188 (4.78)	0.787 (19.99)	0.372 (9.45)
	201–400	2.750 (69.85)	1.844 (46.84)	0.860 (21.84)	1.000 (25.40)	0.406 (10.31)	1.344 (34.14)	0.250 (6.35)	0.250 (6.35)	0.926 (23.52)	0.453 (11.51)
	401–600	3.063 (77.80)	2.031 (51.59)	0.880 (22.35)	1.250 (31.75)	0.484 (12.29)	1.625 (41.28)	0.250 (6.35)	0.312 (7.92)	1.074 (27.28)	0.516 (13.11)
	601–800	3.375 (85.73)	2.219 (56.36)	0.891 (22.63)	1.750 (44.45)	0.547 (13.89)	2.078 (52.78)	0.250 (6.35)	0.375 (9.53)	1.222 (31.04)	0.578 (14.68)
	801–1200	4.000 (101.6)	2.531 (64.29)	1.078 (27.38)	2.000 (50.8)	0.609 (15.48)	2.516 (63.90)	0.250 (6.35)	0.438 (11.11)	1.441 (36.60)	0.735 (18.67)
600	61–100	2.953 (75.01)	2.352 (59.74)	1.640 (41.66)	0.750 (19.05)	0.281 (7.14)	0.828 (21.03)	0.500 (12.70)	0.125 (3.18)	0.646 (16.41)	0.300 (7.62)
	101–200	3.250 (82.55)	2.507 (63.67)	1.660 (42.16)	0.875 (22.22)	0.344 (8.74)	1.078 (27.38)	0.500 (12.70)	0.188 (4.78)	0.787 (19.99)	0.372 (9.45)
	201–400	3.625 (92.08)	2.719 (69.06)	1.730 (43.94)	1.000 (25.40)	0.406 (10.31)	1.625 (41.28)	0.500 (12.70)	0.250 (6.35)	0.926 (23.52)	0.453 (11.51)
	401–600	3.984 (101.19)	2.953 (75.01)	1.780 (45.21)	1.250 (31.75)	0.484 (12.29)	2.094 (53.19)	0.500 (12.70)	0.312 (7.92)	1.074 (27.28)	0.516 (13.11)
	601–800	4.328 (109.93)	3.172 (80.57)	1.875 (47.63)	1.750 (44.45)	0.547 (13.89)	2.516 (63.91)	0.500 (12.70)	0.375 (9.53)	1.207 (30.66)	0.578 (14.68)

[a]Tolerances: 61–200 A, 300 V, ±0.020 (±0.51 mm); 201–1200 A, 300 V and 61–800 A, 600 V, ±0.40 in (±1.02 mm).
[b]Column B tolerance: ±0.015 in (±0.38 mm).
[c]Columns D and J tolerance: ±0.020 in (±0.51 mm), except J tolerance ±0.028 (±0.71 mm) for 801–1200 A, 300 V.
[d]Tolerances: minus 0.000 in (minus 0.00 mm); 61–100 A, plus 0.005 in (plus 0.13 mm); 101–200 A, plus 0.006 in (plus 0.15 mm); 201–400 A, plus 0.007 in (plus 0.18 mm); 401–1200 A, plus 0.008 in (plus 0.20 mm).
[e]Column H tolerance: ±0.006 in (±0.15 mm).

**Fig 77
Class T Fuse Dimensions**

CLASS CC FUSE

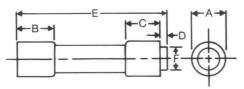

DIMENSIONS OF CLASS CC FUSES

Rating		Dimensions in Inches (mm)					
Volts	Amperes	Ferrule Diameter A[a]	Ferrule Length B[a]	Ferrule Length C[a]	Rejection Length D[a]	Overall Length E[b]	Rejection Diameter F[a]
600	0–30	0.405 (10.29)	0.375 (9.53)	0.375 (9.53)	0.125 (3.18)	1.500 (38.10)	0.250 (6.35)

[a]Tolerance: ±0.005 in (±0.13 mm)

[b]Tolerance: ±0.31 in (±0.79 mm)

Fig 78
Class CC Fuse Dimensions

CLASS R FUSES
Ferrule Type 0–60 A

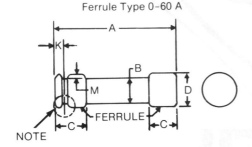

NOTE

Knife-Blade Type 61–600 A

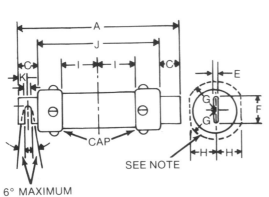

6° MAXIMUM

Fig 79
Class R Fuse Dimensions

CLASS R FUSES — FERRULE TYPE REJECTION
GROOVE DETAILS

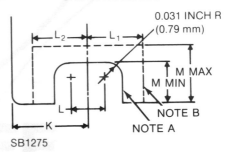

SB1275

Note A — Solid line indicates location, shape and dimensions for minimum rejection groove.

Note B — Dashed line indicates location, shape and dimensions for maximum rejection groove.

Table 25
Dimensions of Ferrule-Type Class R Fuses in Inches (mm)

Rating		Overall Length of Fuse[a]	Maximum Outside Diameter of Tube	Minimum Length of Ferrule	Outside Diameter of Ferrule[b]	Distance of Rejection Feature From End[c]	Minimum Width of Rejection Feature[d]	Minimum-Maximum Depth of Rejection Feature	Maximum Width Toward End	Maximum Width Toward Body
Volts	Amperes	A	B	C	D	K	L	M	L₂	L₁
									L_2	L_1
250	0–30	2.000 (50.80)	0.531 (13.49)	0.500 (12.70)	0.562 (14.27)	0.156 (3.96)	0.070 (1.78)	0.085–0.130 (2.16–3.30)	0.115 (2.92)	0.150 (3.81)
	31–60	3.000 (76.20)	0.781 (19.84)	0.625 (15.88)	0.812 (20.62)	0.188 (4.78)	0.086 (2.18)	0.085–0.130 (2.16–3.30)	0.123 (3.12)	0.170 (4.32)
600	0–30	5.000 (127.00)	0.781 (19.84)	0.500 (12.70)	0.812 (20.62)	0.188 (4.78)	0.086 (2.18)	0.085–0.130 (2.16–3.30)	0.123 (3.12)	0.170 (4.32)
	31–60	5.500 (139.70)	1.031 (26.19)	0.625 (15.88)	1.062 (26.97)	0.250 (6.35)	0.086 (2.18)	0.085–0.130 (2.16–3.30)	0.154 (3.91)	0.180 (4.57)

[a] Column A tolerance: ±0.031 inch (±0.79 mm).
[b] Column D tolerance: ±0.008 inch (±0.20 mm). To provide proper contact, the diameter of rejection ferrule end shall be equal to or not more than 0.050 inch (1.27 mm), smaller than actual diameter of main contact area for any fuse, and no part of rejection ferrule end shall protrude beyond the diameter of the main part of the ferrule.
[c] Column K tolerance: + 0.008, − 0.016 inch (+ 0.20, − 0.41 mm).
[d] Dimension column L: distance between centers of 0.031 inch (0.79 mm) radius fillets. Shape of rejection groove is not specified but shall be completely within solid and dashed lines regardless of shape.

Table 26
Dimensions of Blade-Type Class R Fuses in Inches (mm)

Rating Volts	Rating Amperes	Overall Length of Fuse[a] A	Minimum Length of Blade[b] C	Thickness of Blade[c] E	Width of Blade[d] F	Maximum Dimensions Over Projections — Measured Parallel to Blade G	Measured at Right Angles To Blade H	Minimum Distance from Midpoint of Fuse to Nearest Live Part I	Minimum-Maximum Overall Length of Cylindrical Body[e] J	Distance of Rejection Feature from End[f] K	Width of Rejection Feature[g] L	Web Width of Blade at Rejection Feature[h] N
250	61–100	5.875 (149.22)	1.000 (25.40)	0.125 (3.18)	0.750 (19.05)	0.656 (16.66)	0.594 (15.09)	1.031 (26.19)	3.375–3.781 (85.72–96.04)	0.500 (12.70)	0.281 (7.14)	0.250 (6.35)
	101–200	7.125 (180.98)	1.375 (34.92)	0.188 (4.78)	1.125 (28.58)	0.938 (23.83)	0.844 (21.44)	1.188 (30.18)	4.125–4.281 (104.78–108.74)	0.688 (17.48)	0.281 (7.14)	0.438 (11.13)
	201–400	8.625 (219.08)	1.875 (47.62)	0.250 (6.35)	1.625 (41.28)	1.203 (30.56)	1.203 (30.56)	1.188 (30.18)	4.625–4.813 (117.48–122.25)	0.938 (23.83)	0.406 (10.31)	0.625 (15.88)
	401–600	10.375 (263.52)	2.250 (57.15)	0.250 (6.35)	2.000 (50.80)	1.453 (36.91)	1.453 (36.91)	1.531 (38.89)	5.188–5.813 (131.78–147.65)	1.125 (28.58)	0.531 (13.49)	0.750 (19.05)
600	61–100	7.875 (200.02)	1.000 (25.40)	0.125 (3.18)	0.750 (19.05)	0.781 (19.84)	0.719 (18.26)	1.750 (44.45)	5.375–5.781 (136.52–146.84)	0.500 (12.70)	0.281 (7.14)	0.250 (6.35)
	101–200	9.625 (244.48)	1.375 (34.92)	0.188 (4.78)	1.125 (28.58)	1.062 (26.97)	0.984 (24.99)	2.250 (57.15)	6.125–6.781 (155.58–172.24)	0.688 (17.48)	0.281 (7.14)	0.438 (11.13)
	201–400	11.625 (295.28)	1.875 (47.62)	0.250 (6.35)	1.625 (41.28)	1.453 (36.91)	1.453 (36.91)	2.500 (63.50)	7.125–7.831 (180.98–198.90)	0.938 (23.83)	0.406 (10.31)	0.625 (15.88)
	401–600	13.375 (339.72)	2.250 (57.15)	0.250 (6.35)	2.000 (50.80)	1.719 (43.66)	1.719 (43.66)	2.688 (68.28)	8.188–8.844 (207.98–224.64)	1.125 (28.58)	0.531 (13.49)	0.750 (19.05)

[a] Tolerances: 61–200 A, ±0.062 inch (±1.57 mm); 201–600 A, ±0.094 inch (±2.39 mm).
[b] The length of one blade shall not be more than 0.062 inch (1.57 mm) longer than the other blade.
[c] Column E tolerance: ±0.003 inch (±0.08 mm).
[d] Column F tolerance: ±0.035 inch (±0.89 mm).
[e] The length of the cylindrical body may be less than the indicated value if other acceptable interference means (pins through the blades, collars, or the like) are provided to prevent mounting the fuse in a fuseholder that will accommodate a fuse rated in the next lower bracket of current rating.
[f] Column K tolerance: ±0.008 inch (±0.20 mm).
[g] Column L tolerance: − 0.005, + 0.025 inch (− 0.13, + 0.64 mm). Dimension is diameter of slot at semicircle. Maximum rounding of corner at end of slot 0.125 inch (3.18 mm) radius.
[h] Column N tolerance: ±0.031 inch (±0.79 mm).

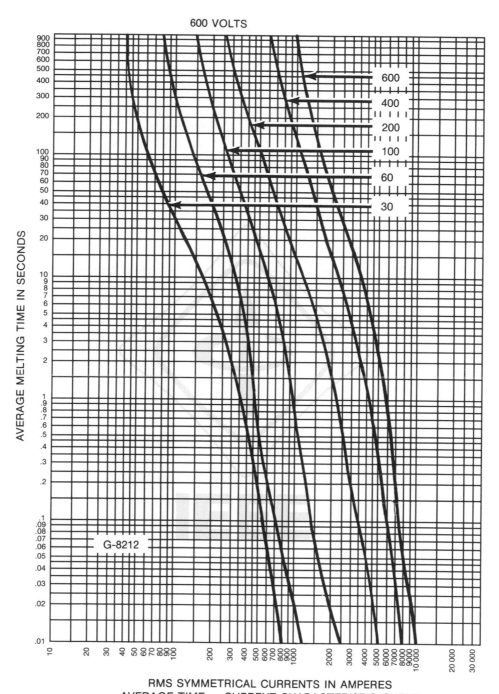

600 VOLTS

AVERAGE MELTING TIME IN SECONDS

RMS SYMMETRICAL CURRENTS IN AMPERES
AVERAGE TIME — CURRENT CHARACTERISTIC CURVE

NOTE: For illustration purposes only. Refer to fuse manufacturer for specific and up-to-date data.

**Fig 80
Time-Delay Fuses, Class RK-5**

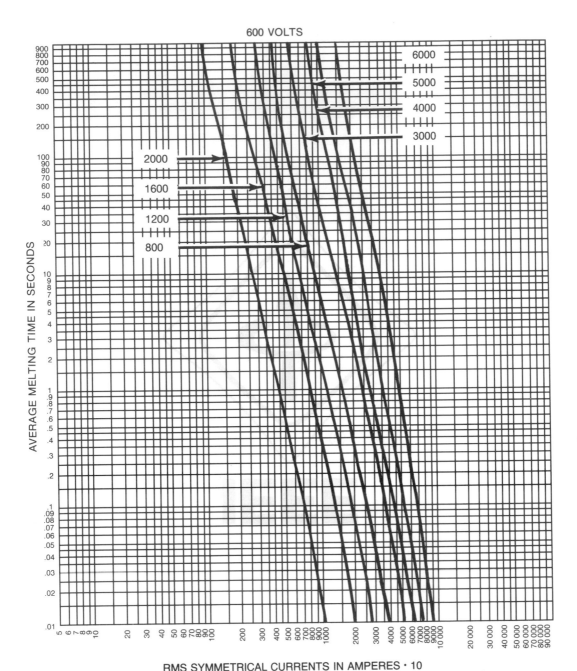

600 VOLTS

RMS SYMMETRICAL CURRENTS IN AMPERES · 10
AVERAGE TIME-CURRENT CHARACTERISTIC CURVE

NOTE: For illustration purposes only. Refer to fuse manufacturer for specific and up-to-date data.

Fig 81
Current-Limiting Fuse, Class L

can be avoided by the proper selection of overcurrent protective devices. Selectivity (often referred to as selective coordination) may be defined as the complete isolation of a faulted circuit to the point of fault without disturbing any of the other protective devices in the system. Figure 82 shows a selective system.

Figure 83 illustrates the general principle by which current-limiting fuses coordinate with one another for any value of short-circuit current sufficient for them to operate in the current-limiting mode. Note that for selectivity, the total clearing I^2t of fuse B must be less than the minimum melting I^2t of fuse A.

In low-voltage fuse applications, coordination may sometimes be achieved through the use of selectivity ratio tables (5.9.1).

5.9.1 Selectivity Ratio Tables. Table 27 shows a typical selectivity schedule for various combinations of fuses. This schedule is very general and does not apply to all lines of fuses. Specific data are available from the fuse manufacturers.

An example of using Table 27 is found in Fig 84, where a 1200 A Class L fuse is to be selectively coordinated with a 400 A Class J5 current-limiting fuse.

Selectivity schedules or tables are used as a simple check for selectivity, assuming that identical or reduced fault currents flow through the circuits in descending order, that is, from main, to feeder, to branch. Where closer fuse sizing than indicated is desired, check with the fuse manufacturer as the ratios may be reduced for lower values of short-circuit current. A coordination study may be desired when the simple check as outlined is not sufficient and can be accomplished by plotting fuse time-current characteristic curves on standard log–log graph paper. If fuse ratios for high- or medium-voltage fuses to low-voltage fuses are not available, it is recommended that the fuse curves in question be plotted on log–log paper. Also, fuse ratios cannot be used with fuses at different voltages or from different manufacturers. Fuse manufacturers can furnish selectivity tables showing actual ampere ratings.

5.9.2 Time-Current Characteristic Curves. Time-current characteristic for fuse curves are available in the form of minimum melting and total clearing curves on transparent paper, which is easily adapted to tracing. A typical example of coordinating high-voltage and low-voltage fuses using graphic analysis is shown in Fig 85. Note that the total clearing curve of the 1200 A fuse is plotted against the minimum melting curve of the 125E rated 5 kV fuse. The curves are referred to 240 V for a study of secondary faults.

5.10 Current-Limiting Characteristics. Due to the speed of response to short-circuit currents, current-limiting fuses have the ability to cut off the current before it reaches its full prospective short-circuit value. Figures 71, 72, and 83 show the current-limiting action of fuses. The available short-circuit current would flow if there were no fuse in the circuit or if a noncurrent-limiting protective device were in the circuit. A current-limiting fuse will limit the peak current to a value less than that available and will open in one half-cycle or less in its current-limiting range, therefore letting through only a portion of the available short-circuit energy. The degree of current limitation is usually represented in the form of peak let-through-current charts.

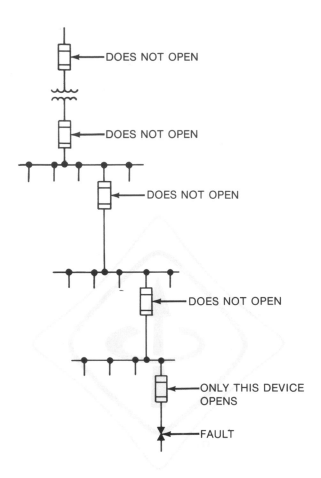

Fig 82
Selective Operation of Overcurrent Protective Devices

5.10.1 Peak Let-Through Current Charts. Peak let-through current charts (sometimes referred to as current-limiting effect curves) are useful from the standpoint of determining the degree of short-circuit protection that a current-limiting fuse provides to the equipment beyond it. These charts show fuse instantaneous peak let-through current as a function of available symmetrical rms current, as shown in Fig 86. The straight line running from the lower left to the upper right shows a 2.3 relationship (based on a 15% power factor) between the instantaneous peak current that would occur without a current-limiting device in the circuit and the available symmetrical rms current. The following data can be determined from the let-through current charts and are useful in relating to equipment withstandability:

    (1) Peak let-through current (magnetic effect)
    (2) Apparent equivalent symmetrical rms let-through current (heating effect)
    (3) Current magnitude for less than one half-cycle clearing time (fuse is operating in current-limiting range)

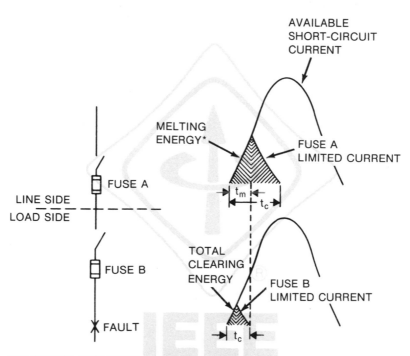

*INDICATES, BUT DOES NOT EQUAL, ENERGY.

**Fig 83
Selectivity of Fuses; Total Clearing Energy of Fuse B Must Be
Less Than Melting Energy of Fuse A**

Table 27
Typical Selectivity Schedule* for UL Listed Fuses

Line Side	Load Side					
	Class L Fuse 601-6000 A	Class K1 Fuse 0–600 A	Class J Fuse 0–600 A	Class K5 Time-Delay Fuse 0–600 A	Class J Time-Delay Fuse 0–600 A	Class G Fuse 0–60 A
Class L Fuse 601–6000 A	2:1	2:1	2:1	6:1	5:1	—
Class K1 Fuse 0–600 A		3:1	3:1	8:1	4:1	4:1
Class J Fuse 0–600 A		3:1	3:1	8:1	4:1	4:1
Class K5 Time-Delay Fuse 0–600 A		1.5:1	1.5:1	2:1	1.5:1	1.3:1
Class K5 Time-Delay Current-Limiting Fuse 0–600 A		1.5:1	1.5:1	4:1	2:1	2:1
Class J Time-Delay Fuse 0–600 A		1.5:1	1.5:1	4:1	2:1	2:1

* Exact ratios vary with ampere ratings, system voltage, and short-circuit current.
NOTE: For illustration purposes only. Refer to fuse manufacturer for specific and up-to-date data.

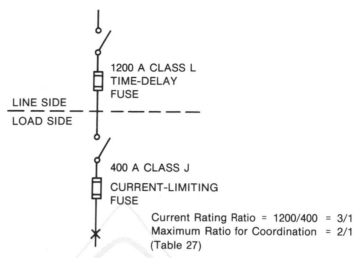

NOTE: Check specific manufacturer for exact data.

Fig 84
**Typical Application Example of Table 4; Selective Coordination is Apparent
as Fuses Meet Coordination-Ratio Requirements**

NOTE: This procedure will yield a conservative value of symmetrical rms let-through current if the component has been given a withstand time rating of one half-cycle or longer under a test circuit power factor of 15% or larger.

These data may be compared to short-circuit withstand data of those components that are static in nature, such as wire, cable, bus, etc, so that engineering decisions can be made concerning short-circuit protection and equipment bracing.

An example showing the application of the let-through-current charts is represented in Fig 87, where the load-side component is protected by an 800 A current-limiting fuse. It is desired to determine the fuse let-through current values with 40 000 A, rms symmetrical, available at the line side of the fuse. Enter the let-through current chart of Fig 86 at an available current of 40 000 A, rms symmetrical, and find a fuse peak let-through current of 40 000 A and an effective rms current of 17 000 A. The clearing time will be less than one half-cycle. The downstream circuit components will not be subjected to an I^2t duty greater than the total clearing let-through I^2t of the fuse.

5.10.2 Maximum Clearing I^2t and Peak Let-Through Current Ip. I^2t is a measure of the energy that a fuse lets through while clearing a fault. Every piece of electrical equipment is limited in its capability to withstand energy. When equipment has an I^2t withstand rating, it can be compared with maximum clearing I^2t values for fuses. These maximum clearing I^2t values are available from fuse manufacturers. (UL does not have any procedures to assign I^2t withstand ratings to equipment.)

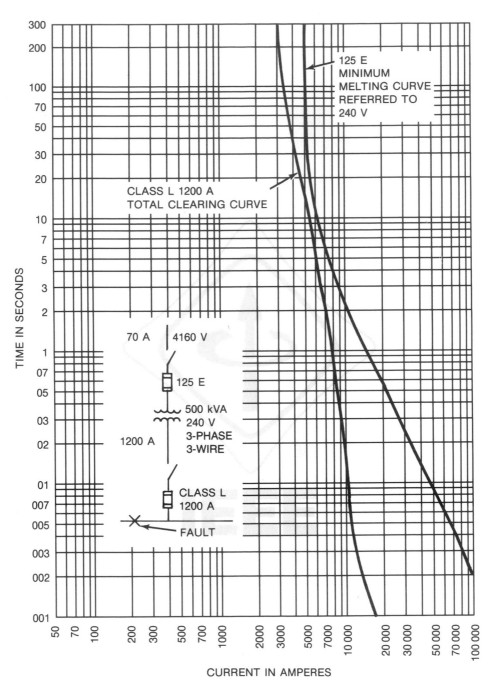

NOTE: For illustration purposes only. Refer to fuse manufacturer for specific and up-to-date data.

Fig 85
Typical Coordination Study of Primary and Secondary Fuses
Showing Selective System

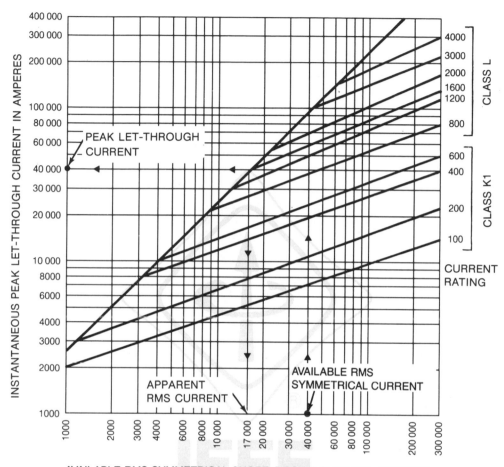

Fig 86
Typical 60 Hz Peak Let-Through Current as a Function of
Available RMS Symmetrical Current—15 Percent Power Factor

Magnetic forces can be substantial under short-circuit conditions and should also be examined. These forces vary with the square of the peak current Ip^2 and can be reduced considerably when current-limiting fuses are used, recognizing that the duration of the short-circuit current is an important factor. Some types of electrical equipment should be examined from the standpoint of peak current withstand, as well as I^2t withstand.

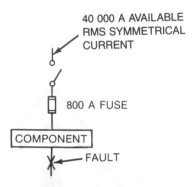

Fig 87
Example for Applying Fuse Let-Through Charts

5.11 Special Applications for Low-Voltage Fuses

5.11.1 Bus-Bracing Requirements. Reduced bus-bracing requirements may be attained when current-limiting fuses are used. Figure 88 shows an 800 A motor-control center being protected by 800 A Class L fuses. The maximum available fault current to the motor-control center (taking into consideration future growth) is 40 000 A rms symmetrical. If a noncurrent-limiting device were used ahead of the motor-control center, the bracing requirement would be a minimum of 40 000 A rms symmetrical, with possible offset giving a peak value of 92 000 A (2.3 · 40 000 A). For this example with current-limiting fuses, the maximum peak current has been reduced from 92–40 kA with a corresponding reduction in effective rms available current from 40–17.4 kA; then bus bracing of 17.4 kA or greater is possible rather than requiring the full 40 kA bracing.

5.11.2 Circuit Breaker Protection. Circuit breakers protected by current-limiting fuses may be applied in circuits where the available short-circuit current exceeds the interrupting rating of the circuit breakers alone. The short-circuit withstand data for circuit breakers can be compared to fuse let-through current values to determine the degree of protection provided. *The circuit breaker and fuse manufacturers should be contacted for proper applications.*

An example of applying fuses to protect molded-case circuit breakers is shown in Fig 89, where a 225 A lighting panel has circuit breakers with an interrupting

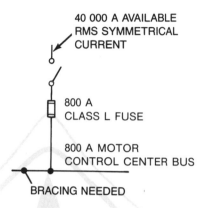

Fig 88
Example for Determining Bracing Requirements of 800 A Motor-Control Center

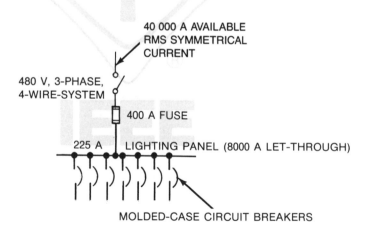

Fig 89
Application of Fuses to Protect Molded-Case Circuit Breakers

capacity of less than 40 000 A rms symmetrical. The available fault current at the line side of the lighting panel is 4000 A rms symmetrical. A 400 ampere RK1 fuse would reduce the current at the breakers to an effective 8000 A rms available. The breakers then need to be able to withstand 8000 A rms or more.

5.11.3 Wire and Cable Protection. Fuses must be sized for conductor protection according to the NEC [9]. Where noncurrent-limiting fuses are used, short-circuit protection for small conductors may not be available and reference should be made to the wire damage charts for short-circuit withstands of copper and aluminum cable in ICEA P32-283 (1969) [19]. (Also, see Chapter 8.)

Due to the current-limiting ability of fuses, small conductors are protected from high-magnitude short-circuit currents even though the fuse may be 300–400% of the conductor rating as allowed by the NEC [9] for nontime-delay fuses for motor branch circuit protection.

5.11.4 Motor Starter Short-Circuit Protection. UL tests motor starters under short-circuit conditions. The short-circuit test performed may be used to establish a withstand rating for starters. UL tests starters of 50 hp and under with 5000 A of available short-circuit current and uses one-time fuses sized at 40% of the maximum continuous-current rating of the starter. Starters over 50 hp in size are tested in similar fashion, except with greater available fault currents. (ANSI/UL 508-1983 [17], 13.6.)

When applying starters in systems with high available fault currents, current-limiting fuses should be used to reduce the let-through energy to a value less than that established by the UL test procedures described. *The starter manufacturer should be contacted for proper applications.*

Figure 90 is a typical one-line diagram of a motor circuit, where the available short-circuit current has been calculated to be 40 000 A rms symmetrical at the motor-control center and the fuse is to be selected such that short-circuit protection is provided. The fuse selected should limit the fault current to 5000 A rms (the withstand rating of the motor starter).

The 1971 NEC originally recognized motor short-circuit protectors, which are fuse-like devices having unique dimensions for use only in combination starters. They provide interrupting ratings up to 100 000 A. Proper selection is made by referring to motor starter manufacturers' application literature.

5.11.5 Transformer Protection. Low-voltage distribution-type transformers are often equipped on the high-voltage side (above 600 V) with medium-voltage fuses sized for short-circuit protection. Transformer overload protection may be provided by fusing the low-voltage secondary with appropriate fuses sized at 100–125% of the transformer secondary full-load amperes. Figure 91 shows a proper size of low-voltage fuse for a 1000 kVA transformer to provide overload protection. For complete protection of transformers, refer to Chapter 10.

Transformers are frequently used in low-voltage electrical distribution systems to transform 480 V to 208Y/120 V. For these types of transformers appropriate time-delay fuses should be provided, sized at 100–125% of the primary full-load current. Some consideration should be given to the magnetizing inrush current since dry-type and oil-filled transformers have inrush currents equivalent to about 12 times full-load rating lasting for 0.1 s—also about 20 to 25 times

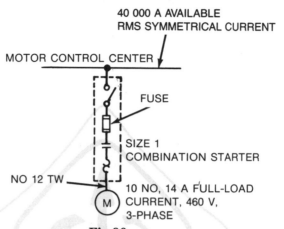

40 000 A AVAILABLE
RMS SYMMETRICAL CURRENT

MOTOR CONTROL CENTER

FUSE

SIZE 1
COMBINATION STARTER

NO 12 TW

10 NO, 14 A FULL-LOAD
CURRENT, 460 V,
3-PHASE

**Fig 90
Application of Fuses to Provide Short-Circuit Protection
As Well As Backup Protection for Motor Starters**

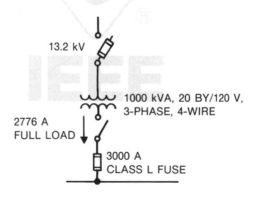

13.2 kV

1000 kVA, 20 BY/120 V,
3-PHASE, 4-WIRE

2776 A
FULL LOAD

3000 A
CLASS L FUSE

**Fig 91
Sizing of Low-Voltage Fuses for Transformer Secondary Protection**

about 20 to 25 times rating for 0.01 s. These inrush currents can be easily checked against the minimum melting curve, so that needless opening may be avoided; if necessary, a larger size time-delay fuse may need to be selected. Figure 92 shows a 225 kVA lighting transformer with time-delay fuses. See Chapter 10 for *transformer protection*.

5.11.6 Motor Overcurrent Protection. Single- and three-phase motors can be protected by specifying time-delay fuses for motor running overload protection according to the NEC [9]. These ratings will depend on service factor or temperature rise, or both. Where overload relays are used in motor starters, a larger size time-delay fuse may be used to coordinate with the overload relays and provide short-circuit protection.

Combination motor starters that employ overload relays sized for motor running protection (100–115%) can incorporate time-delay fuses sized at 125% or the next larger standard size to serve as backup protection. (Time-delay fuses sized up to 175% may be used for branch circuit protection only.) A combination motor starter with backup fuses will provide excellent protection, motor control, and flexibility. Figure 93 illustrates the use of fuses for protection of a motor circuit.

During the period that motors are operated near full-load single-phasing protection may be provided by time-delay fuses sized at approximately 125% of motor full-load current. Loss of one phase, either primary or secondary, will result in an increase in the line current to the motor. This will be sensed by the motor fuses as they are sized at 125% and the single-phasing current will open the fuses before damage to the windings results. If the motors are operated at less than full load, the overload relays and dual-element fuses should be sized to the actual running amperes of the motor. For example, if a motor with a full-load rating of 10 A is being used in a situation where it is drawing only 8 A, the dual-element fuses should be sized at 10 A instead of 12 A.

Medium- and High-Voltage Fuses—2.3–138 kV

5.12 ANSI Standards—Medium Voltage and High Voltage (2.4–138 kV)

(1) ANSI/IEEE C37.40-1981 [6]
(2) ANSI/IEEE C37.41-1981 [7]
(3) ANSI C37.42-1981 [1]
(4) ANSI C37.46-1981 [2]
(5) ANSI C37.47-1981 [3]
(6) ANSI C37.48-1969 (R1974) [4]

5.13 Types of Medium- and High-Voltage Fuses. Medium- and high-voltage fuses are used extensively in industrial, commercial, and institutional power distribution systems; they are available in a wide range of voltages (2.3–138 kV). Medium-voltage fuses fall into two general categories: distribution fuse cutouts and power fuses. The applicable standards, ANSI C37.46-1981 [2] and NEMA SG2-1981 [21], differentiate between two categories on the basis of dielectric strength—power fuses having greater dielectric strength.

Medium- and high-voltage fuses are also classified according to their use as follows:

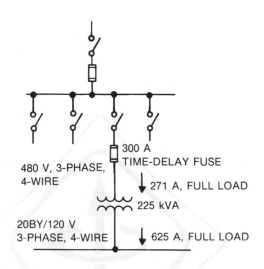

Fig 92
Application of Time-Delay Fuses for Transformer Protection

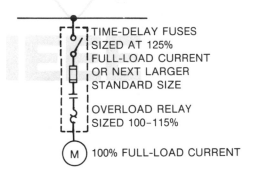

Fig 93
Application of Time-Delay Fuses for Typical Motor Circuit

(1) Medium-voltage fuses suitable for application within buildings, vaults, or enclosures. This classification is comprised of solid-material (boric acid) fuses rated 4.16–34.5 kV and current-limiting fuses rated 2.4–34.5 kV. These represent the majority of medium-voltage-range fuses used in industrial, institutional, and commercial power system protective schemes.

(2) Medium- and high-voltage fuses suitable for outdoor application only. This classification is comprised of solid-material (boric acid) fuses rated 4.16–138 kV, fiberlined expulsion fuses rated 7.2–161 kV, and distribution fuse cutouts rated 4.8–138 kV. This classification of power fuse is typically applied in outdoor substations supplying service from a utility distribution or subtransmission system to an industrial power system (at voltages ranging from 2.4–138 kV), or in industrial distribution systems (at voltages ranging from 2.4–34.5 kV).

5.14 Types of High-Voltage Fuses

5.14.1 Distribution Fuse Cutouts. According to ANSI/IEEE C37.100-1981 [8], the distribution cutout is identified by the following characteristics:

(1) Dielectric withstand (basic impulse insulation level) strengths at distribution levels

(2) Application primarily on distribution feeders and circuits

(3) Mechanical construction basically adapted to pole or crossarm mounting, except for the distribution oil cutout

(4) Operating voltage limits corresponding to distribution system voltage

Characteristically, a distribution fuse cutout consists of a special insulating support and fuseholder. The fuseholder, normally a disconnecting type, engages contacts supported on the insulating support and is fitted with a simple inexpensive fuse link (Fig 94). The fuseholder is lined with an organic material,

Fig 94
Open Distribution Fuse Cutout, Rated 100 A

usually horn fiber. Interruption of an overcurrent takes place within the fuseholder by the action of deionizing gases liberated when the liner is exposed to the heat of the arc established when the fuse link melts in response to the overcurrent.

Distribution fuse cutouts were developed for use in overhead distribution circuits. They are commonly applied on such circuits, where their principal application is in connection with distribution transformers supplying a residential area or a small commercial or industrial plant. Cutouts provide protection to the distribution circuit by de-energizing and isolating a faulted transformer. Another application is the fault protection of small pole-mounted capacitor banks used for power-factor correction or voltage regulation.

Distribution fuse cutouts are available for use on outdoor overhead distribution circuits with operating voltages up to 14.4 kV in maximum continuous-current ratings of 100 and 200 A, and 25 and 34.5 kV in a maximum rating of 100 A. The maximum interrupting ratings, expressed in rms symmetrical amperes, are given in Table 28.

5.14.2 Power Fuses. According to ANSI/IEEE C37.100-1981 [8], the power fuse is identified by the following characteristics:

(1) Dielectric withstand (basic impulse insulation level) strengths at power levels

(2) Application primarily in stations and substations

(3) Mechanical construction basically adapted to station and substation mountings

Power fuses have other characteristics that differentiate them from distribution fuse cutouts in that they are available in higher voltage ratings, higher continuous-current ratings, higher interrupting current ratings, and in forms suitable for indoor and enclosure application, as well as all types of outdoor applications.

A power fuse consists of a fuse support plus a fuse unit, or alternately a fuseholder that accepts a refill unit or fuse link.

The two basic types of power fuses, expulsion type and current-limiting type, effect interruption of overcurrents in a radically different manner. The expulsion type, like the distribution cutout, interrupts overcurrents through the deionizing action of the gases liberated from the lining of the interrupting chamber of the

Table 28
Maximum Short-Circuit Interrupting Ratings for Distribution
Fuse Cutouts

Nominal Rating, kV	Short-Circuit Interrupting Rating, Amperes, rms Symmetrical
4.8	12 500
7.2	15 000
14.4	13 200
25.0	8000
34.5	5000

fuse by the heat of the arc established when the fusible element melts. The current-limiting type interrupts overcurrents when the arc established by the melting of the fusible element is subjected to the mechanical restriction and cooling action of a powder or sand filler surrounding the fusible element.

The earliest form of high-voltage power fuse was the fiber-lined expulsion type, an outgrowth of the distribution fuse cutout, employing longer and heavier fuseholder tubes to cope with higher circuit voltages and short-circuit interrupting requirements. The expulsion type power fuse possesses operating characteristics similar to those of a distribution cutout, except that the noise and emission of gases and flame are greatly magnified as the holder tubes are lengthened and strengthened to handle higher voltages and fault currents. Therefore, this type of fuse has been restricted to outdoor usage only and generally in substations that are remotely located from human habitation. Fiber-lined expulsion type power fuses are still used for the fault protection of small- and medium-size power transformers or substation capacitor banks.

The limited interrupting capacity of these early expulsion type power fuses, coupled with their inability to be used within buildings or enclosures, led to the development in the US during the 1930's of a new form of expulsion fuse known as the boric acid or solid-material fuse. In this fuse the material used to obtain the deionizing action necessary to interrupt overcurrents was not organic, but solid boric acid powder molded into a dense lining for the interrupting chamber of the fuse. The benefits gained by the use of boric acid are listed below.

(1) For identical physical dimensions of interrupting chambers, a fuse lined with boric acid can interrupt (a) a higher voltage circuit, (b) a higher value of overcurrent in a structure of the same strength, (c) the full range of currents from minimum melting to the interrupting rating, with (d) lower arc energy and reduced emission of gases and flame than is possible in a fuse where organic fiber is used as the liner.

(2) The gas liberated from the boric acid is noncombustible and highly deionized, and thus reduces the amount of flame discharged. As a result, the clearances in the path of the exhaust gases necessary to prevent an electrical flashover of circuits or devices are often reduced to virtually the clearance distances required in air.

(3) The most significant feature obtained by the use of boric acid is the ability to control or completely eliminate the gas discharge from the fuse on operation. When exposed to the heat of the arc, boric acid liberates steam, which can be condensed to the liquid stage by venting the gas into a suitably designed cooling device. This allows use indoors or in small size enclosures.

(4) The solid-material type power fuse permits wider use of power fuses. It can operate without objectionable noise or emission of flame and gases in a greatly expanded range of current and interrupting ratings.

(5) Virtually simultaneously with the development of the solid-material boric acid power fuse, an American version of the European current-limiting or "silver-sand" fuse was introduced. Two forms evolved. One was a line of current-limiting fuses to be used with, and coordinated with, high-voltage motor starters for high capacity 2400 and 4160 V distribution circuits. The second was a

line of current-limiting fuses suitable for use with 2400–34 500 V voltage, distribution, and small power transformers.

The medium-voltage current-limiting power fuse has three features that have led to its extensive usage on high-capacity medium-voltage power-distribution circuits.

(1) Interruption of overcurrents is accomplished quickly without the expulsion of arc products or gases, as all the arc energy of operation is absorbed by the sand filler of the fuse and subsequently released as heat at relatively low temperatures. This enables the current-limiting fuse to be used indoors or in enclosures of small size, as there is no noise accompanying operation. Furthermore, since there is no discharge of hot gases or flame, only normal electrical clearances need be provided.

(2) Current-limiting action or reduction of current through the fuse to a value less than that available from the power-distribution system at the location of the fuse occurs if the value of overcurrent greatly exceeds the continuous-current rating of the fuse. Such a current reduction reduces the stresses and possible damage to the circuit up to the fault or to the faulted equipment itself. In the case of current-limiting fuses used with a medium-voltage motor starter, the contactor is only required to have momentary current and to make current capabilities equal to the maximum let-through current of the largest current rating of fuse that will be used in the starter.

(3) Very high interrupting ratings are achieved by virtue of current-limiting action so that current-limiting power fuses can be applied on medium-voltage distribution circuits of very high short-circuit capacity.

Current-limiting fuses by nature generate transient overvoltage when they function. This overvoltage should be coordinated with the spark-over voltage of surge arresters. Fuse manufacturers should be consulted if this is a question. The transient overvoltage created by current-limiting fuses is impressed on the equipment on the line side to the fuses. Motors are on the load side of the fuses.

5.15 Fuse Ratings

5.15.1 Medium- and High-Voltage Fiber-Lined Expulsion Type Power Fuses.
This category has its principal usage in outdoor applications at the subtransmission voltage level. The available ratings of this fuse are given in Table 29.

5.15.2 Medium- and High-Voltage Solid-Material Boric Acid Fuses.
These fuses are available in two styles: (1) the fuse-unit style in which the fusible element, interrupting element, and operating element are all combined in an insulating tube structure, with the entire unit being replaceable, and (2) the fuseholder and refill-unit style (for system voltages through 34.5 kV) of which only the refill unit is replaced after operation.

Fuses of style (1) are principally used outdoors at subtransmission voltages (Fig 95). However, fuses in this style are also available for use at distribution voltages up to 34.5 kV, in current ratings up to 200 A. The fuse units are specifically designed for use in outdoor pole-top or station-style mountings, as well as indoor mountings installed in metal-enclosed interrupter switchgear,

Table 29
Maximum Continuous-Current and Short-Circuit Interrupting Ratings
for Horn Fiber-Lined Expulsion Type Fuses

Rated Maximum Voltage, kV	Continuous-Current Ratings, Amperes (Maximum)	Maximum Interrupting Rating,* kA, rms Symmetrical
8.3	100, 200, 300, 400	12.5
15.5	100, 200, 300, 400	16.0
25.8	100, 200, 300, 400	20.0
38.0	100, 200, 300, 400	20.0
48.3	100, 200, 300, 400	25.0
72.5	100, 200, 300, 400	20.0
121.0	100, 200	16.0
145.0	100, 200	12.5
169.0	100, 200	12.5

*Applies to all continuous-current ratings.

Fig 95
Substation Serving Large Industrial Plant at
69 kV; Combination of Interrupter Switch
with Shunt Trip Device and Solid-Material
Boric-Acid Power Fuses Provides Protection
for Power Transformer

indoor vaults, and pad-mounted gear. Indoor mountings incorporate an exhaust control device that contains most of the arc interruption products and virtually eliminates noise accompanying a fuse operation. These exhaust control devices do not require a reduction of the fuse's interrupting rating.

Indoor mountings for use with fuse units up to 25 kV can be furnished with an integral hookstick-operated load-current-interrupting device, thus providing for single-pole live switching in addition to the fault interrupting function provided by the fuse.

The ratings of fuses of style (1) are given in Table 30.

The fuses of style (2) are used either indoors or outdoors at medium distribution voltages (Fig 96) and their ratings are given in Table 31. These fuses are also available with integral load-current-interrupting devices for single-pole live switching.

5.15.3 Current-Limiting Fuses.

Current-limiting power fuses suitable for the protection of voltage transformers, auxiliary power transformers, power transformers, and capacitor banks are available in the ratings given in Table 32.

The interrupting rating of fuses for voltage transformers is about 2000 MVA, three-phase symmetrical, for voltages of 14.4–34.5 kV.

Some current-limiting power fuses are E-rated. Fuses rated 100E and less open in 300 s at currents between 200 and 240% of their E ratings. Fuses rated above 100E open in 600 s at currents between 220 and 264% of their E ratings. E-rated fuses are classified as general-purpose fuses.

Some current-limiting distribution fuses are C rated. These fuses open in 1000 s at currents between 170 and 240% of their C ratings.

Current-limiting fuses suitable for use with medium-voltage motor starters are designated by an R rating. These R-rated fuses are available in ratings from 2R–36R up to 5500 V and have an asymmetrical interrupting rating up to 80 000 A rms at 5000 V. Standards for these fuses assign maximum continuous-current ratings of 70 A for the 2R fuse and up to 650 A for the 36R fuse. These ratings are at an ambient temperature of 55 °C. The R rating typically indicates a 20 s melting point at 100 times the R rating of the fuse. Thus, a 12R fuse opens in 20 s at 1200 A. These fuses are selected by their characteristic to coordinate with

Table 30
Maximum Continuous-Current and Short-Circuit Interrupting Ratings
for Solid-Material Boric Acid Fuses (Fuse Units)

Rated Maximum Voltage, kV	Continuous-Current Ratings, Amperes (Maximum)	Short-Circuit Maximum Interrupting Ratings, kA, rms Symmetrical
17.0	200	14.0
27.0	200	12.5
38.0	100, 200, 300	6.7, 17.5, 33.5
48.3	100, 200, 300	5.0, 13.1, 31.5
72.5	100, 200, 300	3.35, 10.0, 25.0
121.0	100, 250	5.0, 10.5
145.0	100, 250	4.2, 8.75

**Fig 96
Solid-Material Boric-Acid Power Fuse
Rated 14.4 kV with Controlled-Venting Device:
200 A Indoor-Disconnecting Type**

**Table 31
Maximum Continuous-Current and Short-Circuit Interrupting Ratings
for Solid-Material Boric Acid Fuses (Refill Units)**

Rated Maximum Voltage, kV	Continuous-Current Ratings, Amperes (Maximum)	Short-Circuit Maximum Interrupting Ratings, kA, rms Symmetrical
2.75	200, 400, 720*	19.0, 37.5, 37.5
4.8	200, 400, 720*	19.0, 37.5, 37.5
8.3	200, 400, 720*	16.6, 29.4, 29.4
15.5	200, 400, 720*	14.4, 34.0, 29.4
25.8	200, 300	12.5, 20.0, 21.0, 21.0
38.0	200, 300	6.9, 17.5, 16.8, 16.8

*Parallel fuses.

the motor and motor controller to provide short-circuit protection. Figure 97 shows how R-rated fuses are mounted in a typical fused medium-voltage drawout-type motor starter.

5.16 Areas of Application of Fuses. Power fuses can be utilized on systems rated up to 138 kV to protect the transmission circuits, power and voltage

Table 32
Maximum Continuous-Current and Short-Circuit Interrupting Ratings for Current-Limiting Fuses

Rated Maximum Voltage, kV	Continuous-Current Ratings, Amperes (Maximum)	Short-Circuit Maximum Interrupting Ratings, kA, rms Symmetrical
2.75	225, 450*, 750*, 1350*	50.0, 50.0, 40.0, 40.0
2.75/4.76	450*	50.0
5.5	225, 400, 750*, 1350*	50.0, 62.5, 40.0, 40.0
8.3	150, 250*	50.0, 50.0
15.5	65, 100, 125*, 200*	85.0, 50.0, 85.0, 50.0
25.8	50, 100*	35.0, 35.0
38.0	50, 100*	35.0, 35.0

*Parallel fuses.

transformers, and capacitor banks. Power fuses can also provide backup protection for transformer secondary faults.

5.16.1 Power Distribution. From the basic concept of overcurrent protection applicable to the medium-voltage distribution systems used in industrial plants and commercial buildings, it becomes apparent that the principal functions of overcurrent protective devices at these voltages are to detect fault conditions in elements of the medium-voltage circuits and to interrupt these high values of overcurrent. Fuses provide fast clearing of high-magnitude fault currents. Their speed of operation for low and medium magnitudes of fault currents should be checked to determine whether supplementary protection (such as ground-fault protection) is needed to clear these arcing type faults. Their secondary function is to act as backup overcurrent protection in the event the next overcurrent device closer to the fault either fails to operate due to a malfunction or operates too slowly due to incorrect (higher) ratings or settings.

Modern medium-voltage power fuses have the capabilities and characteristics to provide this vital overcurrent protection for virtually all types and sizes of distribution systems ranging from a simple radial circuit where power is purchased at medium voltage with a transformer to provide utilization voltage, up to extensive primary-selective circuits supplying several transformer substations. Such fuses used with load-interrupter switches may be applied outdoors, in vaults, or in metal-enclosed switchgear. (See Fig 96.)

Use of relatively inexpensive fuses allows segmentation of radial systems and protection of individual loads. This can improve system reliability by interrupting the least amount of load in order to remove a faulted segment. Fuses are widely used on main switchgear busses, where their fast operating times (less than one half-cycle for current-limiting fuses when operating in their current-limiting range of 1–2 cycles for power fuses on high-current faults) reduce damage to faulted equipment.

5.16.2 Fuses for Medium-Voltage Motor Circuits. The power requirements for large motors driving compressors for air-conditioning systems, pumps, and numerous other loads in industrial plants are frequently too great for low-voltage distribution circuits, so motors operating at 2400 and 4160 V and higher are now commonly used. Since the distribution circuits supplying such a motor usually have many parallel loads, their load-current and short-circuit current capacity is high. The latter condition imposes a severe duty on the motor starter. A coordinated current-limiting fuse/contactor combination is available for motor starting and control devices for motors up to about 2500 hp on systems rated up to 13 800 V.

The current-limiting fuses for this service fall into a special performance category and are designated by an R rating. In addition to carrying normal motor load current plus harmless transient overloads, they must repeatedly withstand starting inrush currents without deteriorating or operating. Such inrush currents occasionally exceed six times the motor full-load current and last for several seconds until the motor and load are brought up to speed.

When an insulation failure occurs in the motor (or in the circuit from the motor starter to the motor) and the current could reach a very high value, the fuse must detect such a fault in a few milliseconds. The fast melting of the fusible element and its subsequent current-limiting action prevent the current from rising to the maximum available magnitude, and then in a few more milliseconds this restricted value of overcurrent is interrupted. This current-limiting action occurs in a total of less than one half-cycle.

In a properly designed and coordinated current-limiting fuse/contactor type motor starter, the overload relays and the contactor should relieve the fuses of all overload and locked-rotor overcurrent interrupting duty and provide protection from undervoltage or single-phasing of the supply. Combinations of this type are generally restricted to motors having less than a 600 A load current as fuses of any higher current rating do not have appreciable current-limiting action at the fault current levels available on these medium-voltage circuits.

5.17 Selection of Fuse Type and Rating. The selection of the most suitable type and rating of medium-voltage power fuses for the various applications that have been enumerated for medium-voltage industrial distribution circuits deserves very careful consideration and study.

The basic rules for the application of a power fuse are that it must be selected for voltage rating, current-carrying capacity, and interrupting capacity. More important, however, is the selection of the type of fuse that will meet the requirements of the distribution system and its components in order to provide the proper balance of primary overcurrent protection for the system, with adequate backup of the next loadside overcurrent protective device, to optimize system protection and continuity of service. No less important is the selection of the fuse type that will also result in a minimum investment of capital, a minimum space requirement, and a minimum of maintenance.

5.17.1 Voltage Rating. The voltage rating should be selected as the next higher standard voltage rating above the maximum operating voltage level of the system.

Solid-material boric acid power fuses are not voltage critical in that they can safely be applied at voltages less than their rated voltage with no detrimental effects. These fuses have a constant current interrupting ability or at best a slightly increased current-interrupting ability when applied to systems operating one or more voltage levels below the fuse rating.

The current-limiting power fuse is voltage critical. This fuse functions by developing a back electromotive force, and in older fuse designs that do not use ribbon type elements, this electromotive force is a function of its maximum rated voltage rather than system voltage. This means that care must be used in selecting the fuse voltage rating to match the system voltage to avoid subjecting system components to excessive voltages. If necessary, the fuse manufacturer should be consulted on the application of specific fuse types.

5.17.2 Current Rating. The current rating of a fuse should be selected considering many factors.

(1) Normal Continuous Load. All modern power fuses are so rated that they will carry their rated current continuously if they are applied in locations where the ambient temperature conditions do not exceed the ANSI standard value of 40 °C. This should be checked where enclosures are required, as temperatures within an enclosure may be 15 °C or more above outside ambient.

(2) Transient Inrush Currents of Transformers. Energizing a transformer from the distribution circuit or subtransmission circuit is accompanied by a short duration inrush of magnetizing current. The integrated heating effect, as seen by a fuse in such a circuit, is customarily represented as that of a current having a magnitude of 8 – 12 times the full-load current of the transformer for a duration of 0.1 s and alternately, 20 – 25 times the full-load current for a duration of 0.01 s. Expulsion power fuses or solid-material boric acid power fuses are available with a selection of time-current characteristics, all of the inverse-time form, and having sufficient time delay such that a fuse having a current rating equal to the transformer rating will withstand such transient inrush currents without operating and without damage to the element. Current-limiting power fuses have an inherently steeper time-current characteristic producing a "faster" fuse. This frequently requires a fuse current rating that is typically three times the transformer full-load current rating.

(3) Motor Starting. The magnitude and duration of the inrush current associated with the starting of a large motor are functions of the motor's characteristics and the load it drives. The current-limiting fuses used with motor starting contactors should be of the type specifically designed for the service. The ratings assigned to such R-rated fuses indicate the maximum starting inrush current and duration of inrush that these fuses can withstand.

(4) Normal Repetitive Overloads. Transformer and distribution circuit components inherently have some short-time overload capability that may be used to advantage in planning maximum economy into an industrial distribution system. These overloads may simply be due to the starting inrush of a large motor on a general-purpose circuit, to a shift startup in an industrial plant, or to peaking air-conditioning or heating loads in abnormal weather. Expulsion fuses and solid-material boric acid fuses have inherent continuous overload capability in all ratings. It is a maximum in the smaller current ratings that are available in a particular physical size of fuse refill or fuse unit and decreases in the larger

231

current ratings in the same fuse refill or fuse unit size. These capabilities are published in tabular form in the ½ – 8 h range.

(5) Emergency Overloads. In the more sophisticated distribution systems, continuity of service is achieved by switching the loads from a faulty circuit or transformer to another circuit or transformer until repairs or replacement can be made. Emergency overload capability data similar to the continuous overload capability data are available for the solid-material boric acid power fuses. It is imperative to consider the effect of such overloads on the melting time-current characteristics of the fuse as it affects selectivity with other overcurrent protective devices.

5.17.3 Interrupting Rating. The interrupting rating of a fuse relates to the value of the maximum rms asymmetrical current available in the first half-cycle after the occurrence of a fault. The rating may be expressed in any of three ways:

(1) Maximum rms symmetrical current
(2) Maximum rms asymmetrical current
(3) Equivalent three-phase symmetrical power in kilovolt-amperes

RMS symmetrical current ratings are particularly useful if a careful system short-circuit study has been made, since the results obtained are expressed in these terms.

RMS asymmetrical current ratings merely represent the maximum current that the fuse may have to interrupt because of its fast-acting characteristic. For power fuses, it is 1.6 times the rms symmetrical current-interrupting rating.

Equivalent three-phase symmetrical interrupting ratings are given as a reference for comparison with circuit breaker capabilities. Also, simplified short-circuit duty requirements can be calculated in these terms.

In selecting a fuse for proper interrupting rating it is, of course, important to make sure that it is adequate for the short-circuit duty required. Generally, however, other requirements automatically ensure that the fuses used have inherently more interrupting capability than the system requires. Current-limiting fuses used for the protection of voltage, distribution, or small power transformers have interrupting ratings that exceed the requirements of the medium-voltage distribution circuits used in industrial systems.

Distribution circuits supplying a number of large medium-voltage motors must be designed so that the short-circuit interrupting duty does not exceed the interrupting rating of the fuse-contactor type motor starters.

5.17.4 Selectivity, General Considerations. A minimum melting time-current characteristic curve indicates the time that a fuse will carry a designated current without operating (assuming no initial load). Figure 102 illustrates a typical family of minimum melting time-current characteristics for current-limiting power fuses. Figure 103 shows the minimum melting time-current characteristics for a representative line of solid-material boric acid fuses. The curve should indicate the tolerance applicable to these times. This curve should further indicate whether the fuse is nondamageable, that is, whether it can carry without damage the designated current for a time that immediately approaches the time indicated by the curve. Since this minimum melting time-current characteristic curve is based on no initial load current through the fuse, it is

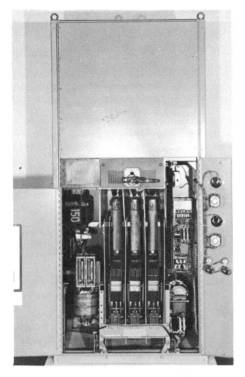

**Fig 97
High-Voltage R-Rated Fuses in
Drawout-Type Motor Starter,
Rated 180 A**

**Fig 98
Metal-Enclosed Switchgear
with Fused Interrupters; 200 A
125 MV A 4.16 kV Outdoor Manual Type**

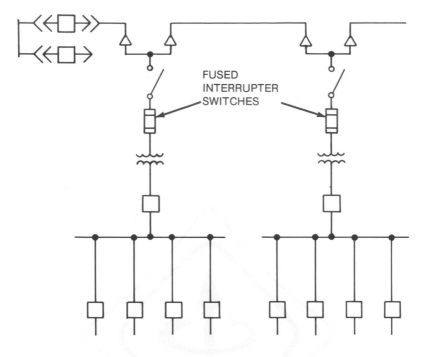

**Fig 99
One-Line Diagram of Simple Radial System**

necessary to modify this curve to recognize the temporary change in melting time due to load current. This modification will permit precise coordination with other overcurrent protective devices nearer the load. Further aids are available in the form of curves that evaluate the temporary change in melting characteristics in the event that the fuse has been carrying a heavy emergency overload or in the case of fuses being applied in a location having an exceptionally high ambient temperature.

If the fuse is susceptible to a permanent change in melting time-current characteristic when exposed to currents for times less than the minimum melting time, the manufacturer's published "safety-zone" allowance or setback curve should be used in any coordination scheme of overcurrent protective devices.

Maximum system protection from primary faults that should cause the fuse to operate, and maximum backup protection to the system and equipment on the load side of the fuse in the event that overcurrent protection malfunctions, as well as optimum transformer protection from secondary arcing faults, require use of the smallest current rating of fuse that meets the requirements stated earlier for the selection of current rating. Fuses having a small tolerance or narrow opening band and utilizing nondamageable elements are best suited to accurate selective application. To assist in this evaluation, total clearing time-current characteristic curves are used. They represent the maximum total operating time of the fuses, taking into account maximum manufacturing

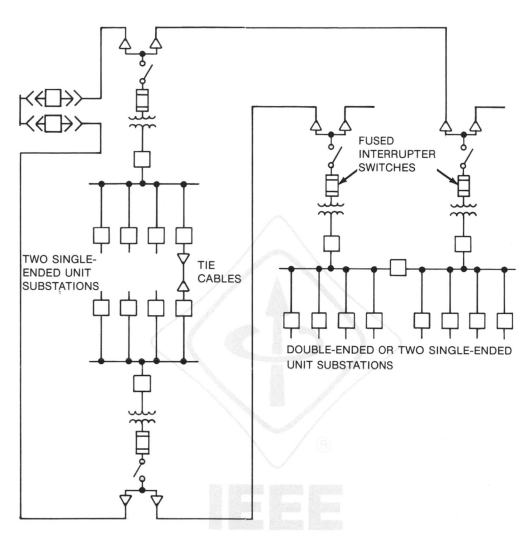

**Fig 100
One-Line Diagram of Typical Secondary-Selective Arrangements
of Load-Center Distribution System**

tolerances of the fusible element, the operating voltage, and recognizing that the
fuse may not be carrying any initial load to reduce the melting time of the fuse
element. Figure 104 indicates the corresponding total clearing time characteristics for the current-limiting fuses of Fig 102. Figure 105 shows the total clearing
time characteristics for the solid-material boric acid fuse of Fig 103.

For the very simple radial system with a single transformer, manufacturers
frequently supply a table of recommended fuse current ratings for transformers
of various voltage and power classes. The bases for such tables should be carefully
scrutinized to see that they are adequate for the necessary requirements

previously outlined, but more particularly, that the recommended rating is not too large, as the system protection may be jeopardized and the full benefits of backup overcurrent protection possible with fuses will not be achieved.

A good axiom in selecting a fuse to fit into an overcurrent protective scheme is to use the very smallest current rating that will carry and not be damaged by any load that should be maintained without interruption. The effort required to make a precise selection of fuse current rating and speed characteristic is small compared to the benefits obtained in overall system overcurrent protection.

5.17.5 Selectivity for Motor Controllers. The application of R-rated fuses in medium-voltage motor controllers is basically one of comparing the minimum melting time-current characteristics of the R-rated fuses with the time-current characteristics of the overload relay curve. The fuse size that is selected should be such that short-circuit protection is provided by the fuse and overload protection is provided by the controller overload relays. The following data are required for proper application:

(1) Motor full-load current rating
(2) Motor locked-rotor current
(3) R-rated fuse minimum melting time-current characteristic curves
(4) Overload relay time-current characteristic curves

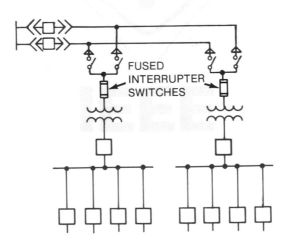

Fig 101
One-Line Diagram of Simple Radial System

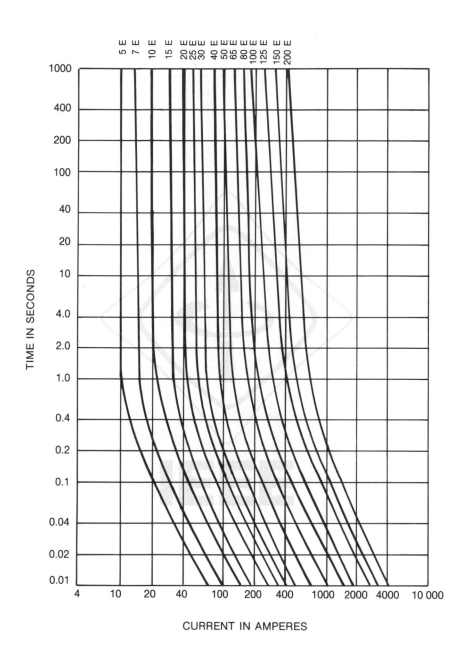

Fig 102
Typical Minimum Melting Time–Current Characteristics for
High-Voltage Current-Limiting Power Fuses

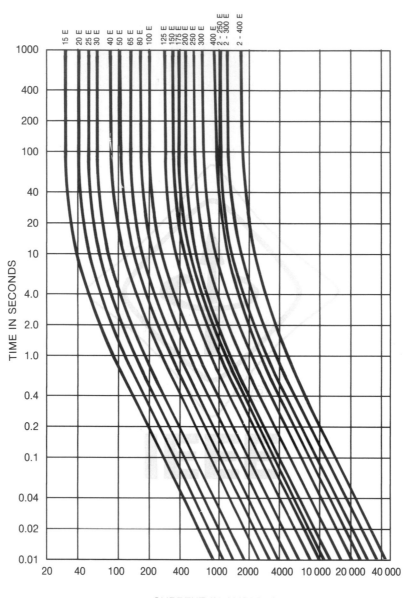

Fig 103
Typical Minimum Melting Time–Current Characteristics for
High-Voltage Solid-Material Boric-Acid Power Fuses

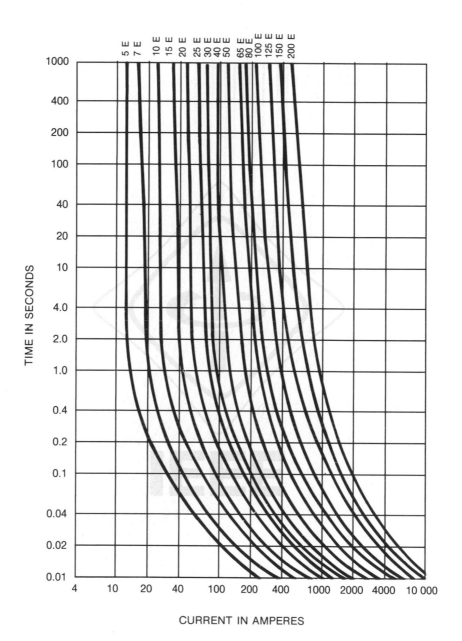

Fig 104
Typical Total Clearing Time – Current Characteristics for
High-Voltage Current-Limiting Power Fuses (See Fig 102)

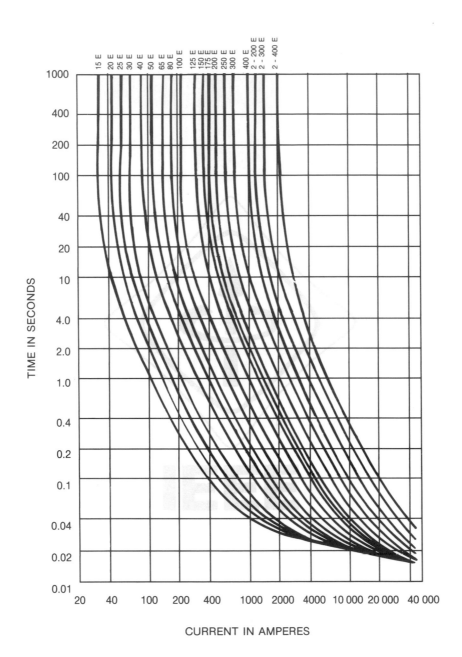

**Fig 105
Typical Total Clearing Time – Current Characteristics for High-Voltage
Solid-Material Boric-Acid Power Fuses (See Fig 103)**

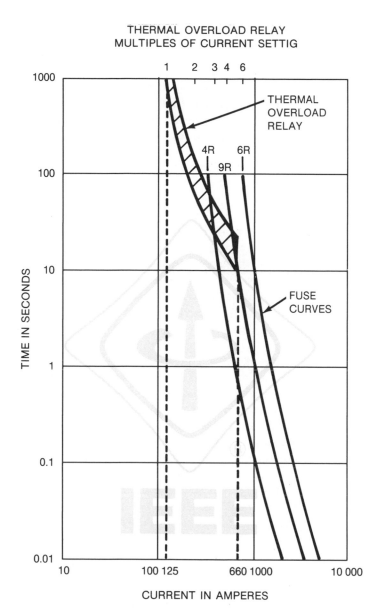

**Fig 106
Time – Current Coordination of R-Rated Fuses and
Thermal Overload Relay in High-Voltage Fused Motor Starter**

For example, a 2300 V motor has a 100 A full-load current rating and a locked-rotor current of 600 A. The overload relay is to be sized at 125% of motor full-load current and system voltages could vary enough to cause a 10% increase in locked-rotor current. Taking these percentages into consideration, the overload relay is sized at 125 A and the locked rotor current would be taken as 660 A.

Figure 106 shows these adjusted values (dashed lines). Also shown is the thermal overload relay curve with the multiple 1 of the relay current setting lining up with the 125 A adjusted full-load current.

The fuse selected for the application is the smallest fuse whose minimum melting time-current characteristic does not cross the relay curve for currents less than the adjusted locked-rotor value. In this example, the 9R rating would be the proper choice. Since the application is at 2300 V, 9R fuses rated 2300 V would be recommended.

The largest fuse size that could be applied is that rating that still protects the controller when the fuse clears the maximum available fault current. Maximum fuse sizes can best be chosen by the motor controller manufacturer.

5.17.6 Selection of Fuse Type. Since there is absolutely no noise, flame, or any expulsion of gases, the current-limiting fuse is especially suited to indoor industrial applications in close-fitting enclosures. The only electrical clearances required are those determined by the insulation class corresponding to the voltage rating of the fuse. These fuses are often used for the protection of voltage transformers, capacitors, distribution-class transformers, and small- to medium-size power transformers. The fault-current-limiting action inherent in the operation of these fuses upon the occurrence of high fault currents reduces the damaging let-through energy to the distribution circuit and simplifies selectivity with overcurrent protective devices that are closer to the source of power. However, their fast fusing action may require selecting a current rating that is several times the transformer full-load current rating. This is necessary in order to accommodate transient inrush current conditions and to coordinate with secondary protective devices. This fuse application is in line with fusing practice that is based on protecting the continuity of the electrical service with the possible loss of some transformer service life. Consideration should be given to the reduced degree of transformer protection from secondary faults, especially during arcing ground faults or when the fuse must back up a secondary device that has malfunctioned. Current-limiting fuses are available in a range of voltage, current, and interrupting ratings (as shown in Table 32) to cover most of the required applications. Medium-voltage current-limiting fuseholders are not of the rejection type, and precautions should be taken to assure that replacement fuses are of the correct size and type.

The solid-material boric acid fuse was developed initially for high load and interrupting capacity in indoor or confined locations. Thus, it is suitable for use in medium-voltage industrial distribution systems. It is available in a wide range of voltage and current ratings with corresponding high short-circuit interrupting ratings (Tables 30 and 31) in order to cover the required applications. An important factor to be considered in making the selection of fuse type is that solid-material boric acid fuses make use of fuse refills, which include the complete fuse element and interrupting element necessary to restore the fuse to original operating condition after it has been called on to function. Their slower speed makes it possible to select a fuse rating equal to or even somewhat less than the transformer full-load current, for maximum protection from through faults.

Working tools for use in selection of each of the fuse types are available from manufacturers for the system engineer who is developing these distribution systems. Time-current characteristic curves, preloading adjustment factors, and

**Fig 107
Peak Let-Through Currents for
Medium-Voltage Current-Limiting Fuses**

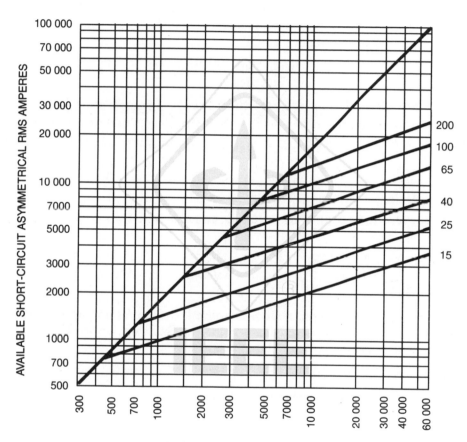

AVAILABLE SHORT-CIRCUIT RMS ASYMMETRICAL AMPERES

ambient temperature adjustment factors are provided for each fuse type. In addition, overload capability data are provided for solid-material boric acid fuses, and peak let-through current curves, minimum melting and maximum total I^2t charts, and peak arc-voltage data are provided for current-limiting fuses. A typical peak let-through current curve is shown in Fig 107.

5.18 References. The following publications shall be used in conjunction with this chapter.

[1] ANSI C37.42-1981, American National Standard Specifications for Distribution Cutouts and Fuse Links.

[2] ANSI C37.46-1981, American National Standard Specifications for Power Fuses and Fuse Disconnecting Switches.

[3] ANSI C37.47-1981, American National Standard Specifications for Distribution Fuse Disconnecting Switches, Fuse Supports, and Current-Limiting Fuses.

[4] ANSI C37.48-1969, American National Standard Guide for Application, Operation, and Maintenance of Distribution Cutouts and Fuse Links, Secondary Fuses, Distribution Enclosed Single-Pole Air Switches, Power Fuses, Fuse Disconnecting Switches, and Accessories.

[5] ANSI C97.1-1972, American National Standard Low-Voltage Cartridge Fuses 600 Volts or Less.

[6] ANSI/IEEE C37.40-1981, IEEE Standard Service Conditions and Definitions for High-Voltage Fuses, Distribution Enclosed Single-Pole Air Switches, Fuse Disconnecting Switches, and Accessories.

[7] ANSI/IEEE C37.41-1981, IEEE Standard Design Tests for High-Voltage Fuses, Distribution Enclosed Single-Pole Air Switches, Fuse Disconnecting Switches, and Accessories.

[8] ANSI/IEEE C37.100-1981, IEEE Standard Definitions for Power Switchgear.

[9] ANSI/NFPA 70-1984, National Electrical Code.

[10] ANSI/UL 198B-1982, Safety Standard for Class H Fuses.[15]

[11] ANSI/UL 198C-1981, Safety Standard for High-Interrupting Capacity Fuses, Current-Limiting Types.

[12] ANSI/UL 198D-1982, Safety Standard for Class K Fuses.

[13] ANSI/UL 198E-1982, Safety Standard for Class R Fuses.

[14] ANSI/UL 198F-1982, Safety Standard for Plug Fuses.

---

[15] ANSI/UL publications can be obtained from the Sales Department, American National Standards Institute, 1430 Broadway, New York, NY 10018, or from Publication Stock, Underwriters Laboratories, Inc, 333 Pfingsten Rd, Northbrook, IL 60020.

[15] ANSI/UL 198G-1981, Fuses for Supplementary Overcurrent Protection.

[16] ANSI/UL 198H-1982, Safety Standard for Class T Fuses.

[17] ANSI/UL 508-1983, Safety Standard for Industrial Control Equipment.

[18] ICEA P32-382 (1969), Short-Circuit Characteristics of Insulated Cable.

[19] IEC Pub 291 (1969), Fuse Definitions.[16]

[20] NEMA BU 1-1983, Busways.

[21] NEMA SG2-1981, High-Voltage Fuses.

[22] UL 198L-1984, DC Fuse for Industrial Use.[17]

---

[16] IEC publications can be obtained in the US from the Sales Department, American National Standards Institute, 1430 Broadway, New York, NY 10018.
    [17] UL publications can be obtained from Publication Stock, Underwriters Laboratories, Inc, 333 Pfingsten Rd, Northbrook, IL 60020.

6. Low-Voltage Circuit Breakers

6.1 General. ANSI/NFPA 70-1984 [6][18] (National Electrical Code [NEC]) defines a circuit breaker as "a device designed to open and close a circuit by nonautomatic means and to open the circuit automatically on a predetermined overcurrent without injury to itself when properly applied within its rating." Low-voltage circuit breakers are then further classified by ANSI C37.100-1981 [5] as (1) molded-case circuit breakers and (2) low-voltage power circuit breakers. Figure 108 illustrates typical constructions:

(1) A molded-case circuit breaker is one that is assembled as an integral unit in a supporting and enclosing housing of insulating material.

(2) A low-voltage power circuit breaker is one for use on circuits rated 1000 V ac and below, or 3000 V dc and below, but not including molded-case circuit breakers.

Underwriters Laboratories Standard UL 489-1980 [11] further specifies that molded-case circuit breakers are specifically intended to provide service entrance, feeder, and branch circuit protection in accordance with the NEC [5]. Present UL requirements cover molded-case circuit breakers rated through 600 V and 6000 A.

ANSI C37.16-1980 [1] includes information on ratings, application, and operating conditions applicable to low-voltage power circuit breakers.

When specifying either molded-case circuit breakers or low-voltage power circuit breakers, it is necessary, from a circuit protection standpoint, to properly apply the circuit breaker within its ratings.

The term *air circuit breaker* is often used in speaking of low-voltage power circuit breakers. Since the arc interruption takes place in air in both molded-case circuit breakers and low-voltage power circuit breakers, this term really applies to both types.

The following discussion is limited to general-purpose applications of circuit breakers that provide overcurrent protection on ac power systems through 600 V. Circuit breakers intended for special-purpose applications, for example, thermal only, instantaneous trip only, etc, are not covered by this discussion. For circuit breaker ratings, requirements, and recommendations on systems, mining applications, transit systems, or other special-purpose applications, consult standards referenced in the bibliography or manufacturers' literature.

---

[18] The numbers in brackets correspond to those of the references listed at the end of this chapter. The numbers in brackets with the prefix "B" correspond to those of the bibliography listed at the end of this chapter.

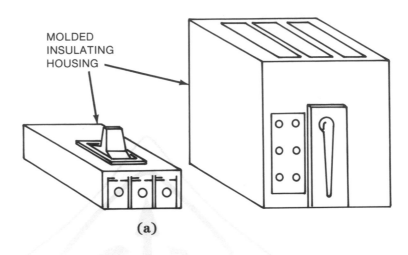

MOLDED
INSULATING
HOUSING

(a)

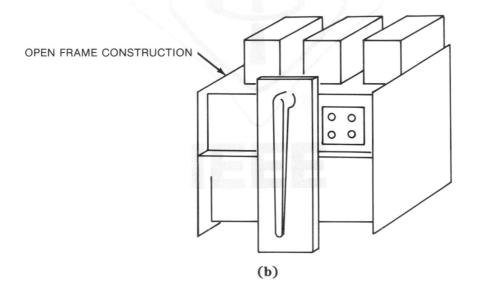

OPEN FRAME CONSTRUCTION

(b)

Fig 108
(a) Molded-Case Circuit Breaker
(b) Low-Voltage Power Circuit Breaker

6.2 Ratings. The ratings that apply to circuit breakers and their actual assigned numerical values reflect the mechanical, electrical, and thermal capabilities of those circuit breakers and generally comply with industry standards published by the National Electrical Manufacturers Association (NEMA), Underwriters Laboratories, Inc (UL), or American National Standards Institute (ANSI). The following is a brief description of the basic ratings.

voltage. Circuit breakers are designed and marked with the maximum voltage at which they can be applied. They can be used on any system where the voltage is lower than the breaker rating.

frequency. Circuit breakers are normally suitable for use in 50 and 60 Hz electrical distribution systems.

continuous current. Standard molded-case circuit breakers are calibrated to carry 100% of their current rating in open air at a given ambient temperature (usually 25 or 40 °C). In accordance with the NEC [5], these breakers, as installed in their enclosures, should not be continuously loaded over 80% of their current rating.

Low-voltage power circuit breakers and certain molded-case circuit breakers are specifically approved for 100% continuous duty. These breakers can be continuously loaded to 100% of their current rating in a 40 °C ambient when installed in their proper enclosures.

interrupting rating. The interrupting rating (or short-circuit current rating, as it is referred to for a low-voltage power circuit breaker) is commonly expressed in rms symmetrical amperes. It may vary with the applied voltage and is established by testing per UL or ANSI standards.

short-time rating. The short-time current rating specifies the maximum capability of a circuit breaker to withstand the effects of short-circuit current flow for a stated period, typically 30 cycles or less, without opening. This provides time for downstream protective devices closer to the fault to operate and isolate the circuit.

The short-time current rating of a low-voltage power circuit breaker without instantaneous trip characteristics is equal to the breaker's short-circuit interrupting rating. Most molded-case circuit breakers are not provided with a short-time current rating; however, some higher ampere rated molded-case circuit breakers are provided with a short-time current rating in addition to the short-circuit interrupting rating.

Circuit breakers with an instantaneous trip function should not be used where continuity of service requires only long-time and short-time delay functions.

control voltage. This is the ac or dc voltage designated to be applied to control devices intended to open or close a circuit breaker. These devices can normally be supplied with a voltage rating needed to meet a particular control system.

Where the interrupting ratings of conventional molded-case circuit breakers or low-voltage power circuit breakers are not sufficient for a particular system application, there are other options available, such as current-limiting circuit breakers, integrally fused circuit breakers, and fuse/breaker coordinated combinations.

Current-limiting circuit breakers not only provide high interrupting capabilities, but also limit let-through current and energy to downstream devices. UL 489-1980 [11] defines a current-limiting circuit breaker as follows: "a circuit breaker that does not employ a fusible element and that when operating within its current-limiting range, limits the let-through I^2t to a value less than the I^2t of a half-cycle wave of the symmetrical prospective current."

I^2t, as used in the definition, is an expression related to the energy resulting from current flow. Specific manufacturers' literature should be consulted for information on the current-limiting characteristics of their circuit breakers. Current-limiting circuit breakers provide the system designer with a means of reducing fault current energy levels at downstream system components while still retaining the advantages of circuit breaker construction, such as common trip and reusability. These breakers can be reset and service restored in the same manner as conventional thermal-magnetic breakers. There is nothing to replace —even after clearing maximum level fault currents. It is recommended after interruption at high fault levels that the breaker be checked and tested before being put back into service. Figure 109 illustrates the current waveform resulting from current-limiting operation.

Integrally fused circuit breakers provide high interrupting capability through the use of current-limiting fuses that are assembled into the housing of the circuit breaker. The fuses in these devices are designed to blow and need replacement only after a high-level fault. The breaker portion is interlocked so that when any fuse opens, the circuit breaker will automatically trip.

Tables 33 and 34 list typical ratings of molded-case circuit breakers for commercial and industrial applications. Table 33 covers ratings of standard and high interrupting capacity units, while Table 34 covers current-limiting and fused breakers. Tables 35 and 36 show standard ratings for low-voltage power circuit breakers with and without instantaneous overcurrent trip characteristics, respectively. Table 37 covers integrally fused low-voltage power circuit breakers.

6.3 Trip Unit. The trip unit considered here is an integral part of the circuit breaker. It may be electromechanical (thermal-magnetic, Figs 110 and 111, or mechanical dashpots) or solid-state electronic (Fig 112). By continually monitoring the current flowing through the circuit breaker, it will sense abnormal current conditions. Depending on the magnitude of the current, the trip unit will initiate an inverse-time response or an instantaneous response. This action will cause a direct acting operating mechanism to open the circuit breaker contacts and interrupt current flow. The following discussion covers those trip unit characteristics commonly available on low-voltage circuit breakers, specifically:

(1) Continuous current rating
(2) Long-time current
(3) Long-time time delay
(4) Short-time current

(5) Short-time time delay

(6) Instantaneous current response

The continuous-current rating may be fixed or adjustable. Some constructions may require replacing all or part of the trip unit to change the continuous rating. Overcurrent trip characteristics are a function, multiples or percentages, of the continuous-current rating.

Most molded-case circuit breakers are not provided with a long-time adjustment or short-time function. The inverse-time response characteristic is strictly related to how long a particular level of current has been flowing. Molded-case circuit breakers with electronic trip units and low-voltage power circuit breakers usually have long-time and short-time current adjustments. These adjustments are used to set the current level (pickup point) at which the circuit breaker trip unit will begin timing toward initiation of tripping action. Also, low-voltage power circuit breakers may be provided with or without an instantaneous trip function.

Not all of these characteristics are provided or are necessary for most applications. Additional functions, such as ground-fault protection, or combinations of functions that further influence circuit breaker time-current response are available from specific manufacturers.

The basic function of the trip unit is to provide the long-time current/time delay and instantaneous current response characteristics necessary for proper circuit protection. The combination of these characteristics provides time delay to

Fig 109
Current Limiting Waveform
(Fault Initiation at t = o and V max.)

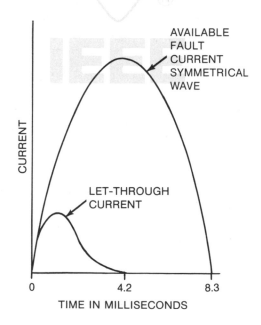

Table 33
**Typical Interrupting Current Ratings of Molded-Case Circuit Breakers
for Commercial and Industrial Applications***

Frame Size (amperes)	Number of Poles	Interrupting Rating in rms Symmetrical Amperes (000's) at AC Voltage				
		120	240	277	480	600
	1	10	–	–	–	–
100	1	–	–	14	–	–
	1	–	–	65	–	–
100, 150	2, 3	–	18	–	14	14
	2, 3	–	65	–	25	18
225, 250	2, 3	–	25	–	22	22
	2, 3	–	65	–	25	22
400, 600	2, 3	–	42	–	30	22
	2, 3	–	65	–	35	25
800, 1000	2, 3	–	42	–	30	22
	2, 3	–	65	–	50	25
1200	2, 3	–	42	–	30	22
	2, 3	–	65	–	50	25
1600, 2000	2, 3	–	65	–	50	42
	2, 3	–	125	–	100	65
3000, 4000	2, 3	–	100	–	100	85
	2, 3	–	200	–	150	100

* Does not include molded-case circuit breakers intended primarily for residential applications. Refer to specific manufacturers for information on those constructions.
NOTE: Ratings shown are typical. Variations among manufacturers or product changes may result in actual ratings that differ from those shown. Consult specific manufacturers' literature for guidance.

override transient overloads; delayed tripping for sustained overloads, low-level short circuits, or ground faults of sufficient magnitude to cause overcurrent response; and instantaneous tripping for higher level short circuits or ground faults.

Trip units of low-voltage power circuit breakers and some molded-case circuit breakers may also be equipped with short-time overcurrent and time-delay characteristics that are useful for providing coordination with downstream or upstream circuit breakers or fuses. The resulting combination of long-time delay and short-time delay characteristics provides delayed tripping for all levels of overcurrent below the instantaneous response. This provides time for downstream circuit breakers to operate and clear the fault. The withstand of other electrical components in the current path should be checked to be certain they can handle the additional stress associated with the longer clearing time.

Table 34
Typical Interrupting Current Ratings of Molded-Case
Circuit Breakers — Current-Limiting and Fused Circuit Breakers

Frame Size (amperes)	Number of Poles	Interrupting Rating in rms Symmetrical Amperes (000's) at AC Voltage		
		240	480	600
Current-Limiting Circuit Breakers				
100	2, 3	200	200	–
225	2, 3	200	200	–
400	2, 3	200	200	–
Fused Circuit Breakers				
100, 150	2, 3	200	200	200
400, 600	2, 3	200	200	200
800	2, 3	200	200	200
1600	2, 3	200	200	200

NOTE: Ratings shown are typical. Variations among manufacturers or product changes may result in actual ratings that differ from those shown. Consult specific manufacturers' literature for guidance.

Low-voltage power circuit breakers are also available without an instantaneous trip characteristic. Molded-case circuit breakers without an instantaneous trip characteristic are not commercially available at this writing.

Solid-state electronic trip units in low-voltage circuit breakers may also provide time-current characteristics for equipment ground-fault protection in accordance with the NEC [6], 230-95, requirements. Maximum current setting is 1200 A as per NEC requirements and time-delay settings are usually adjustable. Coordination is enhanced by the presence of trip unit adjustments that provide inverse ground-fault time-current characteristics. While this chapter will make reference to ground-fault protection with regard to circuit breaker application, the reader is referred to Chapter 7 for a more complete treatment of the subject.

6.3.1 Time-Current Characteristic Curves. Circuit breaker time-current characteristic curves are principally a function of the type of trip unit and its settings. A typical time-current characteristic curve is shown in Fig 113 for a 600 A frame size molded-case circuit breaker in open air at rated temperature, usually 40 °C. This particular curve applies for trip current ratings from 125–600 A and indicates total operational time (from fault current initiation to clearing). The shaded band covers manufacturing tolerance and other variables. To be noted is the long-time delay portion where delays in the order of seconds and minutes vary inversely with current. At the current level setting of the instantaneous element, that is, with no intentional time delay, the operating time drops abruptly to the time required by the circuit breaker mechanism to open the

Table 35
Preferred Ratings for Low-Voltage AC Power Circuit Breakers with Instantaneous Direct-Acting Phase Trip Elements

System Nominal Voltage (volts)	Rated Maximum Voltage (volts)	Insulation (Dielectric Withstand) (volts)	Three-Phase Short-Circuit Current Rating (symmetrical amperes)*	Frame Size (amperes)	Range of Trip-Device Current Ratings (amperes)†
600	635	2200	14 000	225	40–225
600	635	2200	22 000	600	40–600
600	635	2200	22 000	800	100–800
600	635	2200	42 000	1600	200–1600
600	635	2200	42 000	2000	200–2000
600	635	2200	65 000	3000	2000–3000
600	635	2200	85 000	4000	4000
480	508	2200	22 000	225	40–225
480	508	2200	30 000	600	100–600
480	508	2200	30 000	800	100–800
480	508	2200	50 000	1600	400–1600
480	508	2200	50 000	2000	400–2000
480	508	2200	65 000	3000	2000–3000
480	508	2200	85 000	4000	4000
240	254	2200	25 000	225	40–225
240	254	2200	42 000	600	150–600
240	254	2200	42 000	800	150–800
240	254	2200	65 000	1600	600–1600
240	254	2200	65 000	2000	600–2000
240	254	2200	85 000	3000	2000–3000
240	254	2200	130 000	4000	4000

*Ratings in this column are rms symmetrical values for single-phase (2-pole) circuit breakers and three-phase average rms symmetrical values of three-phase (3-pole) circuit breakers. When applied on systems where rated maximum voltage may appear across a single pole, the short-circuit current ratings are 87% of these values. See ANSI/IEEE C37.13-1981 [4], 4.6.
†Note that the continuous-current-carrying capability of some circuit-breaker-trip-device combinations may be higher than the trip-device current rating. See ANSI/IEEE C37.13-1981 [4], 9.1.3.

Table 36
Preferred Ratings for Low-Voltage AC Power Circuit Breakers without Instantaneous Direct-Acting Phase Trip Elements (Short-Time-Delay Element or Remote Relay)

System Nominal Voltage (volts)	Rated Maximum Voltage (volts)	Insulation (Dielectric) Withstand (volts)	Three-Phase Short-Circuit Current Rating or Short-Time Current Rating (symmetrical amperes)*†	Frame Size (amperes)	Range of Trip-Device Current Ratings (amperes)† Setting of Short-Time-Delay Trip Element		
					Minimum Time Band	Intermediate Time Band	Maximum Time Band
600	635	2200	14 000	225	100–225	125–225	150–225
600	635	2200	22 000	600	175–600	200–600	250–600
600	635	2200	22 000	800	175–800	200–800	250–800
600	635	2200	42 000	1600	350–1600	400–1600	500–1600
600	635	2200	42 000	2000	350–2000	400–2000	500–2000
600	635	2200	65 000	3000	2000–3000	2000–3000	2000–3000
600	635	2200	85 000	4000	4000	4000	4000
480	508	2200	14 000	225	100–225	125–225	150–225
480	508	2200	22 000	600	175–600	200–600	250–600
480	508	2200	22 000	800	175–800	200–800	250–800
480	508	2200	42 000	1600	350–1600	400–1600	500–1600
480	508	2200	50 000	2000	350–2000	400–2000	500–2000
480	508	2200	65 000	3000	2000–3000	2000–3000	2000–3000
480	508	2200	85 000	4000	4000	4000	4000
240	254	2200	14 000	225	100–225	125–225	150–225
240	254	2200	22 000	600	175–600	200–600	250–600
240	254	2200	22 000	800	175–800	200–800	250–800
240	254	2200	42 000	1600	350–1600	400–1600	500–1600
240	254	2200	50 000	2000	350–2000	400–2000	500–2000
240	254	2200	65 000	3000	2000–3000	2000–3000	2000–3000
240	254	2200	85 000	4000	4000	4000	4000

*Short-circuit current ratings for breakers without direct-acting trip devices, opened by a remote relay, are the same as those listed here.

†Ratings in this column are rms symmetrical values for single-phase (2-pole) circuit breakers and three-phase average rms symmetrical values of three-phase (3 pole) circuit breakers. When applied on systems where rated maximum voltage may appear across a single pole, the short-circuit current ratings are 87% of these values. See ANSI/IEEE C37.13-1981 [4], 4.6.

*†Note that the continuous-current-carrying capability of some circuit-breaker-trip-device combinations may be higher than the trip-device current rating. See ANSI/IEEE C37.13-1981 [4], 9.1.3.

Table 37
Preferred Ratings for Integrally Fused Low-Voltage AC Power Circuit Breakers with Instantaneous Direct-Acting Phase Trip Elements

Circuit-Braker Frame Size (amperes)*	Rated Maximum Voltage (volts)†	Insulation (Dielectric) Withstand (volts)	Three-Phase Short-Circuit Current Rating (symmetrical amperes)‡	Range of Continuous-Current Rating (amperes)	
				Range of Trip-Device Current Ratings (amperes)§	Maximum Fuse Rating⁼⁼
600	600	2200	200 000	125–600	††
800	600	2200	200 000	125–800	††
1600	600	2200	200 000	200–1600	††

*Two circuit-breaker frame ratings are used for integrally fused circuit breakers. The continuous-current rating of the integrally fused circuit breaker is determined by the rating of either the direct-acting trip device or the current-limiting fuse applied to a particular circuit-breaker frame rating, whichever is smaller.

†Listed values are limited by the standard voltage rating of the fuse.

‡Ratings in this column are rms symmetrical values for single-phase (2-pole) circuit breakers and three-phase average rms symmetrical values of three-phase (3-pole) circuit breakers. When applied on systems where rated maximum voltage may appear across a single pole, the short-circuit current ratings are 87% of these values. See ANSI/IEEE C37.13-1981 [4], 4.6.

§Note that the continuous-current-carrying capability of some circuit-breaker-trip-device combinations may be higher than the trip-device current rating. See ANSI/IEEE C37.13-1981 [4], 9.1.3. Lower rated trip-device current ratings may be used when the fuse size is small or the available current is low, or both. Consult the manufacturer.

⁼⁼Fuse current ratings may be 300, 400, 600, 800, 1000, 1200, 1600, 2000, 2500, and 3000 amperes. Fuses are of the current-limiting type having threshold ratio, peak let-through current, and I^2t characteristics as defined for these ratings in ANSI C97.1-1972 [3].

††Values have not yet been determined; consult the manufacturer.

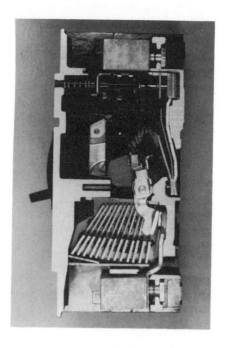

Fig 110
Thermal-Magnetic Trip Molded-Case Circuit Breaker

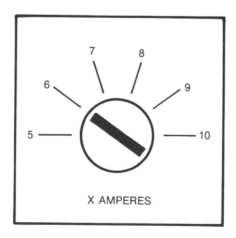

Fig 111
Instantaneous (Magnetic) Current Adjustment Provided on
Thermal-Magnetic Circuit Breakers

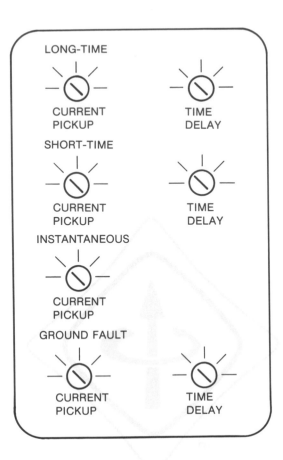

Fig 112
Solid-State Electronic Trip Unit with Long-Time, Short-Time, Instantaneous and Ground Fault Adjustments

contacts plus arcing time. The fault is cleared at the horizontal line labeled "maximum instantaneous opening time." Figure 113 is a typical characteristic curve based on a 40 °C ambient for a molded-case circuit breaker where long-time current sensing is achieved through a thermal element (a bimetal) and where instantaneous tripping is achieved magnetically. Usually, the long-time setting is nonadjustable, whereas instantaneous operation is adjustable (in frame sizes larger than 100 A) from approximately 5–10 times the continuous ampere rating.

Molded-case circuit breakers with more complex trip characteristics and most low-voltage power circuit breakers are equipped with solid-state electronic trip units. The continuous-current rating may be adjustable and long-time and

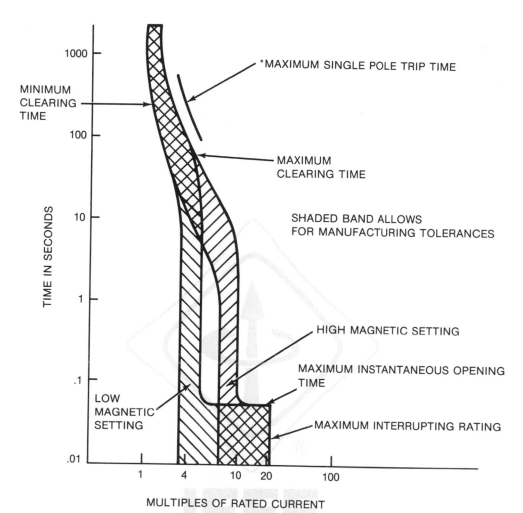

*Single pole test data at 25 °C based on NEMA AB2-1980.

Fig 113
Typical Time-Current Characteristic Curve for a 3-Pole 600 A Frame Size
Molded-Case Circuit Breaker in Open Air at Rated Temperature, Usually 40 °C

short-time functions may be adjustable for current pickup point and time delay. The instantaneous response may also be adjustable. Figure 114 illustrates an overcurrent time-current characteristic curve for a solid-state electronic trip circuit breaker with all of these adjustments.

Circuit breakers with electronic trip units may also include integral ground-fault protection with current pickup and time-delay adjustments. Figure 115

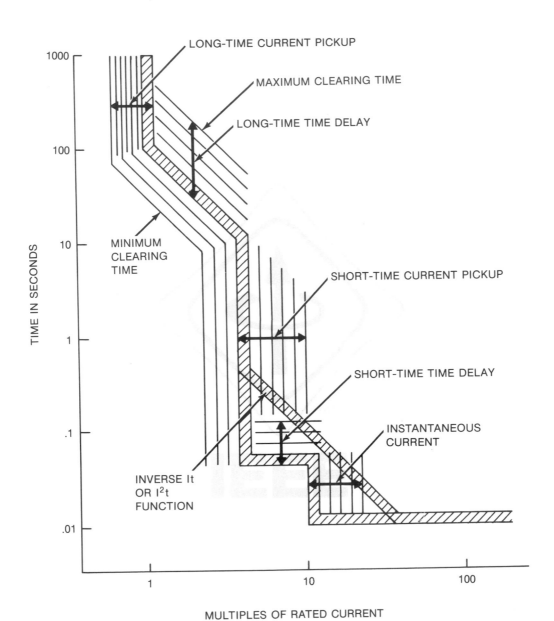

**Fig 114
Time-Current Characteristic Curve for Solid-State Trip Circuit Breaker
(Shown with Maximum Adjustable Functions)**

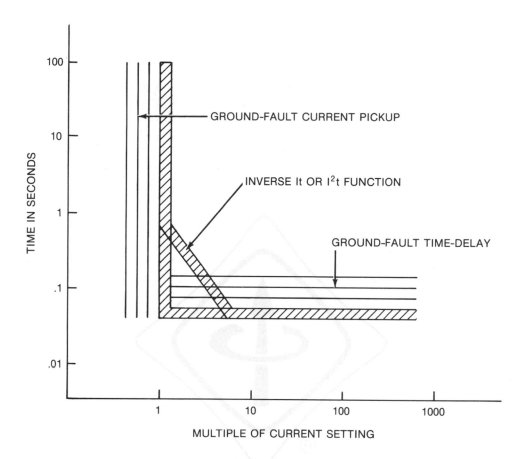

Fig 115
Ground-Fault Time-Current Characteristic Curve
for Solid-State Trip Circuit Breaker

illustrates the time-current characteristic curve for the optional equipment ground-fault protection function.

It should be noted that electronic trip units are available with various functions. Specifications should call out those functions that are necessary on a particular installation. Specific manufacturers' literature should be consulted for guidance on individual units.

Further refinement of the electronic trip unit short-time delay characteristic for overcurrent protection or the ground-fault time-delay characteristic may be obtained by shaping part of the response curve as an inverse function of the product of time and current, "It," or ampere-squared seconds, "I^2t," illustrated by the sloping portion of Figs 114 and 115. This response curve more readily coordinates with downstream thermal-magnetic circuit breakers or fuses that have a similarly sloping response curve.

The time-current characteristic curves of low-voltage power circuit breakers of the simplest type, combining long-time delay and instantaneous characteristics, are similar to those of molded-case circuit breakers. Even the simplest low-voltage power circuit breakers include adjustments for long-time current pickup, long-time time delay, and instantaneous current pickup. Formerly, the tripping elements of low-voltage power circuit breakers were usually mechanical dash-pots; however, today nearly all circuit breakers of this type include solid-state electronic trip units; Figs 114 and 115 also illustrate the possible characteristic curves of these circuit breakers.

6.4 Application. Consideration of all the factors related to proper application of a low-voltage circuit breaker goes beyond voltage, current, and interrupting rating. The performance of a circuit breaker may be influenced by nonelectrical factors related to the installation environment, such as ambient temperature, humidity, elevation, or presence of contaminants. Enclosure type and size, service conditions, loads and their characteristics, outgoing conductors, characteristics of the electrical distribution system, other protective devices upstream and downstream from the circuit breaker under consideration, and even frequency of operation and maintenance, must all be taken into account. For purposes of this chapter, application considerations will be limited to those involving abnormal current conditions and to providing protection and coordination under those conditions.

6.4.1 Protection. The function of system protection may be defined as the detection and prompt isolation of the affected portion of the system whenever a short circuit or other abnormality occurs that might cause damage to, or adversely affect, the operation of any portion of the system or the load that it supplies (1.3).

Treatment of the overall problem of system protection and coordination of electrical power systems will be restricted to the selection, application, and coordination of devices and equipment whose primary function is the isolation and removal of short circuits from the system. Short circuits may be phase-to-ground, phase-to-phase, phase-to-phase-to-ground, three-phase, or three-phase-to-ground. Short circuits may range in magnitude from extremely low-current faults having high-impedance paths to extremely high-current faults having very low-impedance paths. However, all short circuits produce abnormal current flow in one or more phase conductors or in the neutral or grounding circuit. Such disturbances can be detected and safely isolated (1.4).

Of the many types described, there are two types of overcurrent protection that are emphasized, phase overcurrent and ground fault. At the present state of the art, phase-overcurrent conditions are detected on the basis of their magnitudes. Response time is dependent upon the particular overcurrent time-current characteristic curve. Ground-fault conditions of a sufficient magnitude are detectable by phase-overcurrent devices. Those below the minimum current sensitivity of phase-overcurrent devices or arcing ground faults of sufficiently short duration to exclude overcurrent time-delay response will not be cleared. Separate means (either internal to the circuit breaker or externally mounted) must be provided to detect low-level arcing ground faults. This detection means commonly consists of current sensors that monitor each phase and the neutral circuit conductor separately, or one current sensor that monitors all phases and the neutral circuit conductor together, or one current sensor that monitors the ground circuit conductor. Circuit breakers have the advantage of providing a convenient means for opening all phase conductors that will respond to a signal from either the overcurrent or ground-fault detection means; plus the additional advantage of mounting the current sensors and logic circuitry internal to the breaker, minimizing the need to make external connections to control components.

The fundamental rules of applying circuit breakers within voltage and continuous-current ratings are well defined in the NEC [6], other ANSI standards, and other sources. A fundamental rule necessary for system protection is to apply circuit breakers within their interrupting or short-circuit current ratings. The determination of available short-circuit current at the various levels throughout the electrical distribution system is a necessary step to be completed prior to selecting circuit breakers for system protection (Chapter 2 discusses methods of calculating available short-circuit current). It should be noted that molded-case circuit breakers are available with different interrupting ratings in the same physical frame size. Selection by frame size or continuous ampere rating alone is not sufficient; the interrupting rating must also be considered.

Current-limiting fuses, fused circuit breakers, or current-limiting circuit breakers may be provided to lower the let-through short-circuit current. Curves depicting let-through current and I^2t are available from manufacturers to assist in the application of these circuit breakers; Fig 116.

An alternate method is *series connection* of molded-case circuit breakers, that is, two molded-case circuit breakers electrically in series sharing fault interruption duties. This is a viable protection scheme, provided testing has verified performance. Underwriters Laboratories, Inc, presently recognizes series connected short-circuit ratings and prescribes test procedures to verify performance. See Fig 117 for an example of a test setup. It should be noted that selectivity will not be provided at any current level where the breaker trip characteristic curves overlap when this protection scheme is used, that is, both circuit breakers will trip. Series connected ratings must be based on tests and are only valid for the specific circuit breaker types listed in the test reports. Individual manufacturer's series-connected ratings may be found in the UL Recognized Component Dictionary [12]. Fuse/breaker coordinated combinations are also tested by UL and are applicable within their established ratings.

Determination of available fault current levels, specification of circuit breakers and associated equipment rated for those levels, and inspection to verify that properly rated equipment has been installed will satisfy the basic requirement of providing adequately rated equipment for system protection.

Selection of desirable trip unit functions and their settings to provide protec-

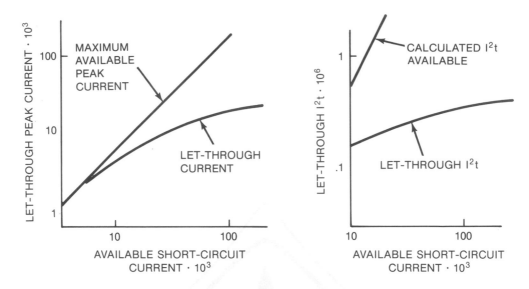

Fig 116
Let-Through Current I^2t for a Current-Limiting Circuit Breaker

tion and coordination is the next consideration. Basic rules applicable to phase overcurrent protection are as follows.

(1) Select continuous-current ratings and pickup settings of long-time-delay characteristics, where adjustable, which are no higher than necessary and which meet applicable standards without causing nuisance tripping. The amount of time delay provided by the long-time-delay characteristics should be selected to be no higher than necessary to override transient overcurrents associated with the energizing of load equipment and to coordinate with downstream protection devices.

(2) Take advantage of the adjustable instantaneous trip characteristic on molded-case circuit breakers and low-voltage power circuit breakers. Set the instantaneous response no higher than necessary to avoid nuisance tripping. Be sure that instantaneous trip settings do not exceed the available short-circuit current at the location of the circuit breaker in the system. This point is frequently overlooked, particularly in service entrance applications.

(3) Provide ground-fault protection in accordance with the NEC [6], as a minimum on systems where required. Ground-fault current settings should be set as low as possible to minimize hazard to personnel and damage to equipment. Time-delay adjustments of ground-fault protective devices should be set so that ground faults are cleared by the device immediately upstream.

6.4.2 Coordination. When protection is being considered, the performance of a circuit breaker with respect to the connected conductors and load is a primary concern. To achieve coordination, consideration is also given to the performance of a circuit breaker with respect to other upstream and downstream

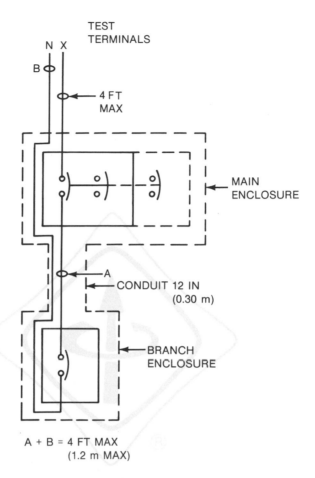

Fig 117
"Series Connection" Test Circuit from UL 489

protective devices. The objective in coordinating protective devices is to make them selective in their operation with respect to each other. In so doing, the effects of short circuits on a system are reduced to a minimum by disconnecting only the affected part of the system. Stated in another way, only the circuit breaker nearest the short circuit should open, leaving the rest of the system intact and able to supply power to the unaffected parts. Chapter 14 covers the general subject of coordination in detail.

Generally, coordination is demonstrated by plotting the time-current characteristic curves of the circuit breakers involved and by making sure that no overlapping occurs between the curves of adjacent circuit breakers; Fig 118. Often coordination is possible only when circuit breakers with short-time time-delay characteristics are used in all circuit positions except the one closest to the load. This is particularly true when there is little or no circuit impedance between successive circuit breakers. This condition often exists in a main

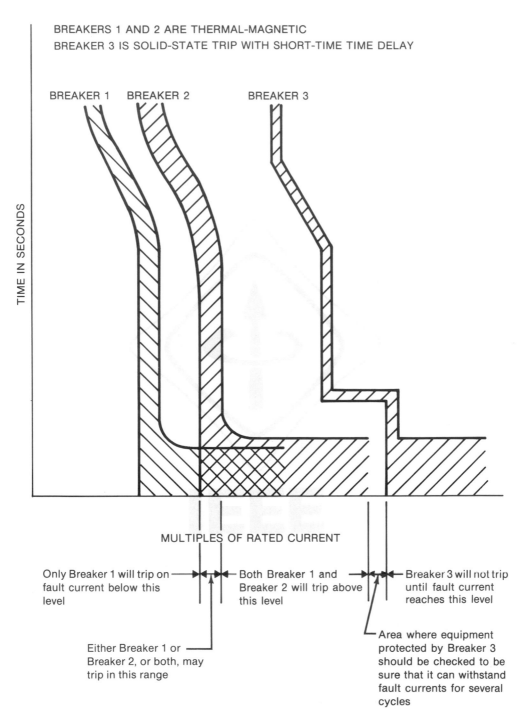

Fig 118
Coordinated Tripping by Overlapping Time-Current Characteristic Curves

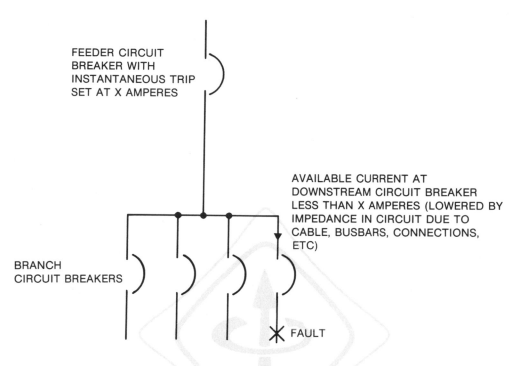

Fig 119
Coordinated Tripping Due to Impedance in Circuit

switchboard or load center unit substation between the main and feeder circuit breakers. Here, to be selective for all levels of possible short-circuit current beyond the load terminals of the feeder circuit breakers requires that the main circuit breaker be equipped with a combination of long-time-delay and short-time-delay trip characteristics. The withstand rating of associated circuit components and assemblies should not be exceeded. Moving downstream, on many feeder circuits there is sufficient impedance in the distribution system to appreciably lower the available short-circuit current at the next downstream circuit breaker. If the available short-circuit current at this next circuit breaker is less than the instantaneous trip setting of the feeder circuit breaker, selectivity is possible; see Fig 119.

The preceding forms the basis for judging selectivity between two circuit breakers in series. If the fault current being interrupted by the downstream circuit breaker flows through the upstream circuit breaker for a period of time equal to or greater than its response time, the upstream circuit breaker will trip. Under these conditions the circuit breakers will not be selective. However, if because of impedance between the circuit breakers the maximum current that can flow during short circuit conditions is insufficient to initiate a tripping response in the upstream circuit breaker, selectivity will exist.

6.5 Conclusions. The following considerations apply to low-voltage circuit breakers for system protection:

(1) They combine a switching means with an overcurrent protective device in a compact, generally self-contained unit.

(2) Little or no exposure to live parts is involved during routine operation.

(3) They are resettable. After tripping (and removal of the fault or overload that caused tripping), service may be restored without replacing any part of the assembly. Inspection of the circuit breaker assembly after fault current interruption is recommended to verify suitability for further use. Inspection of the circuit breakers may require replacement of fuses or fuse assemblies after interruption of high-level fault currents.

(4) They provide simultaneous disconnection of all phase conductors.

(5) High short-circuit interrupting ratings, the availability of current-limiting circuit breakers, and series-connected interrupting ratings permit application on systems with high available fault currents.

(6) The advent of highly sophisticated solid-state trip units has increased circuit breaker versatility and made coordination easier.

(7) Selection of molded-case circuit breakers should consider interrupting rating since different interrupting ratings are available in the same frame size.

(8) Coordination of ground-fault protective devices requires time-delay adjustments and is enhanced by the presence of adjustments that provide inverse-time-current characteristics.

6.6 References. The following publications shall be used in conjunction with this chapter.

[1] ANSI C37.16-1980, American National Standard Preferred Ratings, Related Requirements, and Application Recommendations for Low-Voltage Power Circuit Breakers and AC Power Circuit Protectors.

[2] ANSI C37.17-1979, American National Standard for Trip Devices for AC and General-Purpose DC Low-Voltage Power Circuit Breakers.

[3] ANSI C97.1-1972, American National Standard for Low-Voltage Cartridge Fuses 600 Volts or Less.

[4] ANSI/IEEE C37.13-1981, IEEE Standard for Low-Voltage AC Power Circuit Breakers Used in Enclosures.

[5] ANSI/IEEE C37.100-1981, IEEE Standard Definitions for Power Switchgear.

[6] ANSI/NFPA 70-1984, National Electrical Code.

[7] IEC Pub 157-1 (1973), Low-Voltage Switchgear and Controlgear, Part 1: Circuit Breakers.

[8] NEMA AB1-1975, Molded-Case Circuit Breakers.

[9] NEMA AB2-1984, Procedures for Verifying the Performance of Molded-Case Circuit Breakers.

[10] NEMA SG-3-1981, Low-Voltage Power Circuit Breakers.

[11] UL 489-1980, Molded-Case Circuit Breakers and Circuit Breaker Enclosures.

[12] UL Recognized Component Directory.

7. Ground-Fault Protection

7.1 General Discussion. In recent years there has been an increasing interest in the use of ground-fault protection in electric distribution circuits. This interest has been intensified by the requirement of ANSI/NFPA 70-1984 [6][19] (National Electrical Code [NEC]) for ground-fault protection on certain service entrance equipments. This is evident when one inspects today's electrical indoor distribution, construction, and consulting engineering press and notes the number of feature articles dealing with this subject. These articles and the unusual interest in ground-fault protection have been brought about by a disturbing number of electric failures. One editor [15] reports the cost of arcing faults as follows: "One five-year estimate places the figure between $1 billion and $3 billion annually for equipment loss, production downtime, and personal liability." This chapter explores the need for better ground-fault protection, pinpoints the areas where that need exists, and discusses the solutions that are being applied today.

Distribution circuits that are solidly grounded or grounded through low impedance require fast clearing of ground faults. This is especially true in low-voltage grounded wye circuits that are connected to busways or long runs of metallic conduit. The problem involves sensitivity in detecting low ground-fault currents as well as coordination between main and feeder circuit protective devices.

The concern for ground-fault protection is based on four factors.

(1) The majority of electric faults involve ground. Even those that are initiated phase-to-phase will spread quickly to any adjacent metallic housing, conduit, or tray that provides a return path to the system grounding point. Ungrounded systems are also subject to ground faults and require careful attention to ground detection and ground-fault protection.

(2) The ground-fault protective sensitivity can be relatively independent of continuous load current values, and thereby have lower pickup settings than phase protective devices.

(3) Since ground-fault currents are not transferred through system power transformers that are connected delta–wye or delta–delta, the ground-fault protection for each system voltage level is independent of the protection at other voltage levels. This permits much faster relaying than can be afforded by phase-protective devices that require coordination using pickup values and time

---

[19] The numbers in brackets correspond to those of the references listed at the end of this chapter. The numbers in brackets with the prefix "B" correspond to those of the bibliography listed at the end of this chapter.

delays that extend from the load to the source generators, often resulting in considerable time delay at some points in the system.

(4) Arcing ground faults that are not promptly detected and cleared can be extremely destructive.

Much of the present emphasis on ground-fault protection centers on low-voltage circuits, 600 V or less. Low-voltage circuit protective devices have usually involved fused switches or circuit breakers with integrally mounted series tripping devices. These protective elements are termed overload or fault overcurrent devices because they carry the current in each phase and clear the circuit only when the current reaches a magnitude greater than full-load current. To match insulation damage curves of conductors and to accommodate motor and transformer inrush currents, phase overcurrent devices are designed with inverse characteristics that are rather slow at overcurrent values up to about five times rating. For example, a 1600 A low-voltage circuit breaker with conventional phase protection will clear a 3200 A fault in about 100 s, although it can be adjusted in a range of roughly 30–200 s at this fault value. A 1600 A fuse may require 10 min or more to clear the same 3200 A fault. These low values of fault currents are associated predominantly with faults to ground and generally have received little attention in the design of low-voltage systems until the occurrence of a number of serious electric failures in recent years. Arcing fault phenomena surfaced in the late 1940's and early 1950's with the advent of 480/277 V systems. In contrast, on grounded systems 2400 V and above, it has long been standard practice to apply some form of ground-fault protection.

It should be noted that this chapter is primarily directed to ground-fault protection of major equipment of circuits and equipment. The discussions relate to various forms of ground-fault protection to prevent excessive damage to electrical equipment with current sensitivity in the order of amperes to hundreds of amperes.

Although its purpose is to disconnect faulty parts of an electric system, to preserve service continuity in other parts, and to limit ejection of gases and molten metal or to localize faults, ground-fault protection may not satisfy rigid requirements regarding shock hazards or touch potentials that are designed for protection of people and necessitate milliampere sensitivity, and may only be feasible for small loads.

The action initiated by ground-fault sensing devices should vary depending on the installation. In some cases, such as services to dwelling, it may be necessary to immediately disconnect the faulted circuit to prevent loss of life or property. However, the opening of some circuits in critical applications may in itself endanger life or property. Therefore, each particular application should be studied carefully before selecting the action to be initiated by the ground-fault protective devices.

7.2 Types of Systems Relative to Ground-Fault Protection. A comprehensive discussion of grounded and ungrounded systems is given in Chapter 1 of ANSI/IEEE Std 142-1982 [4]. When considering the choice of grounding, it is important to determine the types of ground-fault protection available and their effect on system performance, operation, and safety.

An ungrounded system has no intentional connection to ground except through potential indicating or measuring devices, or through surge protective devices. While it is called *ungrounded,* it is actually coupled to ground through the distributed capacitance of its phase windings and conductors.

A grounded system is intentionally grounded by connecting its neutral or one conductor to ground, either solidly or through a current-limiting impedance. Various degrees of grounding are used ranging from solid to high impedance, usually resistance.

Figure 120 shows ungrounded and grounded systems and their voltage relationships. The term *solidly grounded* and *direct grounded* have the same meaning, that is, no intentional impedance is inserted in the neutral-to-ground connection.

7.2.1 Classification of System Grounding. The types of system grounding normally used in industrial and commercial power systems are

(1) Solid grounding
(2) Low-resistance grounding
(3) High-resistance grounding
(4) Ungrounded

Each type of grounding has advantages and disadvantages, and there is no general acceptance of any one method. Factors that influence the choice include:

(1) Voltage level of power system
(2) Transient overvoltage possibilities
(3) Type of equipment on the system
(4) Required continuity of service
(5) Caliber and training of operating and maintenance personnel
(6) Methods used on existing systems
(7) Availability of convenient grounding point
(8) National Electrical Code (ANSI/NFPA 70-1984 [6])
(9) Cost of equipment, including protective devices and maintenance
(10) Safety, including fire and shock hazard
(11) Tolerable fault damage levels
(12) Effect of voltage dips during faults

There are many factors involved in selecting grounding methods for the different voltage levels found in power distribution systems. ANSI/IEEE Std 142-1982 [4] discusses many of these factors in detail, while the following discussion mentions only the reasons that relate to ground-fault protection.

7.2.2 Solid Grounding (Fig 121). Most industrial and commercial power systems are supplied from electric utility systems that are solidly grounded. If the user must immediately convert to lower voltage, the power transformers typically have a delta-connected primary and a wye-connected secondary that can again be connected solidly to ground. This results in a system that can be conveniently protected against overvoltages and ground faults. The system has flexibility since the neutral can be carried with the phase conductors, which permits connecting loads from phase-to-phase and from phase-to-neutral.

(1) Systems Above 600 V. Ground relaying of medium-voltage and high-voltage systems that are solidly grounded has been successfully accomplished for

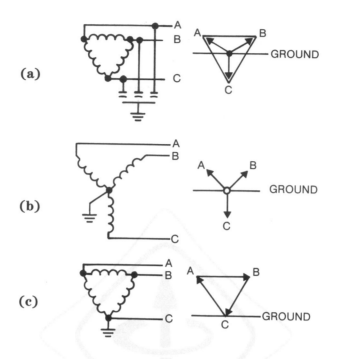

Fig 120
Voltages to Ground under Steady-State Conditions
(a) Ungrounded System, Showing System Capacitance to Ground
(b) Grounded Wye-Connected System
(c) Grounded Delta-Connected System

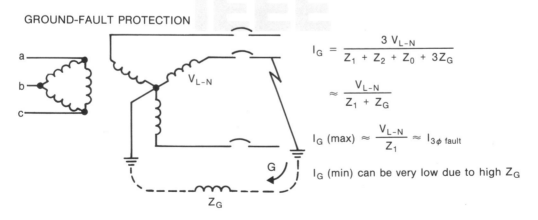

GROUND-FAULT PROTECTION

$$I_G = \frac{3\,V_{L-N}}{Z_1 + Z_2 + Z_0 + 3Z_G}$$

$$\approx \frac{V_{L-N}}{Z_1 + Z_G}$$

$$I_G\,(max) \approx \frac{V_{L-N}}{Z_1} \approx I_{3\phi\,fault}$$

$I_G\,(min)$ can be very low due to high Z_G

Fig 121
Direct or Solid Grounding; Use Ground Relays to Trip

many years using residually connected ground relays or zero-sequence sensing. The circuit breakers normally have current transformers to provide the signal for the phase overcurrent relays and the ground overcurrent relay is connected in the wye point ("residual") to provide increased sensitivity for ground faults. Ground-fault magnitudes usually are comparable to phase-fault magnitudes and are therefore easily detected by relays or fuses unless they occur in equipment windings near the neutral point.

(2) Systems 600 V and Below. All 208 V systems are solidly grounded so that loads can be connected from line-to-neutral to provide 120 V service. Similarly, all 480 V systems that are to serve 277 V lighting must also be solidly grounded. This results in most commercial building and many industrial plant 480 V systems being solidly grounded. Even where 277 V lighting is not used, many industrial plant 480 V systems are solidly grounded to limit overvoltages and to facilitate clearing ground faults.

While higher voltage systems normally use relays, and the ground relay with increased sensitivity is easily provided, low-voltage systems usually use circuit breakers with integrally mounted trip devices in the phases. Until recently, it was not felt that the additional cost of supplementary ground-fault relaying was justified. However, it is now recognized that even solidly grounded low-voltage systems can experience ground faults with a relatively low fault current level for the reasons explained in 7.3. Because of this, sensitive ground-fault relays and trip devices have been developed for use with low-voltage circuit breakers and bolted pressure switches.

One disadvantage of the solidly grounded 480 V system involves the high magnitude of ground-fault currents that can occur, and the destructive nature of arcing ground faults. However, if these are promptly interrupted, the damage is kept to acceptable levels. While it is possible to resistance-ground low-voltage systems, resistance grounding restricts the use of line-to-neutral loads.

Another characteristic of solidly grounded 480 V systems is that the ground fault may cause immediate forced outages. If this cannot be tolerated, then either the high-resistance grounded system (without ground-fault tripping) or ungrounded systems are used to delay the required outage for repairs. On the other hand, immediate removal of a faulty circuit is usually desirable.

7.2.3 Low-Resistance Grounding (Fig 122). The low-resistance grounded system is similar to the solidly grounded system in that transient overvoltages are not a problem. The resistor limits ground-fault current magnitudes to reduce the rate of damage during ground faults. Ground-fault current levels vary, but levels of 100–200 A can be detected by many relaying schemes currently in use. In multisource systems ground fault as high as 800–1600 A can be anticipated. The magnitude of the grounding resistance is selected to allow sufficient current for ground-fault relays to detect and clear the faulted circuit. This type of grounding is used mainly in 2.4–13.8 kV systems, which often have motors directly connected.

The value of resistance also relates to the type of relaying and the amount of motor windings that can be protected. Ground faults in wye-connected motors have reduced driving voltage as the neutral of the motor winding is approached;

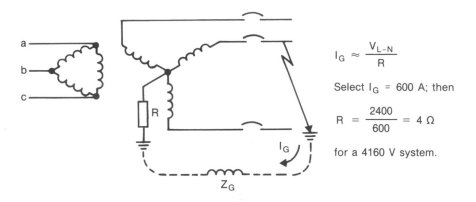

$$I_G \approx \frac{V_{L-N}}{R}$$

Select I_G = 600 A; then

$$R = \frac{2400}{600} = 4\ \Omega$$

for a 4160 V system.

**Fig 122
Low-Resistance Grounding; Use Ground Relays to Trip**

thus, ground-fault current magnitudes are reduced.

7.2.4 High-Resistance Grounding [Fig 123(a) and (b)]. High-resistance grounding limits first fault-to-ground currents to very low values. The fault-current magnitude is predictable regardless of the location of the fault, since the grounding resistor inserted in the neutral is very large compared to the impedance of the remainder of the ground-fault path. Some high-resistance grounding schemes, usually at medium voltages, connect a distribution transformer between the neutral and ground, with a resistor on the transformer secondary. The transformer primary is rated for line-to-line voltage, and a 240 V secondary will limit the secondary voltage to a convenient 139 V maximum.

High-resistance grounding helps insure a ground-fault current of known magnitude, helpful for relaying purposes. This makes it possible to identify the faulted feeder with sensitive ground-fault relays, which are available with fault sensitivity in the range of small fractions of an ampere. If the resistance is chosen so that the fault current is equal to or slightly larger than the charging current of the system, transient overvoltages will be reduced. Charging current can be calculated or measured, and is usually under 2 A for low-voltage systems.

Ground-fault currents of this magnitude seldom require immediate tripping, and thus high-resistance grounding can often maintain continuity of service under first ground-fault conditions until a favorable time for an outage to clear the fault. This is provided so that the cable carrying the fault is rated 173% of the voltage level. If a second ground-fault occurs on another phase before the first fault is cleared, a phase-to-ground-to-phase fault occurs that is not limited by the neutral grounding resistor [see Fig 123(b)]. This may be an arcing fault, whose magnitude is limited by the ground-path impedance to a value high enough to cause severe arcing damage, but too low to activate the overcurrent devices

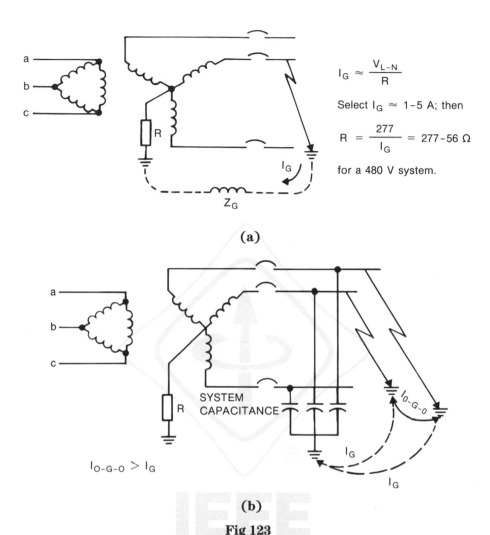

$$I_G \approx \frac{V_{L-N}}{R}$$

Select $I_G \approx$ 1–5 A; then

$$R = \frac{277}{I_G} = 277\text{–}56 \ \Omega$$

for a 480 V system.

(a)

$I_{O-G-O} > I_G$

(b)

Fig 123
(a) High-Resistance Grounding; May Use Ground Relays to Alarm on First Fault, Trip on Second Fault
(b) High-Resistance Grounding with Phase-to-Ground-to-Phase Fault

quickly enough to prevent or limit this damage. Two-level relays are available that will alarm on first (low) fault, but will trip on second (high) fault in time to prevent arcing burndowns.

7.2.5 Ungrounded Systems [Fig 124(a) and (b)]. Ungrounded systems employ ground detectors (lamps or voltmeters connected from each phase to ground) to indicate a ground fault. These detectors show the existence of a ground on the system and identify the faulted phase, but do not locate the ground, which could be anywhere on the entire system. The system will operate

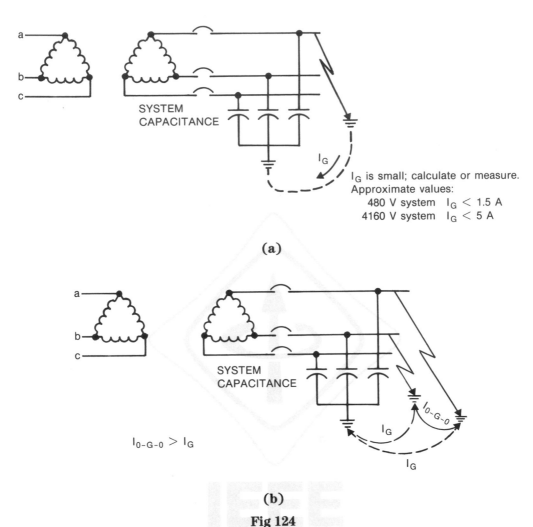

Fig 124
(a) Ungrounded System; Use Bus Ground Detector to Alarm
(Watch for Possible Overvoltage Problem)
(b) Ungrounded System with Phase-to-Ground-to-Phase Fault

with the ground fault acting as the system ground point. The ground-fault current that flows is the capacitive charging current of the system, generally only a few amperes.

If this ground fault is intermittent or allowed to continue, the system could be subjected to possible severe overvoltages to ground, which can be as high as six or eight times phase voltage. This can puncture insulation and result in additional ground faults. These overvoltages are caused by repetitive charging of the system capacitance, or by resonance between the system capacitance and the inductances of equipment in the system.

A second ground fault occurring before the first fault is cleared will result in a phase-to-ground-to-phase fault, usually arcing, with current magnitude large enough to do damage, but sometimes too small to activate the overcurrent devices in time to prevent or minimize damage.

Ungrounded systems offer no advantage over high-resistance grounded systems in terms of continuity of service, and have the disadvantages of transient overvoltages, locating the first ground fault, and burndowns from a second ground fault. For these reasons, they are being used less frequently today than high-resistance grounded systems, and existing ungrounded systems are often converted to high-resistance grounded systems by resistance-grounding the neutral if it exists, or, if the system is from a delta source, by creating a neutral point with a zig-zag or other transformer and then resistance-grounding it.

Once the system is high-resistance grounded, overvoltages are reduced, and modern, highly sensitive ground-fault protective equipment can identify the faulted feeder on first fault and trip one or both feeders on second fault before an arcing burndown does serious damage.

7.3 Ground Faults—Nature, Magnitudes, and Damage. Ground faults on electric systems can (1) originate in many ways, (2) have a wide range of magnitudes, and (3) cause varying amounts of damage. The most serious faults from the standpoint of rate of eroded material are acing faults, both phase-to-phase and phase-to-ground.

7.3.1 Origin of Ground Faults. Ground faults originating from insulation breakdown can be classified, roughly, as follows:

(1) Reduced insulation due to moisture, atmospheric contamination, foreign objects, insulation deterioration, etc

(2) Physical damage to insulation system due to mechanical stresses, insulation punctures, etc

(3) Excessive transient or steady-state voltage stresses on insulation

Good installation and maintenance practices ensuring adequate connections and the integrity of the insulation of the equipment have a significant effect in reducing the probability of ground faults. However, insulation breakdowns to ground can occur at any point in the system where phase conductors are in close proximity to a grounded reference. The contact between the phase conductor and ground is usually not a firm metallic contact, but rather usually includes an arcing path in air or across an insulating surface, or a combination of both. In addition to these arcing ground faults certain bolted type faults occur, usually during installation or maintenance, when there is an inadvertent firm metallic connection from phase-to-ground.

7.3.2 Magnitude of Ground-Fault Currents. Ground-fault current magnitudes can vary greatly. Using the method of symmetrical components (see Chapter 2), the single line-to-ground fault current I_{GF} is calculated by the formula

$$I_{GF} = \frac{3V_{L+N}}{Z_1 + Z_2 + Z_0 + 3Z_G}$$

where Z_1, Z_2, and Z_0 are the positive, negative, and zero-sequence impedances. The term Z_G represents the impedances of the ground return circuit, including those of the fault arc, the grounding circuit, and the intentional neutral impedance, when present.

To illustrate how ground-fault currents can vary greatly in magnitude, consider a solidly grounded system with a bolted ground fault very close to the generator terminals. In this example Z_G could approach zero, and if we assume that $Z_1 = Z_2 = Z_0$, then

$$I_{GF} = \frac{V_{L+N}}{Z_1}$$

which is actually the formula for a bolted three-phase fault. In fact, with many generators, since Z_0 is smaller than Z_1, it is necessary to add an intentional neutral impedance Z_N to reduce the bolted ground-fault current to the magnitude of the bolted three-phase fault current.

For a ground fault in a high-resistance grounded system, the neutral resistance R_N is very large compared to Z_1, Z_2, Z_0, and the remainder of Z_G. Then I_{GF} is approximately equal to E_{L-N} divided by R_N.

For example, in a high-resistance grounded 480 V system with a neutral resistance of 20 Ω, the ground-fault current will be

$$I_{GF} = \frac{480/\sqrt{3}}{20} = 14 \text{ A}$$

This is true since the fault arc impedances and the ground-circuit impedances are negligible when compared to 20 Ω.

Precise calculation of low-voltage ground-fault current magnitudes in solidly grounded systems is much more difficult than the previous example. The reason for this is that the circuit impedances, including the fault arc impedance, that were negligible in the high-resistance example play an important part in reducing ground-fault current magnitudes. This applies even in most cases where a sizable grounding conductor is carried along with phase conductors.

The primary consideration in applying ground-fault protection is whether a selectively coordinated system can be achieved and, if not, to establish the extent to which lack of selectivity is acceptable.

The two main setting characteristics that need to be determined for ground-fault relays are (1) minimum operating current, and (2) speed of operation. Selection of the minimum operating current (pickup) setting is based primarily on the characteristics of the circuit being protected. If the circuit serves an individual load, such as a motor, transformer, or a heater circuit, then the pickup setting can be quite low, such as 5–10 A. If the protected circuit feeds multiple loads, each with individual overcurrent protection, for example, a panelboard, feeder duct, motor control center, etc, the pickup settings will be higher. These higher settings in the order of 200–1200 A are selected to allow the branch phase overcurrent devices to clear low-magnitude ground faults in their respective circuits, if coordination is possible. Furthermore, low-level faults in some parts of the system may be self-extinguishing, thus allowing uninterrupted operation of other equipment.

7.3.3 Damage Due to Arcing Faults. The arcing fault causes a large amount of energy to be released in the arcing area. The ionized products of the arc spread rapidly. Vaporization at both arc terminals occurs, and the erosion at the electrodes is concentrated when the arc does not travel. While there is a tendency of the arc to travel away from the source, this does not necessarily occur at low levels of fault current or at higher levels of current in circuits with insulated conductors. An arcing fault, if allowed to persist indefinitely, (1) is a potential fire hazard, (2) causes considerable damage, and (3) may result in a more extended power outage.

Single-pole interrupters without antisingle-phasing provisions are especially ineffective against low-voltage arcing faults [12]. The reason for this is that often an initial ground fault or phase-to-phase fault can cause one or two fuses to clear faulted phases, but the fault continues to be fed through the load impedances from the uncleared phases. The fault is thus of a diminished magnitude and may never be cleared by the remaining fuses until other circuits are involved. This sequence has caused extended outages of many physically adjacent circuits. In many applications, the probability of extended damage can be reduced by careful design including compatible ground-fault protection.

Some fusible switches can be equipped with antisingle-phasing provisions that consist of installing small actuator type fuses in parallel with the line fuses in the switch. When a line fuse blows, these fuses will also blow and subsequently close a contact to actuate a signal or switch-opening circuit to open all three poles of the switch. Figure 125 shows this particular scheme, which can also be used in conjunction with ground-fault protective relaying. Other equipment utilizes voltage relays in place of the actuator fuses to trip interrupter switches.

Fused circuit breakers and service protectors, as well as circuit breakers, have antisingle-phasing devices incorporated in their basic designs.

The basic need for ground-fault protection in low-voltage grounded systems is illustrated in Fig 126. A 1000 kVA transformer, with a 1600 A main circuit breaker and typical long-time and short-time characteristics, optionally with a fused switch, is shown.

A 1500 A ground fault (point I) on the 480Y/277 V grounded neutral system would not be detected by the main circuit breaker or fuse. A ground relay set at 0.2 s time delay would cause the circuit breaker to clear the fault in about 0.33 s. A 4000 A ground fault (point II) could persist for about 33 s, even if the circuit breaker minimum long-time band were used. The fuse would require up to 5 min to clear this fault. The ground relayed circuit breaker would clear the fault in about 0.25 s. An 8000 A ground fault (point III) would be cleared within about 0.2–0.4 s by the circuit breaker short-time device, assuming it is present; otherwise, between 8–20 s would elapse before the long-time device clears the fault.

Arc energies for these assumed faults are tabulated in Table 38. Arc voltages are assumed to be 100 V. Since the arc voltage tends to have a flat top characteristic (nonlinear arc resistance), the energy of the arc in W/s can be estimated by obtaining the product of the current in rms A, the arc voltage in V, and the clearing time in s. Approximate calculation of the energy required to erode a certain amount of electrode material shows that 50 kW/s of energy divided equally between conductor and enclosure will vaporize about ⅛ in³ of aluminum or ½₀ in³ of copper. The calculation assumes that most of the arc energy goes into the electrodes, while the energy lost to the surrounding air is neglected. Comparisons were made from several arcing fault tests [10], [11], and

Fig 125
Antisingle-Phasing Provisions for Fusible Switches

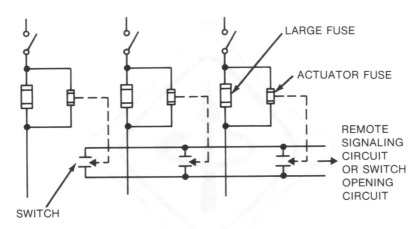

Table 38
Arc Energies for Assumed Faults of Fig 126

Fig 126 Points	Fault (A rms)	Main Device	Clearing Time (s)	Arc Energy (kW · s)
I	1500	Relay	0.33	50
		Circuit breaker	∞	∞
		Fuse	∞	∞
II	4000	Relay	0.25	100
		Circuit breaker	33.00	13 200
		Fuse	300.00	120 000
III	8000	Relay	0.25	200
		Circuit breaker	0.4	320
		Fuse	10.00	8000
IV	20 000	Relay	0.25	500
		Circuit breaker	0.20	400
		Fuse	0.01	20

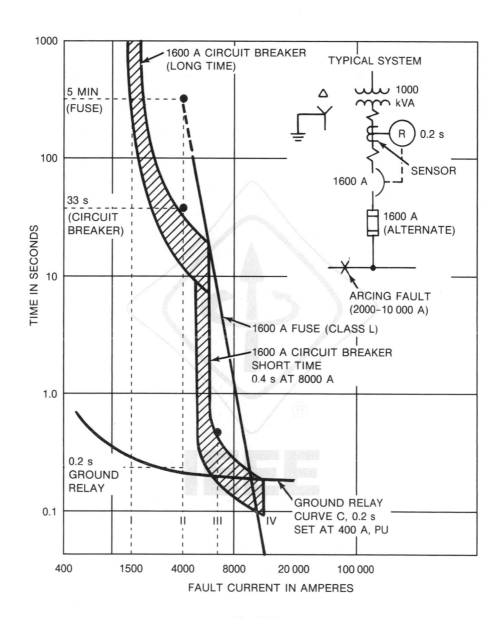

Fig 126
(a) Time-Current Plot Showing Slow Protection Provided by Phase Devices
for Low-Magnitude Arcing Ground Faults

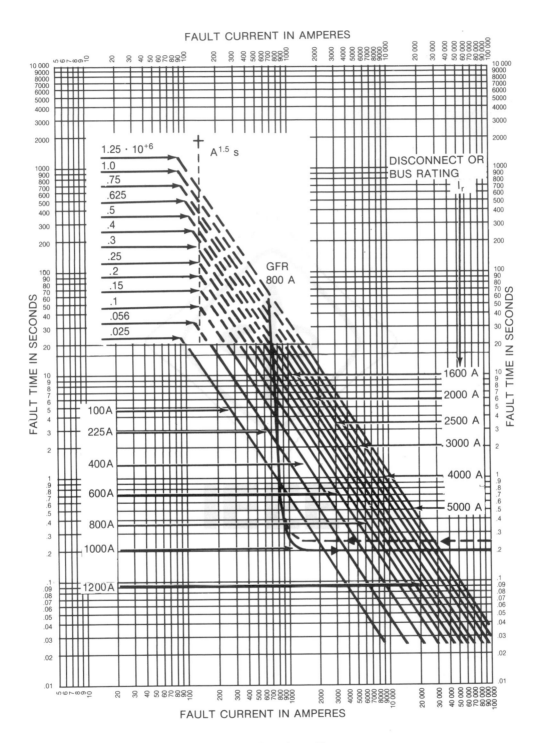

FAULT CURRENT IN AMPERES

Fig 126 *(Continued)*
(b) Assumed Tolerable Damage Levels

good correlation was obtained between calculated energy from test data and measured conductor material eroded.

For the assumed 8000 A fault, even though the current values are the calculated result using all source, circuit, and arc impedances, the actual rms current values passing through the circuit breaker can be considerably lower. This is because of the spasmodic nature of the fault caused by (1) arc elongating blowouts effects, (2) physical flexing of cables and some bus structures due to mechanical stresses, (3) self-clearing attempts and arc reignition, and (4) shifting of the arc terminals from point to point on the grounded enclosures (as well as on the faulted conductors for noninsulated construction). All of these effects tend to reduce the rms value of arcing fault currents. Therefore, a ground fault that would normally produce 8000 A under stablized conditions might well result in an effective value of only 4000 A, and would have the arc energies associated with point II in Table 38.

Expressing acceptable damage in terms of kW · s, or kW cycle units with an assumption of 100 V arc drop in 480Y/277 V circuits has been proposed.

Recent investigations [16] show that damage in standard switchboards at normal arc lengths is proportional to time and 1.5 power of ground-fault current magnitude. Thus, arc voltage magnitude question at varying and unpredictable fault currents may be excluded and damage prediction simplified.

According to the study, specific damage or burning rate:

$$k_s = \frac{\text{damaged volume } V_D}{\text{current}^{1.5} \cdot \text{time}} \, [\text{in}^3/\text{A}^{1.5} \text{ s}]$$

with

$k_s = 0.72 \cdot 10^{-6}$ for copper

$k_s = 1.52 \cdot 10^{-6}$ for aluminum

$k_s = 0.66 \cdot 10^{-6}$ for steel

Since selection of conductors is often based on nearly uniform current densities (say 800–1000 A/in^2), acceptable damage could then be based on conductor or disconnect ratings, or cross sectional area.

Thus, if based on

$$I_F^{1.5}t + k_e I_R$$

where

I_F = fault current

I_R = disconnect or bus rating

the acceptable damage $V_D = k_s k_e I_R$

can be used as a constant for a given system and disconnect rating. Acceptable damage could then be held by appropriate selection of current and time settings for ground-fault protective devices.

For example, if $I_R = 1000$ A and $k_e = 250$ [A$^{0.5}$ s] (as assumed in NEMA PB1.2-1977 [7]) acceptable damage

$I^{1.5} t = 250 \cdot 1000 = 0.25 \cdot 10^{+6}$ [A$^{1.5}$ s], or

$V_D = 0.72 \cdot 10^{-6} \cdot 0.25 \cdot 10^6 = 0.18$ in^3 for copper, conductors will not be exceeded for faults between 800 and 10 000 A, with relay settings as shown in Fig 126(b) if clearing time of the circuit breaker does not exceed 50 ms.

The above computations are based on 277 V single-phase test results and the assumption that the damage would be proportional to the arcing fault current. Therefore, some discretion should be used when referencing the example in Fig 126(b) [13].

7.3.4 Selection of Low-Voltage Protective Device Settings. Maximum protection against ground faults can be obtained by applying ground protection on every feeder circuit from source to load. The minimum operating current for all series devices may be set at about the same pickup setting, but the time curves are selected so that each circuit protective device is opened progressively faster, moving from the source to the load. The load switching device can be opened instantaneously or with brief delay upon occurrence of a ground fault.

The delay required between devices is determined by the addition of (1) the trip operating time of the overcurrent device, (2) the clearing time of the overcurrent device, and (3) a margin of safety. The trip operating time of today's molded-case breakers, service protectors, air, power circuit breakers is usually three cycles or less. Shunt tripped switches may take somewhat longer. Modern current limiting fuses will clear in less than half-cycle when operating in their current limiting range.

This coordination by time delay is similar to other overcurrent coordination. However, a new method of coordination, called zone interlocking, is available for ground-fault protection using solid-state relays. Ground faults, for minimum damage, must be cleared as quickly as possible regardless of their magnitude. Zone coordination assures instantaneous tripping of all ground-fault relays for faults within their zone of protection, with upstream devices restrained to a time delay in response to ground faults outside their zone. This restraining signal requires only one pair of wires from the downstream zone to the upstream relay to carry the interlocking signal. Zone interlocking provides the fastest tripping, for minimum damage, with full coordination so that only the affected part of the system is shut down on ground fault.

Recent developments in bolted-pressure fused switch design has made possible the use of most of the ground-fault protection schemes presently available. Now available are stored energy switching mechanisms that can be shunt tripped to cause the switch to open very quickly. In order to avoid the situation that occurs when the switch might open on ground faults above its nominal interrupting rating but before the fuses can operate, Underwriters Laboratories, Inc (UL) listed switches must incorporate one of two features: (1) a lock-out feature activated by a fault detector on each phase, which allows the fuses to clear on faults above the interrupting rating of the switch (usually 6–7.5 times switch rating), or (2) the switch interrupting rating must equal 12 times the switch continuous-current rating. However, this is not usually the case when the fuses are current limiting and are properly coordinated with the ground-fault relay.

From the standpoint of damage alone, speed of clearing is paramount. However, there are situations where some delay is desirable. This is primarily to obtain coordination between main and feeder circuits and branch currents. Consider a typical 480Y/277 V application consisting of a 3000 A main, an 800 A feeder, and a 100 A branch. If the branch circuits do not have ground-fault protection, then the feeder ground-fault protection must be set with a time delay to allow the branch circuit instantaneous units to clear moderately high-magnitude ground-fault currents without tripping the feeder circuit breaker. When full coordination is essential, it is desirable to set the feeder ground-fault pickup equal to the instantaneous setting of the branch circuits. While infrequent loss of coordination may be acceptable between feeders and branch circuits, it is recommended that full coordination be maintained between main and feeder overcurrent protective devices. It is generally not recommended to set main service ground-fault protection at less than 0.1 s (6 cycle) response time. Proper settings will reduce effects of inrush, startups and switching currents, and prevent nuisance trips.

Another reason for delayed clearing of ground faults on main or large feeder circuits is the threat of a trip where the power outage itself is of greater consequence than the incremental difference in fault damage.

In summary, the sensitivity (minimum operating current setting) of ground-fault protection in solidly grounded low-voltage systems is determined by the following considerations.

(1) When the ground-fault protection is used on devices protecting individual loads, such as motors, instantaneous devices with the lowest available settings can be used providing the devices will not cause false tripping from inrush currents.

(2) For mains and feeders, the setting for ground-fault protective devices is normally in the range of 10–100% of the circuit trip rating or fuse rating. If downstream devices do not have sensitive ground-fault protection, then the circuit ground-fault protection may have to be set higher than the downstream phase-protective device tripping characteristics to ensure full coordination.

For full protection the setting should be somewhat lower than the minimum estimated ground-fault current in the zone of protection for which the circuit protective device is responsible.

7.3.5 Sensing, Relaying, and Trip Devices. The signal for ground-fault protective devices may be derived from the residual of phase current transformers, window type current transformers, or sensors. The CT's or sensors provide isolation between main busses and relaying equipment and should be located in a specific path to detect proper ground-fault currents under all operating conditions.

Sensors are often designed with other than 5 or 1 A nominal secondary rating and for use with specific relays or trip devices as a system. If part of such a system, the relays normally have dials marked in terms of primary ground-fault current amperes.

Ground-fault relays or trip devices may be either self-powered (fault current) or externally powered (operation or trip power), or incorporate both methods. Outputs may be of contact or solid-state type (for example, thyristors).

AC control power, derived from an auxiliary transformer of proper capacity, is popular in systems of 600 V and below, sometimes supplemented by capacitor trips. In these cases, the primaries of control power transformers should be connected line-to-line to reduce effects of voltage dips during ground faults. The need for overcurrent protection and transfer to an alternate control power source should be evaluated.

Supplementary or backup ground-fault protection may be accomplished by monitoring the equipment environment. Such systems detect ionized gases and other fault current by-products, such as abnormal light and heat. By early detection of one or more of the by-products of a ground-fault current and prompt tripping of the interrupting device serving the fault, the magnitude of the damage may be reduced. Supplementary sensing is particularly desirable when the primary ground-fault sensing means are set relatively high to prevent nuisance tripping or to satisfy coordination requirements. To maximize the effectiveness of environmental detectors, care should be exercised in the selection, proper installation within the equipment enclosure, and setting of the detectors [14].

7.4 Frequently Used Ground-Fault Protective Schemes. While ground-fault protective schemes may be elaborately developed, depending on the ingenuity of the relaying engineer, nearly all schemes in common practice are based on one or more of the following methods of ground-fault detection: (1) residually connected overcurrent relays, (2) zero sequence of feeder conductors, (3) detection of ground-return current in the equipment grounding circuit, and (4) differential relaying.

7.4.1 Residual Connection. A residually connected ground relay is widely used to protect medium-voltage systems. The actual ground current is measured by current transformers that are interconnected in such a way that the ground relay responds to a current proportional to the ground-fault current. This scheme, using individual relays and current transformers, is not often applied to low-voltage systems. However, there are available low-voltage circuit breakers with three current transformers built into them and connected residually with the solid-state trip devices of the circuit breakers to provide ground-fault protection.

The term "residual" in common usage is normally reserved for three-phase system connections and seldom applied to single-phase or multiple-signal mixing.

The basic residual scheme is shown in Fig 127. Each phase relay is connected in the output circuit of its respective current transformer, while a ground relay connected in the common or residual circuit will measure the ground-fault current. In three-phase three-wire systems, such as shown in Fig 127, no current flows in the residual leg under normal conditions since the resultant of the three current transformers is zero. This is true for phase-to-phase short circuits also. When a ground-fault occurs, current bypasses the phase conductors and their current transformers and the resultant current flows in the residual leg, operating its relay.

On four-wire circuits a fourth current transformer should be connected in the neutral circuit as shown. The neutral conductor carries both 60 Hz single-phase

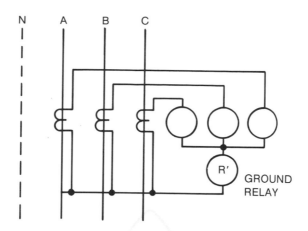

Fig 127
Residually Connected Ground Relay

load unbalance current as well as zero-sequence harmonic currents caused by the nonlinear inductance of single-phase loads, such as fluorescent lighting. Without the neutral-conductor current transformer the current in that conductor will appear to the ground relay as ground-fault current, and the ground relay would have to be desensitized sufficiently to prevent tripping under unbalanced load conditions.

The sensitivity of residually connected relays is determined by the CT ratio and relay pickup setting. If greater sensitive ground-fault protection is needed, consider the core-balance method (7.42).

7.4.2 Core Balance. This method is based on primary current vector addition or flux summation. The remaining zero-sequence component, if any, is then transformed to the secondary. The core-balance current transformer or sensor is the basis of several low-voltage ground-fault protective systems introduced in recent years. (The core-balance current transformer is frequently called a "zero-sequence sensor" or "window" current transformer, but the term "core balance" is preferable since it more specifically describes the function of the current transformer). The principle of the core-balance current transformer circuit is shown in Fig 128. The main conductors pass through the same opening in the current transformer and are surrounded by the same magnetic core. Core-balance current transformers are available in several convenient shapes and sizes, including rectangular designs for use over bus bars.

Under normal conditions, that is, balanced, unbalanced, or single-phase load currents or short circuits not involving ground (if all conductors are properly enclosed), all current flows out and returns through the current transformer. The net flux produced in the current transformer core will be zero and no current will flow in the ground relay. When a ground-fault occurs, the ground fault current

GROUND-FAULT PROTECTION

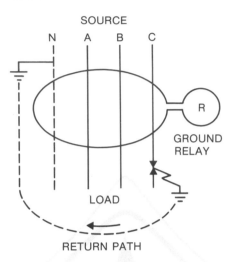

Fig 128
Core-Balance Current Transformer Encircles All Phase and Neutral Conductors

returns through the equipment grounding circuit conductor (and possibly other ground paths) bypassing the current transformer. The flux produced in the current transformer core is proportional to the ground-fault current, and a proportional current flows in the relay circuit. Relays connected to core-balance current transformers can be made quite sensitive. However, care is necessary to prevent false tripping from unbalanced inrush currents that may saturate the current transformer core, or through faults not involving ground. If only phase conductors are enclosed and neutral current is not zero, the transformed current will be proportional to the load zero-sequence or neutral current. Systems with grounded conductors, such as cable shielding, should have the current transformer surround only the phase conductors and not the grounded conductor.

By properly matching the current transformer and relay, ground-fault detection can be made as sensitive as the application requires. The relays are fast to limit damage and may be adjustable (for current, time, or both) in order to obtain selectivity. Many ground protective systems now have solid-state relays specially designed to operate with core-balance current transformers. The relays in turn trip the circuit protective device. Power circuit breakers, molded-case circuit breakers with shunt trips, or electrically operated fused switches can be used. The latter includes service protectors, which use circuit breaker contacts and mechanisms but depend on current-limiting fuses to interrupt the high available short-circuit currents. Fused contactors and combination motor starters may be used where the device interrupting capability equals or exceeds the available ground-fault current.

Figure 129(a) is a typical termination of a medium-voltage shielded cable. After the cable is pulled up through the core-balance CT, the cable jacket is

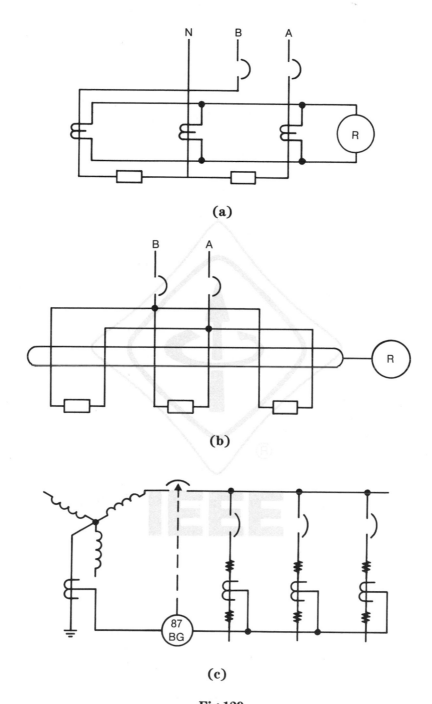

Fig 129
(a) Differential Protection Center-Tapped Loads
(b) Differential Protection Multiple Loads
(c) Ground Differential Scheme Bus and Transformer

removed to expose the shielding tape or braid. Jumpering the shields together, the connection to the ground is made after this shield lead is brought back through the CT. This precaution would have been necessary only if the shield had been pulled through the CT.

Between multiple shield ground connections on a single conductor cable, a potential exists that drives a circulating current, often of such a magnitude as to require derating of the cable ampacity. When applying the core-balance CT, the effects of this circulating current should be subtracted from the measuring circuit.

7.4.3 Ground Return. Ground return relaying is illustrated in Fig 129. The ground-fault current returns through the current transformer in the neutral-bus to ground-bus connection. For feeder circuits, an insulating segment may be introduced in busway or conduit, as shown in Fig 129, and a bonding jumper connected across the insulator to carry the ground-fault current. A current transformer enclosing this jumper will then detect a ground fault. This method is not recommended for feeder circuits due to the likelihood of multiple ground-current return paths and the difficulty of maintaining an insulated joint.

7.4.4 Ground Differential. This generic term is used for a variety of schemes that utilize vector or algebraic subtraction or addition of signals. The currents may be produced by any of the previously discussed methods.

Figure 130(a), for example, shows a ground differential protection for single-phase center-tapped loads and is similar to the residual method.

One core-balance sensor could detect ground faults in a plurality of loads [Fig 130(b)].

Ground differential relaying is effective for main bus protection since it has inherent selectivity. With the differential scheme [Fig 130(c)], core-balance current transformers are installed on each of the outgoing feeders and another smaller current transformer is placed in the transformer neutral connection to ground. This arrangement can be made sensitive to low ground-fault currents without incurring tripping for ground faults beyond the feeder current transformers. All current transformers must be very carefully matched to prevent improper tripping for high-magnitude faults occuring outside the differential zone.

Bus differential protection protects only the zone between current transformers and does not provide backup protections against feeder faults.

7.4.5 Circuit Sensitivity. Factors that affect the sensitivity of ground-fault sensing include:

(1) Circuit charging current drawn by surge arrestors, shielded cables, and motor windings

(2) The number of coordination steps between the branch circuit and the supply

(3) Primary rating and accuracy of the largest current transformer used to supply residually connected relays in the coordination

(4) How well matched are the current transformers used for residually connected relays

(5) Burden on the current transformers, in particular, that of the residually

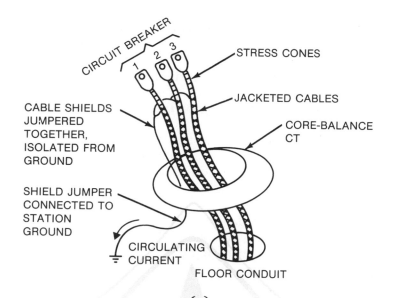

(a)

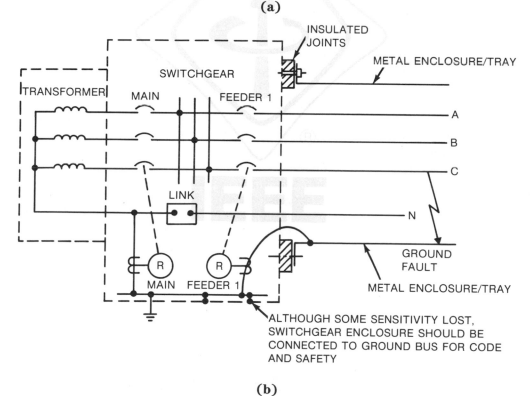

(b)

Fig 130
(a) Common Method for Shielded Cable Termination
(b) Ground-Return Relay Scheme

connected relay; solid-state and some induction disk relays have burdens of 40 Ω for a 0.1 A tap

(6) Maximum through-fault current and its effect upon the current transformers, with selected relays for phase and residual connections

(7) Fault contact resistance

(8) Location of conductors within core-balance transformers

7.5 Typical Applications. The application of ground-fault protection to typical low-voltage power distribution systems is illustrated by Figs 131–141 for the following types of power distribution systems:

The one-line diagrams show the locations for the ground-fault sensing devices as well as the locations for the protective devices. Additional considerations in the application of ground-fault protection follow.

(1) A common economy in system design is to use the simple radial system without transformer secondary main overcurrent protective devices. This results in a particular hazard when low-magnitude ground faults are considered, as there is no protective device to open should a ground fault occur between the transformer secondary winding and the feeder overcurrent protective device. Some systems are designed so that a transformer primary protective device can be opened, which is ideal. However, this is not practical in many systems, and the use of secondary mains should be considered.

(2) Even with a secondary main overcurrent protective device, the zone from the transformer secondary to the main overcurrent protective device is not protected. Even though the ground-fault sensing device may detect the fault and trip the secondary main, the fault is not removed. Thus, to minimize the possibility of trouble in this zone, it should be kept small by locating the main overcurrent protective device as close to the transformer as possible, and it should be designed with extra care to reduce the possibility of faults. The primary overcurrent protective device should be relayed or selected to detect arcing ground faults in this zone.

(3) The use of sensitive ground-fault protection makes the coordination of protective devices extremely important. The first consideration is where to apply the ground-fault protection.

In any evaluation of whether or not to apply certain protective devices, one eventually arrives at a comparison of application cost versus probable consequence of omission. In considering the application of ground-fault protection, there are several alternatives, each varying in cost of application. This discussion will consider two basic approaches: (1) ground-fault protection on the mains only, and (2) ground-fault protection on mains and feeders. In this short treatment, protection versus coordination will be explored in only the most fundamental manner.

7.5.1 Ground-Fault Protection on Mains Only. An example of this approach is shown in Fig 131. Here there is a 3000 A main with long-time and short-time trip devices, a 1200 A feeder with long-time and instantaneous trip devices, and a molded-case circuit breaker in a branch circuit with thermal and

**Fig 131
Ground Relays on Main Circuit Only**

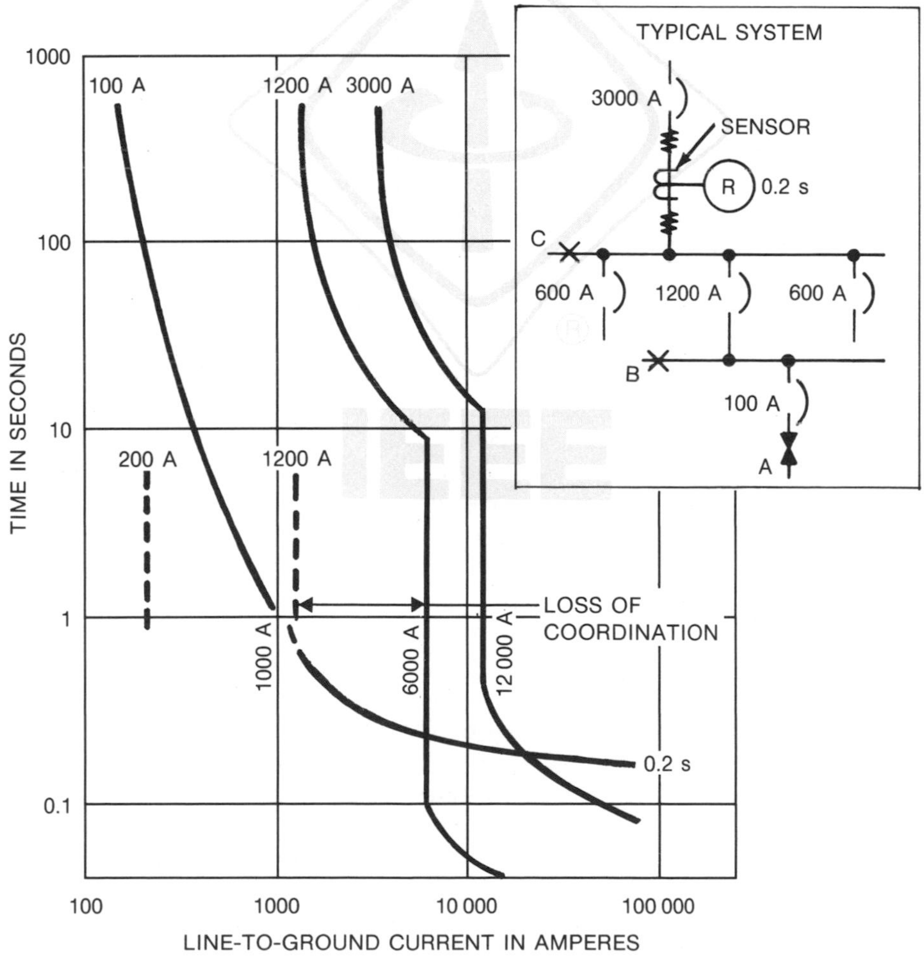

instantaneous trip devices. The ground-fault protection on the main will coordinate with both instantaneous trip devices if given about 0.2 s time delay with a relatively flat characteristic.

The problem arises, where do we set the minimum ground pickup? For full coordination with all feeders, the setting would have to be above 6000 A (above the instantaneous setting of the largest feeder). Obviously, this is too high and violates code requirements. For excellent protection of circuit ground faults, the pickup setting should be about 200 A. This, however, produces loss of coordination for ground faults at A of magnitude between 200 and 1000 A and loss of coordination for faults at B of magnitude between 200 and 6000 A. Thus, while the 200 A setting on the main will provide excellent arcing fault protection, we can expect the main circuit breaker to trip for certain feeder faults where heretofore we were accustomed to having them handled by the feeder or branch circuit breakers. In short, we have lost a rather substantial degree of coordination. In a few applications, this loss of coordination can be tolerated.

Under the circumstances, the best setting is approximately a 1200 A minimum pickup. Here we have protection against the most severe arcing faults and we have only lost coordination on ground-faults between 1200 and 6000 A. The above scheme is fairly common, but it is still clearly a compromise, which should be noted.

7.5.2 Ground-Fault Protection on Mains and Feeders. An example of this approach is shown by Fig 132. Here we have included ground-fault protection on the 3000 A main and also on all feeders above roughly 400–800 A. This application shows a 200 A minimum pickup with a time delay of 0.1 s on each feeder in addition to a 400 A minimum pickup and a 0.3 s time delay on the main.

In this example the main circuit breaker is fully coordinated with each feeder circuit breaker. Also, both main and feeders have sufficiently low settings to provide reasonable arcing fault protection. There is some loss of coordination between the feeder and branch devices, but this is felt to be acceptable in many applications. Better protection can be obtained using zone-selective relay interlocking, with less damage occurring because of instantaneous tripping in the faulted zone.

7.5.3 Ground-Fault Protection, Mains Only—Fused System. Figure 133(a) shows a situation similar to that of Fig 131, involving a fused system with 1200 A ground pickup. This setting will coordinate with the 200 A fuses and branch circuit lighting circuit breakers. However, coordination of the 1200 A pickup with the 800 A feeder fuses is sacrificed.

7.5.4 Ground-Fault Coordination. There are various means of coordinating ground-fault protection. One approach is zone-selective interlocking, which not only allows operation at the minimum desirable time for units in every zone when they are responding to ground faults in their own zone, but also establishes positive coordination between mains, feeders, and branches so that the smallest possible segment of the system is opened in the event of a ground fault.

Refer to the diagram in Fig 133(b) showing a typical coordination arrangement. Relay no 3 will restrain in a time-delay mode relay no 2, which in turn

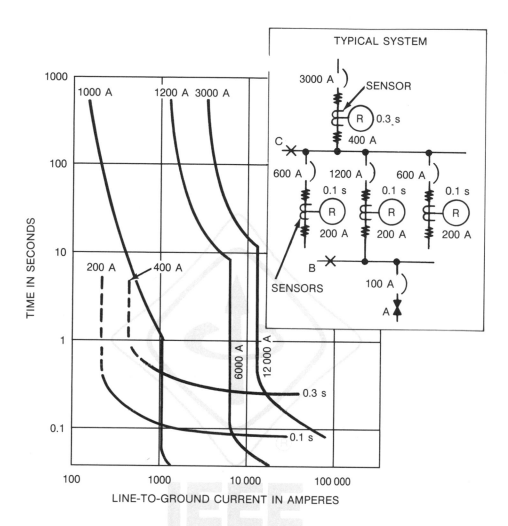

Fig 132
Ground Relays in Both Main and Feeder Circuits

restrains relay no 1; relay no 2 restrains relay no 1. For example, with a fault of 1500 A as location no 3, relay no 3 will initiate a signal in 0.03 s to the branch device to open the circuit. At the time of the fault, relay no 3 also sends a restraining signal to relay no 2, which in turns sends a restraining signal to relay no 1, which causes these two relays to start timing the duration of the fault. Relay no 2 will initiate a signal to the feeder device only if the branch device fails to open the circuit 0.2 s after the ground fault occurred. Relay no 1 will initiate a signal to the main device only if the branch device and feeder device fail to open the circuit 0.5 s after the ground fault occurred. If a 1500 A ground fault occurs at location no 2, relay no 2 will initiate a signal in 0.03 s to the feeder device to open the circuit. At the time of the fault, relay no 2 also sends a restraining signal to

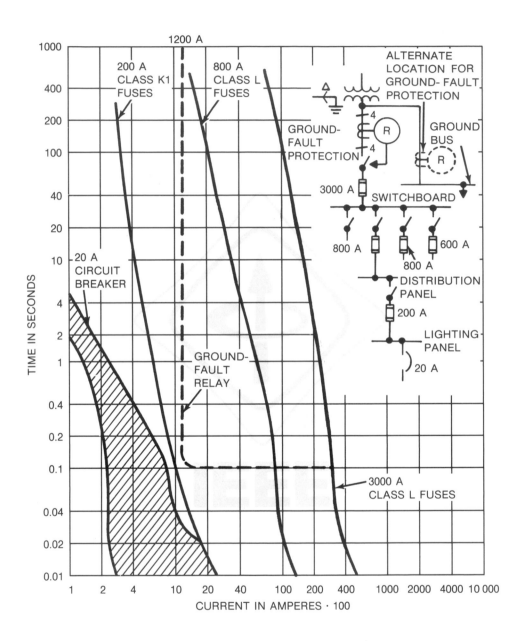

Fig 133
(a) Fused System Using Ground-Fault Relays on Main Circuit Only Shows
Coordination with 200 A Fuses and Lack of Coordination with 800 A Feeder Fuses

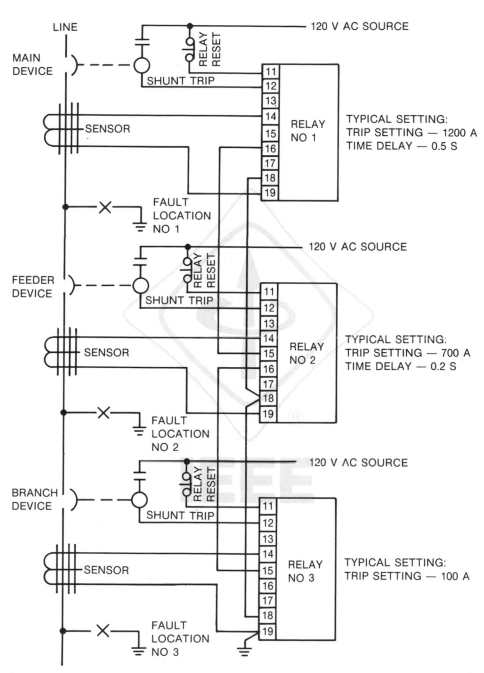

NOTE: For clarity, all power supply and total circuit connections are not shown.

Fig 133 *(Continued)*
(b) Typical Zone Selective Interlocking System

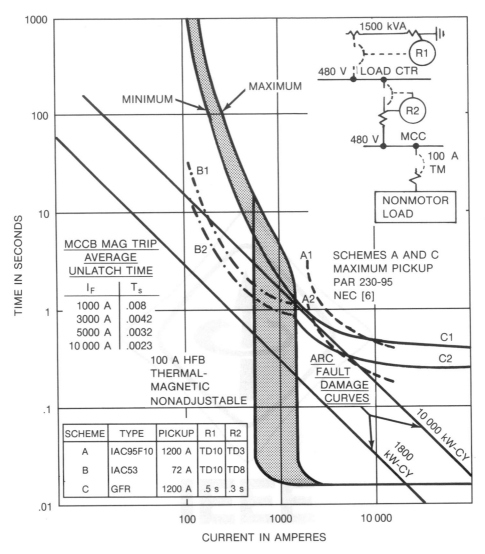

Fig 133 *(Continued)*
(c) Minimum GFP — Nonmotor Circuit

relay no 1. Relay no 1 will initiate a signal to the main device only if the feeder device fails to open the circuit 0.5 s after the ground fault occurred. If a 1500 A ground fault occurs at location no 1, relay no 1 will initiate a signal in 0.03 s to the main device to open the circuit. This type of system coordination allows the circuit interrupting devices nearest to the ground fault to receive an operation signal instantaneously (0.03 s) and provide time delay in only the backup devices. Clearing time of the circuit interrupting device should also be considered in complete system coordination.

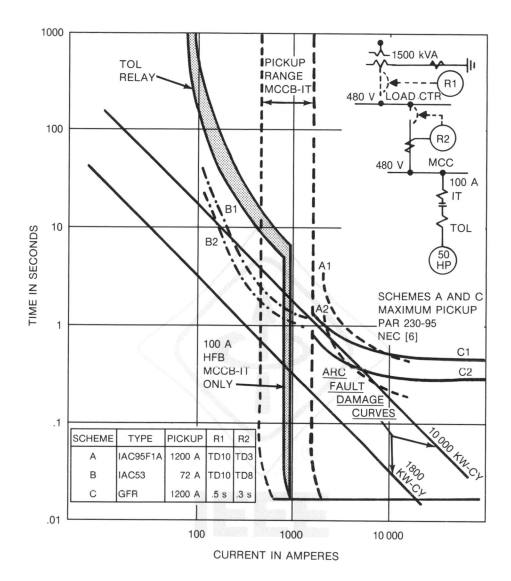

Fig 133 *(Continued)*
(d) Minimum GFP — Motor Circuit

This system requires a pair of control wires between relays of each successive coordination step. The control leads add exposure to possible faults on these control leads. This should be considered when zone-selective interlocking schemes are applied.

Another approach to a coordinated ground-fault protection scheme is based upon protection against low-level arcing faults in which the arc has seriously reduced the fault current. Backup protection utilizes a standard time overcurrent relay. This scheme is illustrated by curves B1 and B2 in Fig 133(c) and (d), where its protection scope is compared against two common GFP schemes.

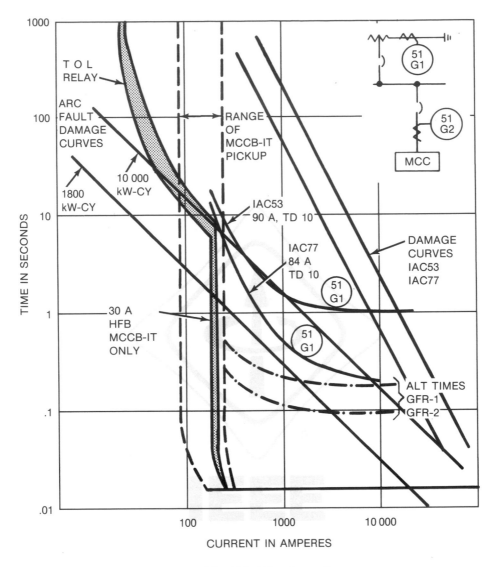

Fig 133 *(Continued)*
(e) GFP — 15 hp Motor Load

Curves C1 and C2 represent the tripping characteristics of a fixed-delay solid-state GFR. Curves A1 and A2 represent a special electromechanical GFR. Both types are set with a minimum pickup of 1200 A. The curves B1 and B2 represent the tripping characteristics of a standard electromechanical overcurrent relay with very inverse tripping characteristics, but set to operate on a minimum pickup of 72 A. A and C GFP relays are not sensitive to low ground-fault currents, will not initiate nuisance trips, and backup phase overcur-

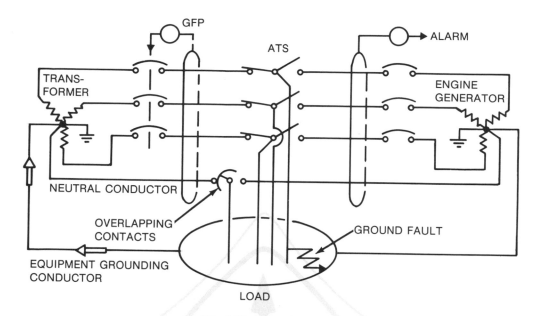

Fig 133 *(Continued)*
**(f) Emergency Power Systems with Ground-Fault Protection on Normal and
Ground-Fault Alarm on Emergency**

rent protection in the 1000–20 000 A range. But they may not necessarily provide
real protection service.

There have been numerous references that have studied the arcing nature of
480 V ground faults. Often a 90–140 V arc has been measured, such an arc
limiting the ground-fault current magnitude. Some articles have equated arcing
fault damage to 1800, 2000, or even 10 000 kW cycle energy limits.

The 1800 kW cycle and 10 000 kW cycle curves shown in Fig 133(c) and (d) are
based upon a 100 V arc and an arcing current of essentially resistive characteristics.

(1) Fault energy $= V_{arc} \cdot I_F \cdot$ time

(2) Fault energy $= \dfrac{100 \cdot I_F}{1000} \cdot 60t$ kW cycles

where

$V_{arc} = 100$ arc V

$t = $ s

$I_F = $ ground-fault current

(3) If fault energy $= 1800$ kW cycles $= 6\,(I_F \cdot t)$ kW cycles

then

$I_F \cdot t = \dfrac{1800}{6} = 300$

Example:

$I_F = 100$ A, $t = 3$ s

$I_F = 1000$ A, $t = 0.3$ s

$I_F = 10\,000$ A, $t = 0.03$ s

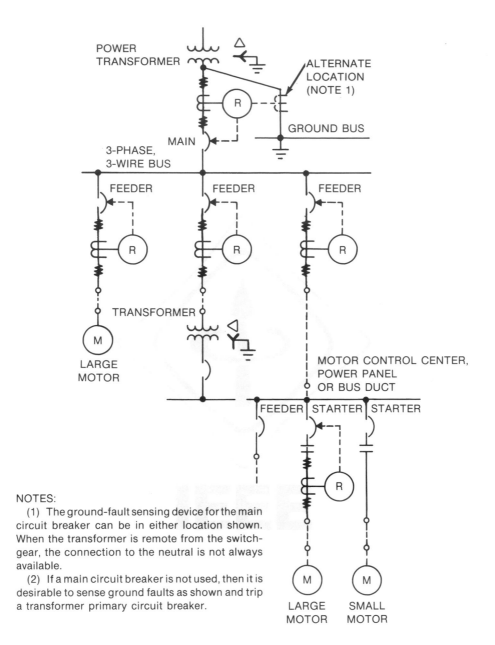

POWER
TRANSFORMER

ALTERNATE
LOCATION
(NOTE 1)

R

GROUND BUS

MAIN

3-PHASE,
3-WIRE BUS

FEEDER   FEEDER   FEEDER

R   R   R

TRANSFORMER

R   R   R

LARGE
MOTOR

MOTOR CONTROL CENTER,
POWER PANEL
OR BUS DUCT

FEEDER STARTER STARTER

R

NOTES:

(1) The ground-fault sensing device for the main
circuit breaker can be in either location shown.
When the transformer is remote from the switchgear, the connection to the neutral is not always
available.

(2) If a main circuit breaker is not used, then it is
desirable to sense ground faults as shown and trip
a transformer primary circuit breaker.

M   M

LARGE   SMALL
MOTOR   MOTOR

Fig 134
Three-Wire Solidly Grounded System; Single Supply;
Ground Relays Must Have Time Coordination

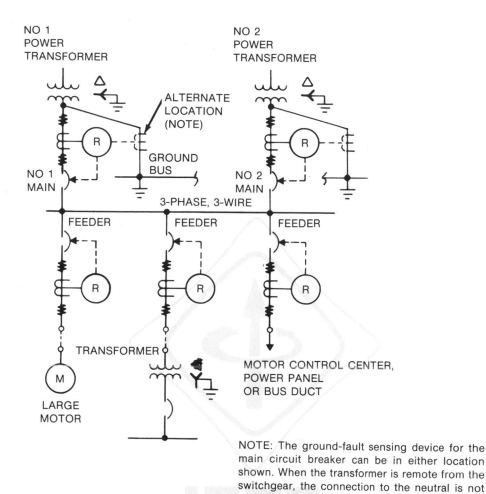

NOTE: The ground-fault sensing device for the main circuit breaker can be in either location shown. When the transformer is remote from the switchgear, the connection to the neutral is not always available.

Fig 135
Three-Wire Solidly Grounded System; Dual Supply;
Ground Relays Must Have Time Coordination

Plotted on a log–log scale, the locus of the boundary points is a straight line, which makes an ideal criteria for coordinating with a very inverse or extremely inverse type GFP relay. The calculations for the 10 000 kW cycle damage limit was developed on the same basis.

Referring again to Fig 133(c) and (d), it can be seen that GFP type A and C settings permit extremely low (< 1000 A) ground-fault currents, whereas the GFP type B settings limit the damage for arcing ground faults in both the motor as well as the nonmotor circuit. However, scheme B loses coordination below 1000 A unless there is a ground-fault relay on MCC branch circuits above 30 A and for motor branch circuits above 15 hp, as illustrated in Fig 133(e).

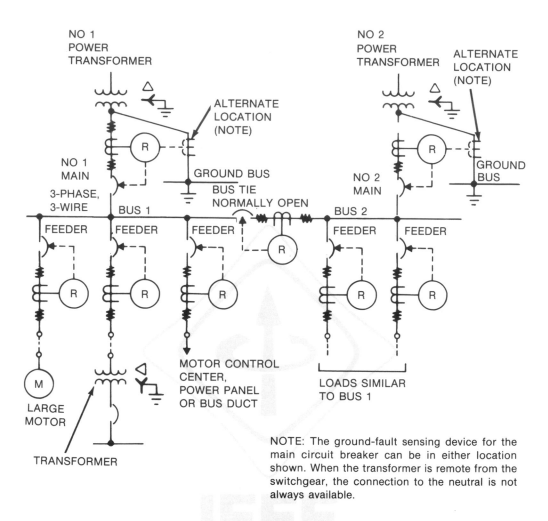

Fig 136
Three-Wire Solidly Grounded Secondary Selective System;
Ground Relays Must Have Time Coordination

7.5.5 Ground-Fault Protection of Systems with an Alternate Power Source. Following usual considerations of reliability or redundancy, it may be advantageous to ground the neutral terminal of an alternate power source, such as a second utility source (service) or an engine–generator set, at its location. However, this may create multiple neutral-to-ground connections, which in turn may cause problems unless appropriate steps are taken.

For example, consider a typical 3-phase, 4-wire 480 V emergency power system in which a 3-pole transfer switch connects the load to either a normal utility source or an engine–generator set. Ground-fault protection is provided at the utility incoming service. The neutral conductors of both the utility incoming

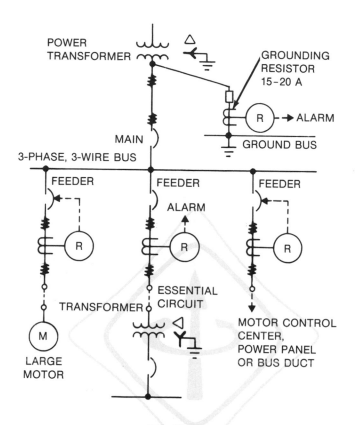

**Fig 137
Three-Wire High-Resistance Grounded System;
Grounded Relays Must Have Time Coordination**

service and the engine–generator set are separately grounded and interconnected (continuous neutral).

Because of the multiple neutral-to-ground connections, two problems can develop. One is if a ground-fault occurs anywhere in the system. Then the fault current has two paths of flow—one directly to the grounded neutral of the incoming utility source via equipment grounding conductors and one to the grounded neutral of the engine–generator set via equipment grounding conductors, and back to the neutral of the incoming utility source via the continuous neutral. The latter path will not tend to actuate the ground-fault sensor, which would cause incomplete sensing of the total fault current.

The other problem can occur in the case where, if the load is unbalanced, the return neutral current has two paths of flow. It can go directly to the service neutral via the neutral conductor or back to the service neutral via the equipment grounding conductors and the continuous neutral. The current through the latter path would have the same effect on the ground-fault sensor as a ground-fault current and will be carried by equipment grounding conductors,

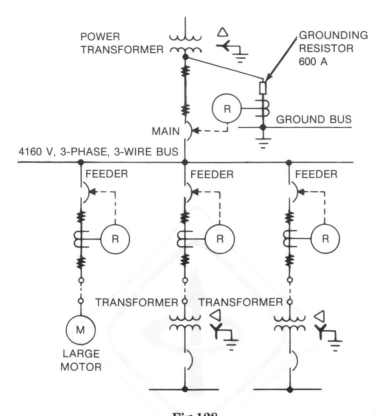

Fig 138
Three-Wire Low-Resistance Grounded System;
Ground Relays Must Have Time Coordination

which are not provided for this purpose. If the second path has sufficiently low impedance (comparable to same of the neutral path), an unbalanced load may cause the ground-fault sensor to trip the breaker even though a ground fault or short circuit does not occur.

Various solutions include single-point grounding, 4-pole transfer switches, overlapping neutral contacts, and transformer isolation. In all cases, consideration should be given to both modes of transfer switch operation, to providing adequate area protection and ground-fault protection, and conformance to the NEC [6].

(1) Single-point grounding eliminates the above identified problems and permits simple relaying methods. The basic scheme is shown in Fig 141 and can be used for multiple services or in utility service-generator source systems with a continuous neutral.

The method can be modified to detect not only phase-to-ground (enclosure) faults (most commonly occurring), but also phase-to-neutral failures in a system. However, consideration should be given to the possibility of power disruption

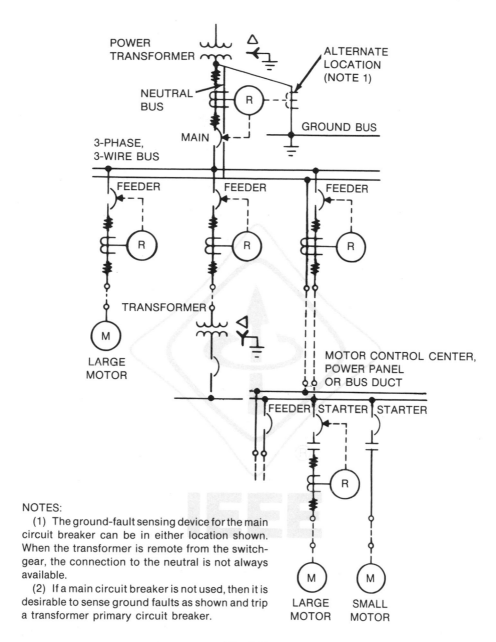

NOTES:

(1) The ground-fault sensing device for the main circuit breaker can be in either location shown. When the transformer is remote from the switchgear, the connection to the neutral is not always available.

(2) If a main circuit breaker is not used, then it is desirable to sense ground faults as shown and trip a transformer primary circuit breaker.

**Fig 139
Four-Wire Solidly Grounded System, Single Supply;
Ground Relays Must Have Time Coordination**

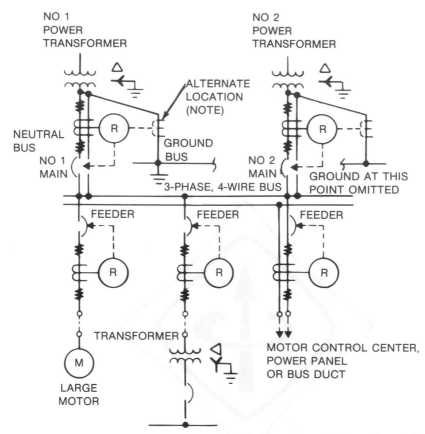

NO 1
POWER
TRANSFORMER

NO 2
POWER
TRANSFORMER

ALTERNATE
LOCATION
(NOTE)

NEUTRAL
BUS

NO 1
MAIN

GROUND
BUS

R

NO 2
MAIN

3-PHASE, 4-WIRE BUS

GROUND AT THIS
POINT OMITTED

FEEDER

FEEDER

FEEDER

R

R

R

TRANSFORMER

M

LARGE
MOTOR

MOTOR CONTROL CENTER,
POWER PANEL
OR BUS DUCT

NOTE: The ground-fault sensing device for the
main circuit breaker can be in either location
shown. When the transformer is remote from the
switchgear, the connection to the neutral is not
always available.

Fig 140
Four-Wire Solidly Grounded System, Dual Supply;
Ground Relays Must Have Time Coordination

within the facility and transferring the load to ungrounded emergency power
source. Also, a ground-fault condition when the transfer switch is in the
emergency position may cause nuisance tripping of the normal source breaker.

(2) Using 4-pole transfer switches throughout the system may allow complete
isolation of service and generator neutral conductors. This eliminates possible
improper ground-fault sensing and nuisance tripping caused by multiple neutral-
to-ground connections. When this is done, the generator will comply with the
NEC [6] definition of a separately derived system. However, momentary opening
of the neutral conductor may cause voltage surges. Furthermore, it should be
remembered that the contacts of the fourth switch pole do interrupt current and

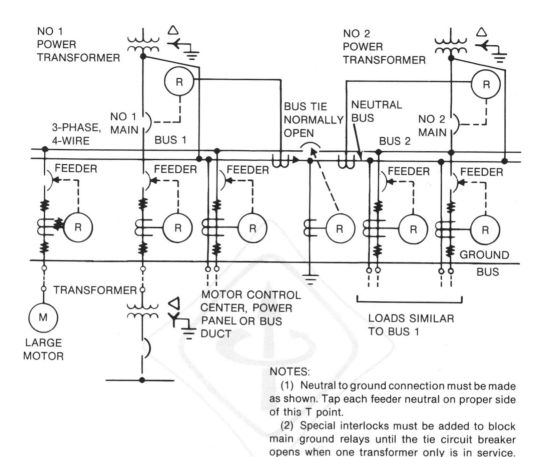

**Fig 141
Four-Wire Solidly Grounded Secondary Selective System;
Ground Relays for Feeders, Tie, and Mains Must Have Time Coordination**

are, therefore, subject to arcing and contact erosion. A good maintenance program should reaffirm at intervals the integrity of the fourth pole as a current-carrying member with sufficiently low impedance.

(3) A variation of the method of isolating the normal and emergency source neutrals is for the automatic transfer switch to include overlapping neutral transfer contacts. This provides the necessary isolation between neutrals and at the same time minimizes abnormal switching voltages. With overlapping contacts, the only time the neutrals of the normal and emergency power sources are connected is during transfer and retransfer. With a conventional double-throw transfer switch, this duration can be less than the operating time of the ground-fault sensor, which is usually set anywhere from 6–24 cycles (100–400 ms). As with 4-pole transfer switches, conventional ground-fault sensing can be readily added to the emergency side. Figure 133(f) shows isolation by

overlapping contacts and ground-fault sensing on the emergency side for actuating an alarm circuit rather than tripping a breaker. If required, emergency source tripping on ground faults can also be provided.

(4) Where a 3-phase, 4-wire critical load is relatively small compared to the rest of the load, an isolating transformer is sometimes used. This requires both power sources connected to the transfer switch to be 3-phase, 3-wire, and the delta-wye isolating transformer must be inserted between the transfer switch and the 4-wire load. An unbalance of the critical load would have no effect on the ground-fault protector at the incoming service. Furthermore, ground-fault currents would not be transmitted through the delta-wye transformer. Any increase in primary current due to ground-faults in the secondary is seen simply as an overload by the primary protective devices. Circuits on the secondary side of the isolating transformer have the advantage of reduced fault current available but may require their own ground fault protection if the transformer secondary is grounded. Because the transfer switch is not located directly ahead of the load, it does not guarantee emergency power supply in the event of isolating transformer failure.

7.6 Special Applications

7.6.1 Ground-Fault Identification and Location. Ground faults in solidly grounded systems usually require opening of disconnect devices to reduce damage. Many relays and trip devices have targets or lamp indicators to identify faulty circuits. The equipment and conductors should be inspected and repaired, if necessary, prior to restoration of service.

The conventional method of ground-fault detection used on ungrounded wye or delta 3-wire systems utilizes three potential transformers supplying (1) ground detector lamps, (2) ground detecting voltmeters, or (3) ground alarm relays. It should be noted that the presence of potential transformers connected to ground from each phase may in itself be the cause of dangerous overvoltages because the detector becomes the grounding mechanism, as well as being a detector.

To reduce the probability of transient overvoltages, high-resistance grounding is often applied, as described in detail in ANSI/IEEE Std 142-1982 [4]. When high-resistance grounding is used, the basic reason is to eliminate a trip operation when the first ground fault occurs. Operation with a ground fault on the system entails a substantial hazard and it becomes important to (1) locate the fault as soon as system operation allows, and (2) provide additional ground-fault protection against a second ground fault until the first one is corrected.

In applying high-resistance grounding, the resistance of the ground circuit should be of a magnitude to pass a ground circuit at least equal to the charging current of the system. Figure 142 shows a typical high-resistance grounded system with a neutral resistor and a ground alarm relay. Most similar applications use a distribution transformer with a low voltage resistor connected to its 240 V secondary.

Each feeder is equipped with a low-ratio core-balance transformer and ammeter or milliammeter, which will indicate the faulted feeder. No automatic tripping occurs with this scheme.

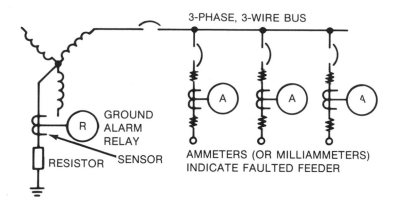

**Fig 142
High-Resistance Grounded System Using Core-Balance Current
Transformers and Fault-Indicating Ammeters**

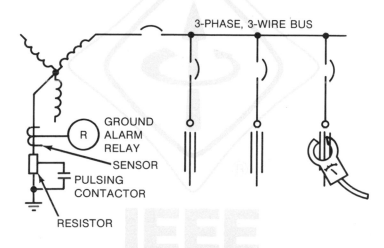

**Fig 143
High-Resistance Grounded System
with Pulsing Contactor to Locate Faulted Feeder**

In lieu of current transformers and ammeters on each feeder, it is possible to
arrange a pulsing ground circuit as shown in Fig 143. The pulsing circuit is
manually initiated and serves to reduce the resistance to about 50–75% of its full
value, about once or twice a second. This causes the ground-fault current to vary
sufficiently to be detected by a clamp-on ammeter, which can be placed in turn
around the conductors of each feeder circuit.

If a ground fault is not cleared immediately, a relatively dangerous condition
may arise upon the occurrence of a second fault. A second fault has a high
probability of occurring because (1) the steady-state voltage on the unfaulted
phases has increased, and (2) the initial fault may be intermittent, which will

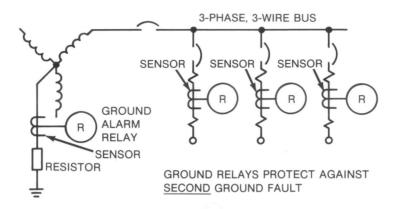

Fig 144
High-Resistance Grounded System Continues Operation on First Ground Fault;
Ground Relays Protect Against Second Ground Fault

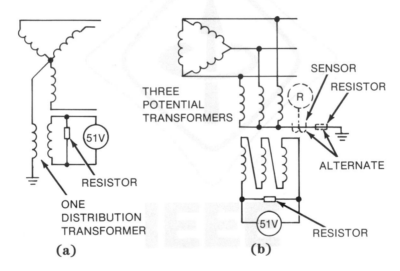

Fig 145
High-Resistance Grounding Using Distribution or Grounding Transformers
(a) Wye-Connected System
(b) Delta-Connected System

cause some transient overvoltages in spite of the resistor grounding. Figure 144 shows how feeder ground relays are applied to trip on the occurrence of the second fault on a different feeder. The feeder ground relays are set to pick up at a value higher than the maximum initial ground-fault current. For example, on a 4.16 kV system a 300 Ω resistor would limit the ground fault current to 8 A. The ground alarm relay would be set to pick up at 5 A or less. The feeder ground relays are set to pick up at a current level higher than that for a single line-to-ground fault, perhaps 10–15 A.

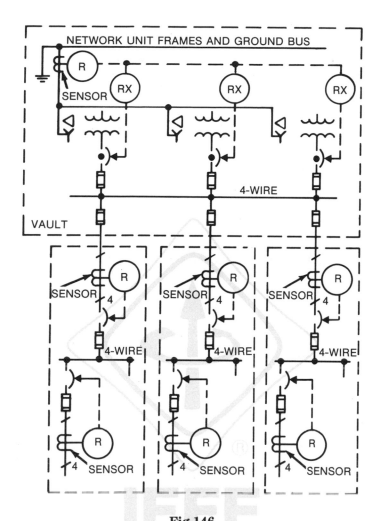

Fig 146
Ground-Fault Protection for Typical Three-Transformer Spot Network

The use of low-voltage grounding resistors coupled with grounding (or standard distribution) transformers is preferred by some engineers. Figure 145(a) and (b) shows how these transformers are utilized for (a) 3-wire grounded wye systems, and (b) 3-wire grounded delta systems.

7.6.2 Spot Network Applications. Spot networks provide continuity of service against the loss of one or more of several utility supply feeders. Each feeder supplies one network transformer at each vault. All transformer secondaries at each vault are paralleled at the network service bus from which the user service switchgear is connected.

Figure 146 shows one method of providing ground-fault protection for a 3-transformer spot network serving three physically separate service switchboards. Ground protection is provided (1) on each switchboard feeder, (2) in each

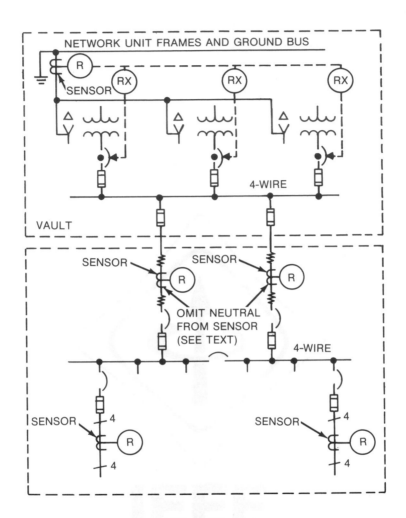

Fig 147
Ground-Fault Protection for Spot Network
When User Switchgear Is Secondary Selective Type

switchboard main, and (3) in the fault system neutral, which will trip out all network protectors in the event of a vault fault. However, not all utilities will allow user relays to trip network protectors. In a typical approved installation the feeder ground relays are set at about 0.1 s, the main relays at about 0.3 s, and the vault relays at about 0.5 s. The minimum operating current settings for the vault relays are set substantially higher than the setting on the user switchgear so that the vault relays will operate only for a fault in the vault area. Though not shown in the figure, each service switchboard must include a connection from neutral bus to ground bus.

Where it is desired to provide secondary selective flexibility in the user service switchgear, the system sometimes takes the form shown in Fig 147. Here one must be careful in applying the ground sensors (current transformers) in the

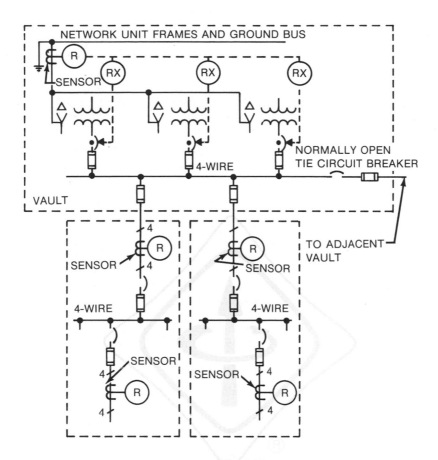

Fig 148
Ground-Fault Protection for Sectionalized Vault Bus Spot Network

main circuits. If the ground sensors are installed over all phases and neutral, then a fault on either user bus will always cause tripping of both main circuit breakers. One method of circumventing this is to install the main sensors over the phases only. This will provide the proper selectivity, except that the ground relay minimum operating current setting must be set at a value above the normal neutral current. This normal neutral current will consist of the single phase-to-neutral load unbalance plus the third harmonic in the neutral caused by nonincandescent lighting systems. Settings of 1000–2000 A may be required. With this type of system it will be difficult to meet the present 1200 A maximum setting of the NEC and still accommodate the secondary selective arrangement.

An alternate vault arrangement that has been used is shown in Fig 148. Here the vault is sectionalized into two halves and one of the advantages is that a vault fault does not shut off all electrical service to the user. Ground-fault protection is essentially identical to that shown in Fig 146. An alternate ground-fault protection scheme consisting of heat sensing systems of the continuous and probe types is described in [16].

7.7 References. The following publications shall be used in conjunction with this chapter.

[1] ANSI C37.16-1980, American National Standard Preferred Ratings, Related Requirements and Application Recommendations for Low-Voltage Power Circuit Breakers and AC Power Circuit Protectors.

[2] ANSI C37.17-1979, American National Standard for Trip Devices for AC and General-Purpose DC Low-Voltage Power Circuit Breakers.

[3] ANSI/IEEE Std 141-1986, IEEE Recommended Practice for Electric Power Distribution for Industrial Plants.

[4] ANSI/IEEE Std 142-1982, IEEE Recommended Practice for Grounding of Industrial and Commercial Power Systems.

[5] ANSI/IEEE Std 446-1980, IEEE Recommended Practice for Emergency and Standby Power Systems for Industrial and Commercial Applications.

[6] ANSI/NFPA 70-1984, National Electrical Code.

[7] NEMA PB1.2-1977, Application Guide for Ground-Fault Protective Devices for Equipment.

[8] UL 1053, Ground Fault Sensing and Relaying Equipment.

[9] CHUMAKOV, W.V. Grounding and Ground-Fault Protection—Part 1, Single-Source Power Systems. *IEEE Conference Record, Industry Applications Society Annual Meeting,* Sept 1980, pp 417–422.

[10] CONRAD, R.R. and DALASTA, D.A. New Ground Fault Protective System for Electrical Distribution Circuits. *IEEE Transactions on Industry and General Applications,* vol IGA-3. May/June, 1967, pp 217–227.

[11] FISHER, L.E. Resistance of Low-Voltage AC Arcs. *IEEE Transactions on Industry and General Applications,* vol IGA-6, Nov/Dec 1970, pp 607–616.

[12] KAUFMANN, R.H. Application Limitations of Single-Pole Interrupters in Polyphase Industrial and Commercial Building Power Systems. *IEEE Transactions on Applications and Industry,* vol 82, Nov 1963, pp 363–368.

[13] LOVE, D.J. Discussion of Predicting Damage from 277 V Single Phase Ground Arcing Faults. *IEEE Transactions on Industry Applications,* vol IA-14, no 1, Jan/Feb 1978, pp 93–95.

[14] NEUHOFF, C.J. Improving Ground-Fault Protection by Monitoring Equipment Environment. *IEEE Conference Record, Industry Applications Society Annual Meeting,* Oct 1979, pp 1204–1208.

[15] O'CONNOR, J.J. The Threat of Arcing Faults. *Power Magazine,* July 1969, p 48.

[16] STANBACK, H.I. Predicting Damage from 277 V Single Phase to Ground Arcing Faults. *IEEE Transactions on Industry Applications,* vol IA-13, no 4, July/Aug 1977, pp 307–314.

7.8 Bibliography

[B1] BAILEY, B.G. Clearing Low-Voltage Ground Faults with Solid-State Trips. *IEEE Transactions on Industry and General Applications,* vol IGA-3, Jan/Feb 1967, pp 60–65.

[B2] BAILEY, B.G. and HEILMANN, G.H. Clear Low-Voltage Ground Faults with Solid-State Trips. *Power Magazine,* Mar 1966, pp 69–71.

[B3] BEEMAN, D.L., Ed. *Industrial Power Systems Handbook.* New York: McGraw-Hill, 1955.

[B4] BISSON, A.J. and ROCHAU, E.A. Iron Conduit Impedance Effects in Ground Circuit Systems. *AIEE Transactions on Applications and Industry,* pt II, vol 73, July 1954, pp 104–107.

[B5] BRINKS, J.W. and WOLFINGER, J.P. System Grounding Protection and Detection for Cement Plants. *IEEE Transactions on Industry Applications,* vol IA-17, no 6, Nov/Dec 1981, pp 587–596.

[B6] CASTENSCHIOLD, R. Ground-Fault Protection of Electrical Systems with Emergency or Standby Power. *IEEE Transactions on Industry Applications,* vol IA-13, no 6, Nov/Dec 1977, pp 517–523.

[B7] DALASTA, D. Ground Fault Protection for Low and Medium Voltage Distribution Circuits. *Proceedings of the American Power Conference,* vol 32, 1970, pp 885–895.

[B8] DUNKI-JACOBS, J.R. The Effects of Arcing Ground Faults on Low-Voltage System Design. *IEEE Transactions on Industry Applications,* vol IA-8, no 3, May/June 1972, pp 223–230.

[B9] EDMUNDS, W.H., SCHWEIZER, J.H., and GRAVES, R.C. Protection Against Ground Fault Hazards in Industry, Hospital, and Home. *Conference Record, 1968 IEEE Industry and General Applications Group Annual Meeting,* IEEE 68C27-IGA, pp 863–874.

[B10] FISHER, L.E. Arcing-Fault Relays for Low-Voltage Systems. *IEEE Transactions on Applications and Industry,* vol 82, Nov 1963, pp 317–321.

[B11] FISHER, L.E. Proper Grounding Can Improve Reliability in Low-Voltage Systems. *IEEE Transactions on Industry and General Applications,* vol IGA-5, July/Aug 1969, pp 374–379.

[B12] FISHER, L.E. and SCHWIEGER, W.L. Tripping Network Protectors to Protect Service Entrance Conductors to Industrial and Commercial Buildings.

IEEE Transactions on Industry and General Applications, vol IGA-5, Sept/Oct 1969, pp 536–539.

[B13] FOX, F.K., GROTTS, H.J., and TIPTON, C.H. High-Resistance Grounding of 2400 V Delta Systems with Ground-Fault Alarm and Traceable Signal to Fault. *IEEE Transactions on Industry and General Applications,* vol IGA-1, Sept/Oct 1965, pp 366–372.

[B14] FREUND, A., Ground-Fault Protection: How It Works. *Electrical Construction and Maintenance,* June 1973, pp 53–60.

[B15] FREUND, A. Ground-Fault Protection for Ungrounded Distribution Systems. *Electrical Construction and Maintenance,* June 1979, pp 80–84.

[B16] GIENGER, J.A., DAVIDSON, O.C., and BRENDEL, R.W. Determination of Ground-Fault Current on Common AC Grounded Neutral Systems in Standard Steel or Aluminum Conduit. *AIEE Transactions on Applications and Industry,* pt II, vol 79, May 1960, pp 84–90.

[B17] GROSS, E.T.B. Sensitive Ground Protection for Transmission Lines and Distribution Feeders. *AIEE Transactions,* vol 60, 1941, pp 968–975.

[B18] KAUFMANN, R.H. Some Fundamentals of Equipment-Grounding Circuit Design. *AIEE Transactions on Applications and Industry,* pt II, vol 73, Nov 1954, pp 227–232.

[B19] KAUFMANN, R.H. Let's Be More Specific about Grounding Equipment. *Proceedings of the American Power Conference,* vol 24, 1962, pp 913–922.

[B20] KAUFMANN, R.H. Ignition and Spread of Arcing Faults. *Conference Record, 1969 IEEE Industrial and Commercial Power Systems and Electric Space Heating and Air Conditioning Joint Technical Conference,* IEEE 69C23-IGA, pp 70–72.

[B21] KAUFMANN, R.H. and PAGE, J.C. Arcing-Fault Protection for Low-Voltage Power Distribution Systems—Nature of the Problem. *AIEE Transactions on Power Apparatus and Systems,* pt III, vol 79, June 1960, pp 160–167.

[B22] KNOBEL, L.V. A 480 V Ground Fault Protection System. *Electrical Construction and Maintenance,* Mar 1970, pp 110–112.

[B23] KREIGER, C.H., LERA, A.P., and CRENSHAW, R.M. Ground Fault Protection for Low-Voltage Systems. *Electrical Construction and Maintenance,* June 1967, pp 88–92.

[B24] LOVE, D.J. Designing a Comprehensive Coordinated 480 Volt Ground-Fault Protection System. *IEEE Conference Record, 1977 Industry Applications Society Annual Meeting,* vol 77CH1246-8-14, pp 533–537.

[B25] LOVE, D.J. Ground-Fault Protection Electric Utility Generating Station Medium Voltage Auxiliary Power Systems. *IEEE Transactions on Power Apparatus and Systems,* vol PAS-97, no 2, Mar/Apr 1978, pp 583–586.

[B26] PEACH, N. Protect Low-Voltage Systems from Arcing Fault Damage. *Power Magazine,* Apr 1964.

[B27] PEACH, N. Get Ground Fault Protection Now, *Power Magazine,* Apr 1968, pp 84–87.

[B28] QUINN, G.C. New Devices End an Old Electrical Hazard—Ground Faults. *Modern Manufacturing Magazine,* Apr 1969.

[B29] READ, E.C. Ground Fault Protection at Buick. *Industrial Power Systems,* vol 2, June 1968.

[B30] SHIELDS, F.J. The Problem of Arcing Faults in Low-Voltage Power Distribution Systems, *IEEE Transactions on Industry and General Applications,* vol IGA-3, Jan/Feb 1967, pp 15–25.

[B31] SOARES, E.C. Designing Safety into Electrical Distribution Systems. *Actual Specifying Engineer,* Aug 1968, p 38.

[B32] VALVODA, F. Protecting Against Arcing and Ground Faults. *Actual Specifying Engineer,* May 1967, p 92.

[B33] WEDDENDORF, W.A. Evidence of Need for Improved Coordination and Protection of Industrial Power Systems. *IEEE Transactions on Industry and General Applications,* vol IGA-1, Nov/Dec 1965, pp 393–396.

8. Conductor Protection

8.1 General Discussion. This chapter deals with power cable protection as well as with busway protection. The primary considerations are presented along with some methods of application.

The proper selection and rating or derating of power cables is as much a part of cable protection as the application of the short-circuit and overcurrent protection devices. The whole scheme of protection is based on a cable rating that is matched to the environment and operating conditions. Methods of assigning these ratings are discussed.

Power cables requires short-circuit current, overload, and physical protection in order to meet the requirements of ANSI/NFPA 70-1984 [3][20] (National Electrical Code [NEC]). A brief description of the phenomena of short-circuit current, overload current, and their temperature rises is presented, followed by a discussion of the time–current characteristics of both cables and protective devices. In addition, a number of illustrations of cable systems and typical selection and correlation of protective devices are included.

Because of their rigid construction, busways provide their own mechanical protection. However, they do require short-circuit current and overload protection. A brief discussion of the types of faults on the busways is presented, followed by a discussion of various methods of fault protection.

The general intent of this chapter is to provide a basis for design, pointing out the problems involved, and providing guidance in the application of cable and busway protection. Each specific case and type of cable or busway requires attention. In most cases, the attention is routine, but the out of the ordinary cable and busway schemes require careful consideration.

8.2 Cable Protection. Cables are the "mortar" that hold together the "bricks" of equipment in an electric system. If the cable system is inadequate due to original unsuitability or from plant expansion, no matter how superb the equipment, unsatisfactory operation will be the inevitable result. Today's cables are vastly superior in performance to those available just a decade or so ago. But even so, they are not unlimited in power capability, and so need protection to prevent possible operation beyond that capability.

---

[20] The numbers in brackets correspond to those of the references listed at the end of this chapter.

Cables are generally classified as either power or control types. Power cables are divided into two voltage classes: 600 V and below, and above 600 V. Control cables include those used in the control of equipment and also for voice communication, metering, and data transmission.

The amount of damage caused by the faulting of power cable has been illustrated many times. As power and voltage levels increase, the potential hazards also increase. High temperature due to continued overload or uncoordinated fault protection is a frequent cause of decreased cable life and failure. Power cables, internally heated as a result of their resistance to the current being carried, can undergo insulation failure if the temperature buildup becomes excessive. Proper selection and rating ensures that the cable is large enough for the expected current. Suitable protection will ensure that cable temperature rising above ambient does not become excessive. Such protection normally is provided by current-time sensitive devices. In addition to insulation breakdown, protection is also required against unexpected overload and short-circuit current. Overcurrent can occur due to an increase in the number of connected loads or due to overloading of existing equipment.

While the extraordinary temperature of the short-circuit arc produces complete destruction of all materials at a fault location, cables carrying energy to (and from) a fault may also incur thermal damage over their entire length if the fault current is not interrupted quickly enough. Depending on conductor size, insulation type, and available fault current, the clearing time of the protection system must be short enough (coordinated) to stop the current flow before damaging temperatures are reached.

Physical conditions can also cause cable damage and failure. Failure due to excessive heat may be caused by high ambient temperature conditions or fire. Mechanical damage may result in short circuits or reduced cable life and may be caused by persons, equipment, animals, insects, or fungus.

Cable protection is required to protect personnel and equipment and to ensure continuous service. From the standpoint of equipment and process, the type of protection selected is generally determined by economics and the engineering requirements. Personnel protection also receives careful engineering attention, as well as special consideration to assure compliance with the various codes that may be applicable to a particular installation.

Protection against overload is generally achieved by means of a device sensitive to current magnitude and duration. Short-circuit protective devices are sensitive to much greater currents and shorter times. Protection against environmental conditions takes on many forms.

Cables may also be damaged by sustained overvoltages such as exist during a ground fault on one phase conductor. Modern cables now bear a rating called "percent insulation level," or % IL. This is described as follows:

(1) 100% IL—Cables that may not be required to operate longer than one minute in case of ground fault

(2) 133% IL—Cables that may not be required to operate longer than one hour in case of a ground fault

(3) 173% IL—Cables that may be required to operate longer than one hour continuously with one phase conductor grounded (consult manufacturers for suitability)

The cable characteristics are not germane to this treatment, but the timing of the permissible protective system should be in accordance with the IL rating of the cables involved.

In general, this chapter covers methods of rating cables and the conditions and problems listed above. It also provides a starting point from which further refinements may be made and other features added for improved power cable protection.

8.3 Glossary. The following symbols are used in cable protection technology. There may be duplication here of some symbols used in other parts of electrical engineering practice, such as fault determination. But there are only a limited number of available symbols, and the ones shown here are deeply rooted in cable technology, so the duplication is something that must be lived with.

(1) Cable Current (A):

I = current flowing in cable
I_O = initial current prior to a current change
I_F = final current after a current change
I_N = normal loading current on base ambient temperature
I_{N1} = normal loading current on nonbase ambient temperature
I_E = emergency loading current on base ambient temperature
I_X = current at values other than normal or emergency loading
I_{sc} = three-phase short-circuit current

(2) Cable Temperature (°C):

T = temperature, in general
T_O = initial temperature prior to a current change
T_f = final temperature after a current change
T_N = normal loading temperature
T_E = emergency loading temperature
T_X = temperature at any loading current
T_t = temperature at time t after a current change
T_a = base ambient temperature
T_{a1} = nonbase ambient temperature

(3) Miscellaneous:

t = time in units as noted
CM = conductor size in circular mils
F_{ac} = skin effect ratio or ac/dc ratio (NEC [3], Table 9)
K = time constant or geometric factor of cable heat flow
K_t = correction factor for initial and final short-circuit temperature

(4) Reactances (%):

X_T = transformer reactance

$X_{d''}$ = subtransient reactance of a rotating machine

$X_{d'}$ = transient reactance of a synchronous machine

8.4 Short-Circuit Current Protection of Cables. A cable must be protected from overheating due to excessive short-circuit current flowing in its conductor. The fault point may be on a section of the protected cable or on any other part of the electric system. The faulted cable section is, of course, to be replaced after the fault has been cleared.

During a phase fault the I^2R losses in the phase conductors elevate first the temperature of the conductor, followed by the insulation materials, protective jacket, raceway, and surroundings. During a ground fault the I^2R losses in both phase conductor and metallic shield or sheath elevate the temperature in a manner similar to that of phase faults.

In most cases, the shield of the cable *beyond* the fault also carries part of the fault return current, which may then return along the shields of other conductors or equipment grounding paths, from common grounding points.

Since the short-circuit current is to be interrupted either instantaneously or in a very short time by the protective device, the amount of heat transferred from the metallic conductors outward to the insulation and other materials is very small. Therefore, the heat from I^2R losses is almost entirely in the conductors, and for practical purposes it can be assumed that 100% of the I^2R losses is consumed to elevate the conductor temperature. During the period that the short-circuit current is flowing, the conductor temperature should not be permitted to rise to the point where it may damage the insulating materials. The task of providing cable protection during a short-circuit condition involves determining the following:

(1) Maximum available short-circuit currents

(2) Maximum conductor temperature that will not damage the insulation

(3) Cable conductor size that affects the I^2R value and its capability to contain the heat

(4) Longest time that the fault will exist and the fault current will flow

8.4.1 Short-Circuit Current

8.4.1.1 Phase-Fault Current and Rates of Decay. The fundamentals of short-circuit current behavior and the calculation of short-circuit currents are described in Chapter 2. The magnitude of short-circuit current should be properly determined. As illustrated in Fig 149, the initial peak current is called asymmetrical current (or current for momentary duty). This then decays in sequence to the subtransient current, transient current, and synchronous current or sustained short-circuit current. The short-circuit current in the subtransient period, transient period, or synchronous period decays exponentially. Figure 149 shows the approximate rate of decay of the total current. Four typical systems are illustrated here to give the reader a general picture of the fault-current behavior. The decay rate in each system depends on the X/R ratio of the system, with the higher X/R ratio on medium voltage and generator voltage level systems.

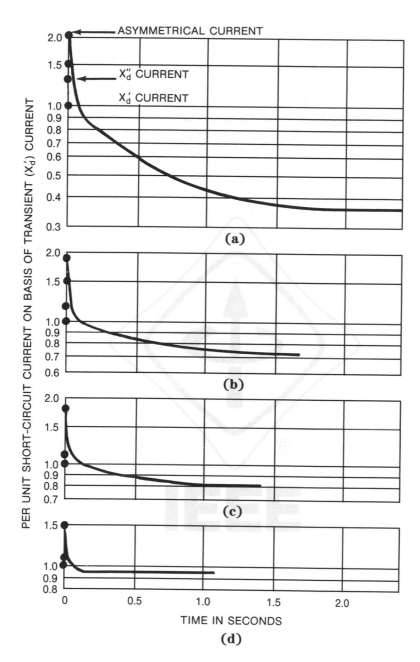

Fig 149
Typical Rate of Short-Circuit Current Decay
(a) Plant Generator System, Medium-Voltage
(b) Utility-Power Supplied System, Medium-Voltage,
with Large Synchronous Motors
(c) Utility-Power Supplied System, Medium-Voltage, No Synchronous Motor
(d) Utility or Plant Generation, Low-Voltage 240 or 480 V Load Centers

327

8.4.1.2 Maximum Short-Circuit Currents. Generally, the subtransient current of a system is used to designate the maximum available short-circuit current in the cables protected by the instantaneous overcurrent relays and medium-voltage switchgear circuit breakers. For cables protected by fuses, cable limiters and protectors (Chapter 5), or low-voltage and instantaneous trip circuit breakers, the asymmetrical current value is used. For delayed tripping at 0.2 s or longer, the rms value of the decayed current over the flow period of fault current should be used. If the latter is not possible, a reduced value of the subtransient current may be estimated. Figure 149 gives the rate of decay from which the subtransient current might be reduced to obtain a realistic value of total fault current over the entire delayed period.

8.4.1.3 Short-Circuit Currents Based on Equipment Ratings. For liberal design margins where economic considerations are not critical, the momentary and interrupting current ratings of the switchgear, circuit breakers, or fuses may be used as the basis for cable selection and protection. This, of course, assumes that the protective devices have been applied within their ratings. Where the available fault currents are substantially less than the ratings of the protective devices, the bottom (abscissa) scale can be conveniently used.

8.4.1.4 Ground-Fault Currents and Rates of Decay. The fundamentals of ground short-circuit current behavior are similar to those of phase-fault current, but the calculations are different, as described in Chapters 2 and 7. For a solidly grounded system, the ground-fault current is of about the same magnitude as the phase-fault current. For a low-resistance grounded system, the magnitude of the ground-fault current is limited to a sustained value determined by the resistor's current rating. The decay of the dc component occurs so rapidly that the asymmetry effect in the current wave shape can be ignored. For a high-resistance grounded or ungrounded system, the ground-fault current is small but should be immediately detected and cleared to prevent persistent arcing and the occurrence of a more serious fault involving other conductors or circuits.

Ground-fault currents of over 3–4 A are likely to ignite organic insulation in the arc path within a few minutes, developing a local fire and subsequent phase-to-phase fault, with extensive damage.

8.4.2 Conductor Temperature

8.4.2.1 Temperature Rise of Phase Conductors. On the basis that all heat is absorbed by the conductor metal and there is no heat transmitted from the conductor to the insulation material, the temperature rise is a function of the size of the metallic conductor, the magnitude of the fault current, and the time of the current flow. These variables are related by the following formula (ICEA P-32-382-1969 [5]):

for copper,

$$\left(\frac{I}{CM}\right)^2 \cdot t \cdot F_{ac} = 0.0297 \log_{10} \frac{T_f + 234}{T_0 + 234}$$

for aluminum,

$$\left(\frac{I}{CM}\right)^2 \cdot t \cdot F_{ac} = 0.0125 \, \log_{10} \frac{T_f + 228}{T_o + 228}$$

If the initial temperature T_O and final temperature T_f are predetermined on the basis of continuous-current rating and insulation material, respectively, the current I versus time t relation of current flow can be plotted for each conductor size (CM).

8.4.2.2 Temperature Rise of Shield and Sheath. On the same basis as for phase conductors, the temperature rise on the shield or sheath due to ground-fault current can be related to the magnitude of the fault current I, the cross section (CM) of the shield and sheath, the spiral cross section and length for the spiral, tape shield, and the time t of current flow, as shown in Table 39 (ICEA P-45-482-1979 [6]).

8.4.2.3 Maximum Short-Circuit Temperature Ratings. ICEA P-32-382-1969 [5] established a guideline for short-circuit temperatures for various types of insulation as shown in Table 40. The short-circuit temperature ratings are considered the maximum temperatures and are not to be exceeded in order to protect the cable insulation from damage.

8.4.2.4 Temperature–Current–Time Curves. For convenience in determining the cable size, the curves depicting the relationship of temperature–current–time are prepared from the *temperature rise* formula and are based on the temperature rise from the continuous to short-circuit temperature presented above. Figures 150 and 151 show the curves for copper and aluminum conductors from 75–200 °C. They also incorporate the total fault clearing times of instantaneous trip devices and the interrupting ratings of various types of switching equipment. For safe design at present and in the future, a cable may be selected on the basis of total clearing time and equipment interrupting rating. For example, AWG No 2/0 copper cable may be selected for connection to 500 MVA,

Table 39
Temperature Rise of Shield and Sheath
Due to Ground-Fault Current

Material	Initial/Final Temperatures 65/200 °C	65/150 °C	90/250 °C
Copper	$I = 0.0694 \dfrac{CM}{\sqrt{t}}$	$I = 0.0568 \dfrac{CM}{\sqrt{t}}$	$I = 0.0779 \dfrac{CM}{\sqrt{t}}$
Aluminum	$I = 0.0453 \dfrac{CM}{\sqrt{t}}$	$I = 0.0371 \dfrac{CM}{\sqrt{t}}$	$I = 0.0509 \dfrac{CM}{\sqrt{t}}$
Lead	$I = 0.0124 \dfrac{CM}{\sqrt{t}}$	$I = 0.0103 \dfrac{CM}{\sqrt{t}}$	$I = 0.0141 \dfrac{CM}{\sqrt{t}}$
Steel	$I = 0.0249 \dfrac{CM}{\sqrt{t}}$	$I = 0.0205 \dfrac{CM}{\sqrt{t}}$	$I = 0.0281 \dfrac{CM}{\sqrt{t}}$

13.8 kV, 8 cycle circuit breakers, and No 4/0 aluminum cable may be selected for connection to the same circuit breaker. However, a smaller cable can be selected if the maximum short-circuit current is less than the equipment interrupting rating.

8.4.2.5 Initial and Final Temperatures. For cables rated at initial (operating) and final (maximum short-circuit) temperatures different from 75 °C and 200 °C, respectively, correction factors for use with Figs 150 and 151 may be determined by use of Fig 152. By this chart, a *correction factor* is obtained by which the actual available fault current is converted to a *virtual available fault current* that is then used with Figs 150 and 151. The actual available fault current is multiplied by the correction factor K_t to obtain the *virtual available fault current*.

Examples (using Fig 152 chart):

(1) Initial temperature = 50 °C
Maximum fault temperature = 200 °C
K_t = 0.899
Actual available fault current = 20 000 A
Virtual available fault current = $0.899 \cdot 20\,000$ = 17 980 A on Figs 149 and 150.

(2) Initial temperature = 90 °C
Maximum fault temperature = 250 °C
K_t = 0.925
Actual available fault current = 20 000 A
Virtual available fault current = $0.925 \cdot 20\,000$ = 18 500 A on Figs 149 and 150.

In both cases, a smaller conductor might be safely used. The interrupting

Table 40
Maximum Short-Circuit Temperatures

Type of Insulation	Continuous Temperature Rating T_0 (°C)	Short-Circuit Current Temperature Rating T_f (°C)
Rubber	75	200
Rubber	90	250
Silicone rubber	125	250
Thermoplastic	60, 75, 90	150
Paper	85	200
Varnished cloth	85	200

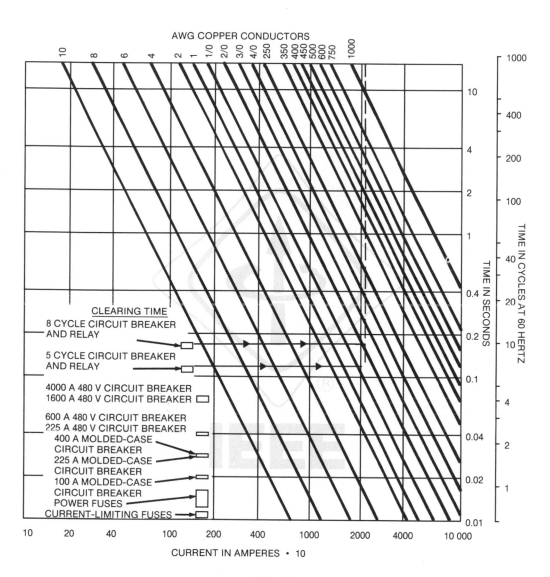

Fig 150
Maximum Short-Circuit Current for Insulated Copper Conductors;
Initial Temperature 75 °C; Final Temperature 200 °C;
for Other Temperatures Use Correction Factors of Fig 152

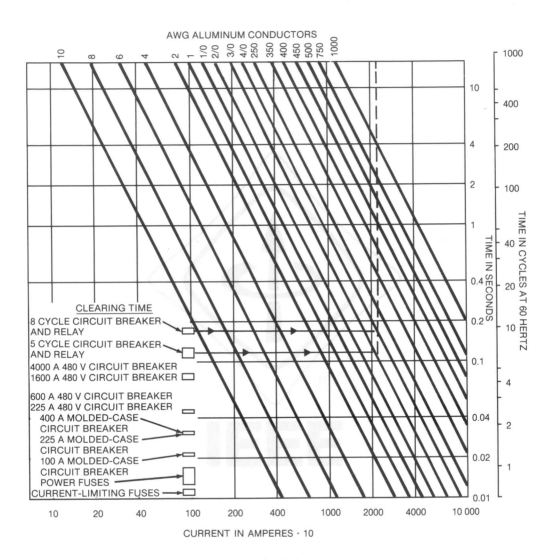

Fig 151
Maximum Short-Circuit Current for Insulated Aluminum Conductors
(Initial Temperature 75 °C; Final Temperature 200 °C;
for Other Temperatures Use Correction Factors of Fig 152)

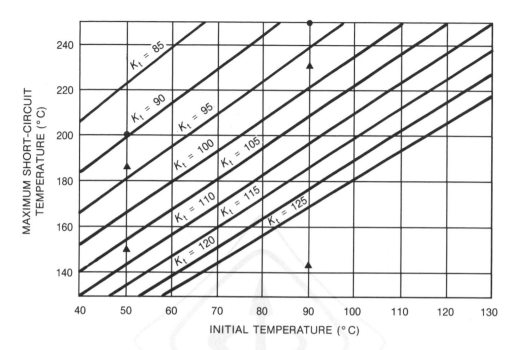

Fig 152
Correction Factors K_t for Initial and
Maximum Short-Circuit Temperatures

equipment ratings at tops of Figs 150 and 151 would need to be "virtually" moved in proportion to the correction factor K_t, as well.

8.4.3 Protective Devices

8.4.3.1 Total Fault-Clearing Time. Devices to protect cables against short-circuit damage should have high reliability and fast tripping speed. In the protective scheme, primary protection is the first line and backup protection the second line of defense. Primary protection normally provides prompt, but not necessarily instantaneous, tripping while backup protection is timed for more delayed tripping. Whether these two levels of protection are to be provided for all cables is a decision to be made in initial design stages. Relay time plus circuit breaker time equals total clearing time t:

(1) Relayed circuit breaker: Total fault clearing time equals current relay time plus auxiliary relay time (if used) plus circuit breaker interrupting time

(2) Direct tripping circuit breaker: Total fault clearing time equals circuit breaker clearing time

(3) Fuses: Total fault clearing time equals fuse total clearing time

8.4.3.2 Protective Devices and Clearing Time. The total clearing time of various types of protective devices depends on the type of relay and circuit breakers or fuses used. Table 41 estimates the total clearing times of various types of protective devices.

Table 41
Estimated Clearing Times of Protective Devices

Relayed Circuit Breakers, 2.4–13.8 kV

	Plunger, Instantaneous	Type of Relay Induction, Instantaneous	Induction, Inverse-Time
Relay times, cycles	0.25–1	0.5–2	6–6000
Circuit breaker inter-rupting time, cycles	3–8	3–8	3–8
Total time, cycles	3.25–9	3.5–10	9–6000

Large Air Power Circuit Breakers, Below 600 V

	Frame Size	
	225–600 A	1600–4000 A
Instantaneous, cycles	2–3	3
Short time, cycles	10–30	10–30
Long time, seconds	over 100 s	
Ground fault, cycles	10–30	10–30

Molded-Case Circuit Breakers, Below 600 V

	Frame Size	
	100 A	225–1200 A
Instantaneous, cycles	1.1	1.5
Long time, seconds	over 100 s	

Medium- and High-Voltage Fuses

High current	0.25 cycles (for current-limiting fuses operating in their current-limiting range)
	1.0 cycle (for power fuses at maximum current)
Low current	600 s (for E-rated fuses operating at 2X nominal rating: other ratings are available with different times at 2X nominal rating)

Low-Voltage Fuses

High current	0.25 cycles (in current-limiting range)
Low current	1000 s (at 1.35 to 1.5 times nominal rating)

For convenience, the total clearing time of instantaneous trip devices is shown in the lower left-hand corner of Figs 150 and 151. These data can be used together with maximum short-circuit current for proper selection of cable sizes.

8.4.3.3 Time–Current Characteristics of Protective Devices. A protective device will provide maximum protection if its time–current characteristics are suitably (20 percent in time) below those of the cable short-circuit current versus time curves shown in Figs 150 and 151. Thus the selection of overcurrent relays or devices is vitally important to the protection of cables. Figures 153, 154, and 155 illustrate the characteristics of relays and devices commonly used in feeder circuits. Shown also are the maximum available short-circuit currents of the system and the maximum short-circuit current curve of the cable.

8.4.3.4 Backup Protection. In some instances, the setting of a given device, rather than just protecting the immediate downstream element, may be selected to protect the second downstream element (cable). This would come into play if a protective device failed; the next upstream device would operate in adequate time to prevent damage to elements such as cable on the load side of the failed device. This is knows as *backup protection*.

Backup protection is almost never applied to industrial or commercial branch circuits, but is occasionally applied to feeder and subfeeder protection. The consequences of failure of a feeder or subfeeder protective device needs to be considered in deciding whether or not backup protection should be provided. Such protection is frequent in utility system practice, but not generally used in industrial or commercial systems.

8.4.4 Application of Short-Circuit Current Protective Devices

8.4.4.1 Typical Cases of Cable Protection. Power cables are used for transmission or distribution, or as feeders to utilization equipment. The following cases are typical in industrial and commercial power systems:

(1) A single cable feeder through a pull box or splice joint or with taps should be protected in the same manner as feeders to panels without transformers [Fig 156 (a), (c), and (d)].

(2) A single or multiple cable feeder without a pull box or taps should be protected from the maximum short-circuit current that can occur at any point in the feeder circuit.

A multiple cable feeder with or without pull box or splice joint should be protected from the maximum short-circuit current caused by a fault on one cable. The short-circuit current in each cable is not equally distributed, since the maximum current on the faulted cable is greater than the total current divided by the number of cables [Fig 156 (b) and (c)]. There are problems in the protection of parallel cables unless individual devices are used for each cable.

(3) A single or multiple cable feeder through a pull box or splice joint, or with taps, should be protected from the maximum short-circuit current caused by a fault on the tapped cables or end section of spliced cable. A cable fault requires the replacement of the faulted cable section only [Fig 156 (d) and (e)].

(4) A single cable feeder with pull box, a splice joint, and transformer tap should be protected from the maximum short-circuit current caused by a fault on the tap or after the splice joint [Fig 156 (f)].

(5) Multiple-feeder circuits should be protected in a manner similar to that of each individual circuit [Fig 156 (g)].

8.4.4.2 Protection and Coordination. The protective device should be selected and coordinated to give the cable sufficient short-circuit protection. This can be done easily by plotting the time–current curves of the protected cable and the protective device on the same log–log graph paper. The time–current curve of the protective device should always be below and to the left of the maximum short-circuit current–time curve (Figs 150 and 151) of the protected cable. Figures 153–157 illustrate that a No 4/0 copper insulated cable may be protected by various protective devices as follows. A 5 kV No 4/0 feeder is protected by a current-limiting fuse [Fig 153 (a)] or a 51 or 49 relay [Fig 153 (b)]. A 600 V No 4/0 feeder is protected by instantaneous tripping [Fig 154 (a)], by instantaneous and short-time tripping [Fig 154 (b)], or by an instantaneous molded-case circuit breaker [Fig 155 (a)]. A 600 V No 4/0 motor circuit is protected by a 400 A NEC [3] fuse [Fig 155 (b)].

8.5 Overload Protection of Cables. Overload protection cannot be applied until the current–time capability of a cable is determined. Protective devices can then be selected to coordinate cable rating and load characteristics.

8.5.1 Normal Current-Carrying Capacity

8.5.1.1 Heat Flow and Thermal Resistance. Heat is generated in conductors by I^2R losses. It must flow outward through the cable insulation, sheath (if any), the air surrounding the cable, the raceway structure, and the surrounding earth in accordance with the following thermal principle [14], [16], [17], [18]:

$$\text{heat flow} = \frac{\substack{\text{difference between conductor} \\ \text{and ambient temperature}}}{\text{thermal resistance from materials}}$$

The conductor temperature resulting from heat generated in the conductor varies with the load. The thermal resistance of the cable insulation may be estimated with a reasonable degree of accuracy, but the thermal resistance of the raceway structure and surrounding earth depends on the size of the raceway, the number of ducts, the number of power cables, the raceway structure material, the coverage of the underground duct, the type of soil, and the amount of moisture in the soil. These are important considerations in the selection of cables.

8.5.1.2 Ampacity. The ampacity of each cable is calculated on the basis of fundamental thermal laws incorporating specific conditions, including (1) type of conductor, (2) ac/dc resistance of the conductor, (3) thermal resistance and dielectric losses of the insulation, (4) thermal resistance and inductive ac losses of sheath and jacket, (5) geometry of the cable, (6) thermal resistance of the surrounding air or earth and duct or conduits, (7) ambient temperature, and (8) load factor. The ampacities of the cable under the jurisdiction of the NEC [3] are tabulated in its current issue or amendments thereto. The current-carrying capacity of cables under general operating conditions that may not come under the jurisdiction of the NEC [3] are published by the Insulated Cable Engineers

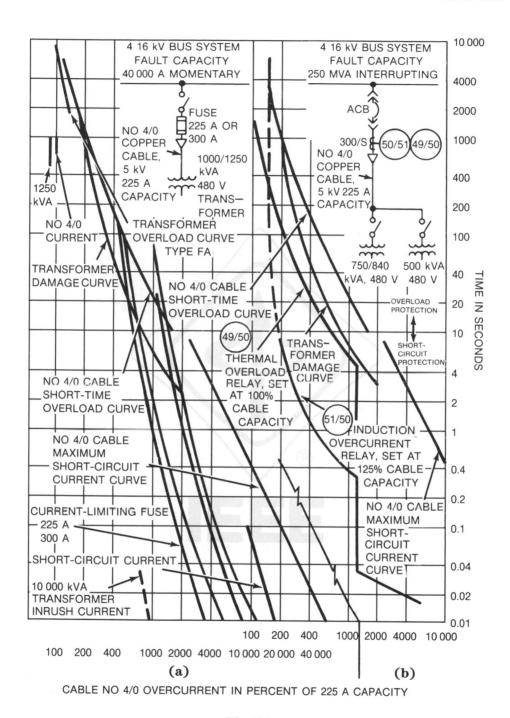

CABLE NO 4/0 OVERCURRENT IN PERCENT OF 225 A CAPACITY

Fig 153
Short-Circuit and Overload Protection of 5 kV Cables
(a) Power or Current-Limiting Fuses
(b) Overcurrent Relays

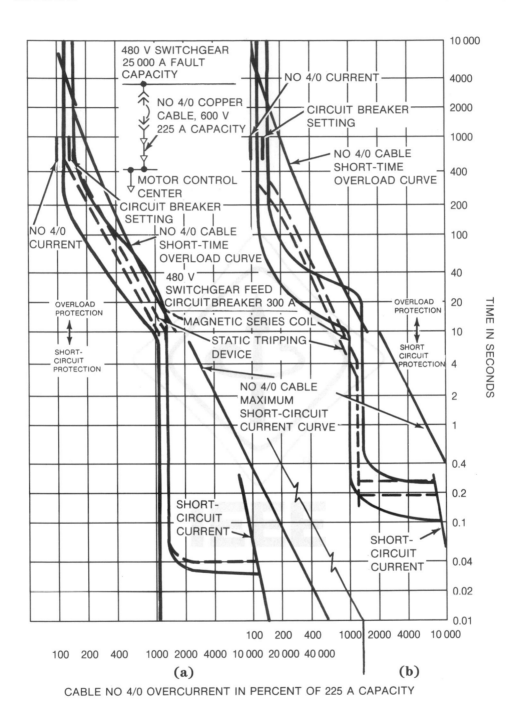

CABLE NO 4/0 OVERCURRENT IN PERCENT OF 225 A CAPACITY

**Fig 154
Short-Circuit and Overload Protection of 600 V Cables
(a) Long-Time and Instantaneous Equipped Circuit Breakers
(b) Long-Time and Short-Time Equipped Circuit Breakers**

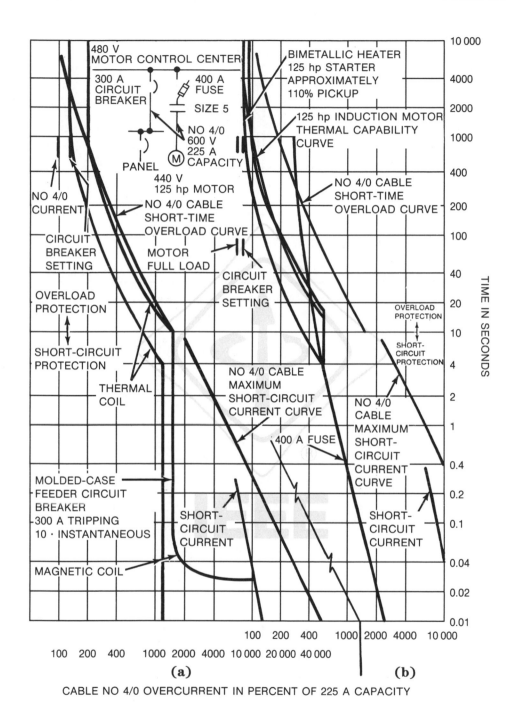

Fig 155
Short-Circuit and Overload Protection of 600 V Cables
(a) Thermal Magnetic Circuit Breakers
(b) Heaters and Fuses

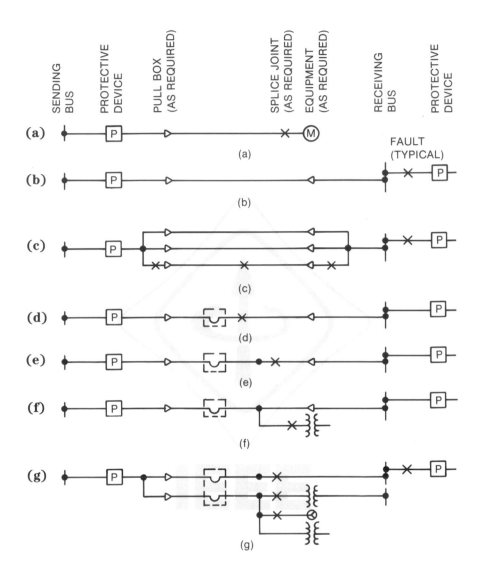

Fig 156
Application of Protective Devices
(a) Single Feeder to Utilization Equipment
(b) Single Feeder to Panel
(c) Multiple Cable Feeder to Panel
(d) Single Feeder with Pull Box
(e) Single Feeder with Pull Box and Splice
(f) Single Feeder with Pull Box and Tap
(g) Multiple Feeder with Pull Box and Taps

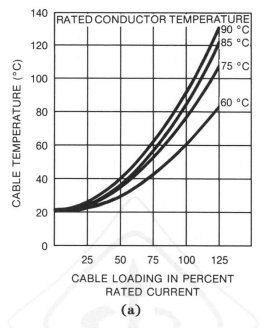

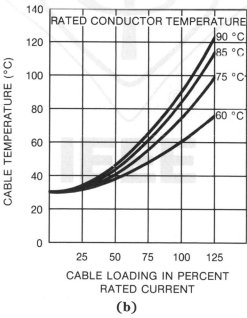

Fig 157
Cable Loading and Temperature Rise
(a) Ambient Temperature at 20 °C
(b) Ambient Temperature at 30 °C

Association (ICEA). In their publications they describe methods of calculation and tabulate the ampacity for 1, 8, 15, and 25 kV cables (ICEA S-19-81/NEMA WC3-1984 [8], ICEA-61-402/NEMA WC5-1984 [9], ICEA S-65-375/NEMA WC4-1983 [10]). The ampacities of specific types of cables are calculated and tabulated by manufacturers. Their methods of calculation generally conform to ICEA P-54-440/NEMA WC51-1979 [7].

8.5.1.3 Temperature Derating Factor. The ampacity of a cable is based on a set of physical and electrical conditions and a base ambient temperature defined as the no-load temperature of a cable, duct, or conduit. The base temperature generally used is 20 °C for underground installation, 30 °C for exposed conduits or trays, and 40 °C for MV cables.

Temperature derating factors (TDF) for ambient temperatures and other than base temperatures are based on the maximum operating temperature of the cable and are proportional to the square root of the ratio of temperature rise, that is,

$$\text{TDF} = \frac{I_N}{I_{N1}} = \frac{\text{current capacity at base ambient temperature}}{\text{current capacity at other ambient temperature}}$$

$$= \sqrt{\frac{T_N - T_a}{T_N - T_{a1}}}$$

$$= \sqrt{\frac{\text{temperature rise above base ambient temperature}}{\text{temperature rise above other ambient temperature}}}$$

8.5.1.4 Grouping Derating Factor. The no-load temperature of a cable in a group of loaded cables is higher than the base ambient temperature. To maintain the same maximum operating temperature, the current-carrying capacity of the cable must be derated by a factor of less than 1. Grouping derating factors (GDF) are different for each installation and environment. Generally, they can be classified as follows:

(1) For cable in free air with maintained space
(2) For cable in free air without maintained space
(3) For cable in exposed conduits
(4) For cable in underground ducts

NEC [3] Tables 318-8 and 318-9 list fill limits for LV cables in cable trays without derating. Greater fills are not permitted, so derating is not required. NEC [3], 318-10, -11, and -12 cover ratings and fills of MV cables in cable trays.

8.5.2 Overload Capacity

8.5.2.1 Normal Loading Temperature. Cable manufacturers specify the normal loading temperature for their products which results in the most economical and useful life of the cables. Based on the normal rate of deterioration, the insulation can be expected to have a useful life of about 20 to 30 years.

Table 42
Typical Normal and Emergency Loading of Insulated Cables

Insulation	Cable Type	Normal Voltage	Normal Loading (°C)	Emergency Loading (°C)
Thermoplastic	T, TW	600 V	60	85
	THW	600 V	75	90
	THH	600 V	90	105
	Polyethylene	0–15 kV	75	95
		>15 kV	75	90
Thermosetting	R, RW, RU	600 V	60	85
	XHHW	600 V	75	90
	RHW, RH-RW	0–2 kV	75	95
	Cross-linked polyethylene	5–15 kV	90	130
	Ethylene-propylene	5–15 kV	90	130
Varnished polyester		15 kV	85	105
Varnished cambric		0–5 kV	85	102
		15 kV	77	85
Paper lead		15 kV	80	95
Silicone rubber		15 kV	125	150

Normal loading temperature of a cable determines its current-carrying capacity under given conditions. In regular service, rated loads or normal loading temperatures are reached only occasionally because cable sizes are generally selected conservatively in order to cover the uncertainties of load variations. Table 42 shows the maximum operating temperatures of various types of insulated cables.

8.5.2.2 Cable Current and Temperature. The temperature of a cable rises as the square of its current. The cable temperature for a given steady load may be expressed as a function of percent full load by the formula

$$T_X = T_a + (T_N - T_a)(I_X/I_N)^2$$

Figure 156 shows this relation for cables rated at normal loading temperatures of 60, 75, 85, and 90 °C.

8.5.2.3 Intermediate and Long Time Zones. Looking at the intermediate and long time ranges from 10 s out to infinity, the definition of temperature versus current versus time is related to the heat dissipation capability of the installation relative to its heat generation, plus the thermal inertias of all parts. The tolerable temperatures are related, as in the previous section, to the thermal degradation characteristics of the insulation. The thermal degradation severity is, however, related inversely to time, so a temperature safely reached during a fault, and maintained for only a few seconds, could cause severe life reduction if it were maintained for even a few minutes. Lower temperatures, above the rated continuous operating temperature, can be tolerated for intermediate times.

The ability of a cable to dissipate heat is a factor of its surface area, while its ability to generate heat is a function of the conductor cross section, for a given current. Thus, the reduction of ampacity per unit cross-section area as the wire sizes increase tends to increase the permissive short-time current for these sizes relative to their ampacities. It may be seen later that the extension of the intermediate characteristic, on a constant I^2t basis, will protect the smallest wire sizes, and will over-protect the largest sizes, as shown in Fig 159. Constant I^2t protection is readily available, and is actually the most common, so a simplification of protection systems is possible.

The continuous current, or ampacity, ratings of cable have been long established and pose no problems for protection. The greatest unknown in the cable thermal characteristic occurs in the intermediate time zone, or the transition from short-time to long-time or continuous state.

8.5.2.4 Development of Intermediate Characteristics. Cable, with its own thermal inertia and that of its surroundings, will take from 1–6 h to change from initial to final temperature as the result of a current change. Consequently, overloads substantially greater than its continuous rating may be placed on a cable for this range of times.

Additionally, all cables except polyethylene (not cross-linked) will withstand, for moderate periods, temperatures substantially greater than their rated operating temperatures. This is a change recently developed from work done within ICEA and published by that organization. (See references in 8.9.) For example, EPR and XLP have emergency ratings of 130 °C, based on maximum time per overload of 36 h, three such periods per year maximum, and an average of one such period per year over the life of the cable. Thermoplastic types degrade in this marginal range by progressive evaporation of the plasticizer, and can operate for several hours at the next higher grade operating temperature 90 °C for 75 °C rating, and so forth) with negligible loss of life. So emergency operating overloads may reasonably be applied to cables within the time and temperature ratings. This should be the basis of application of protection of the cables.

The complete relationship for determination of intermediate overload rating is as follows:

percent overload capability

$$\frac{I_E}{I_N}\% = \sqrt{\frac{\dfrac{T_E - T_O}{T_N - T_O} - \left(\dfrac{I_O}{I_N}\right)^2 \cdot e^{-t/K}}{1 - e^{-t/K}}} \cdot \frac{230 + T_N}{230 + T_E} \cdot 100$$

where
$\quad I_E$ = emergency operating current rating
$\quad I_N$ = normal current rating
$\quad I_O$ = operating current prior to emergency
$\quad T_E$ = conductor emergency operating temperature
$\quad T_N$ = conductor normal operating temperature

T_O = ambient temperature
t = time after start of emergency loading in hours
K = a constant, dependent on cable size and installation type (Table 43)
230 = zero resistance temperature value (234 for copper, 228 for aluminum)
e = 2.7183 (base for natural logarithms)

If the cable has been operated at its rated current prior to the excursion, then I_O/I_N = 1 and its square = 1, so the relation is simplified to:

$$\frac{I_E}{I_N}\% = \sqrt{\frac{\dfrac{T_E - T_O}{T_N - T_O} - e^{-t/K}}{1 - e^{-t/K}} \cdot \frac{230 + T_N}{230 + T_E}} \cdot 100$$

This is the basic equation used in this chapter as representing the maximum safe capability of the cable.

While many medium voltage cables *are* operated at substantially less than full rated capacity, most low-voltage cables *are* operated near their rated ampacity. Even for medium-voltage cable, there are times when full loading is impressed. Regardless of preloading, protection should be coordinated with cable characteristics, not loading. Therefore, data presented here is based on 100% preloading, by the preceding equation. Factors are developed for approximating the characteristic for lower preloadings. For such preloadings, data presented here will be even more conservative.

Intermediate zone characteristics of medium voltage cables and 75 and 90 °C thermoplastic cables are tabulated in Table 44 with the characteristics of medium voltage cable illustrated graphically in Fig 159. These all apply to preloading at rated ampacity at 40 °C ambient temperature. For lower ambient temperatures and when cable ampacities have been increased to take this into account, it will be necessary to *reduce* the intermediate overload current percent by the following factors for each decrease in ambient temperature below 40 °C:

EPR-XLP   0.004
THH   0.002
THW   0.0037

For preloading less than 100% of rating, emergency overload percentages can be increased by the following factors:

	Preloading		
	75%	80%	90%
All insulation types	1.33	1.25	1.11

NOTE: This may safely be done only for permanent preloadings of these percentages.

Table 43
K Factors for Eqs in 8.5.2.4

Cable Size	Air No Cond	Air In Cond	UG Duct	Direct Buried
<#2	0.33	0.67	1.00	1.25
#2−#4/0	1.00	1.50	2.50	3.00
⩾250 MCM	1.50	2.50	4.00	6.00

Table 44
Emergency Overload Current
I_E—Percent of Continuous Rating
at 40 °C Ambient

Time s	Time h	Values of K 0.5	1	1.5	2.5	4	6
		EPR−XLP	$T_N = 90$ °C		$T_E = 130$ °C		
10	0.00278	1136	1602	1963	2533	3200	3916
100	0.0278	374	518	629	807	1018	1244
1000	0.278	160	195	226	277	339	407
10 000	2.78	126	128	132	140	152	168
18 000	5.0	126	127	128	131	137	147
		THH	$T_N = 90$ °C		$T_E = 105$ °C		
10	0.00278	725	1020	1248	1610	2033	2487
100	0.0278	250	338	407	518	651	794
1000	0.278	127	146	163	192	229	270
10 000	2.78	111	112	114	118	124	131
18 000	5.0	111	111	112	113	116	121
		THW	$T_N = 75$ °C		$T_E = 95$ °C		
10	0.00278	987	1390	1703	2197	2775	3396
100	0.0278	329	452	548	702	884	1080
1000	0.278	148	177	202	245	298	357
10 000	2.78	121	123	125	132	142	154
18 000	5.0	121	121	122	125	130	137

8.5.2.5 Direct Buried Cables. With direct buried cables, the conductor operating temperature needs to be kept at no more than 65 °C to keep the outside surface temperature below 60 °C, unless there is an ample supply of moisture in the soil. For higher surface temperature, moisture in the normal soil migrates

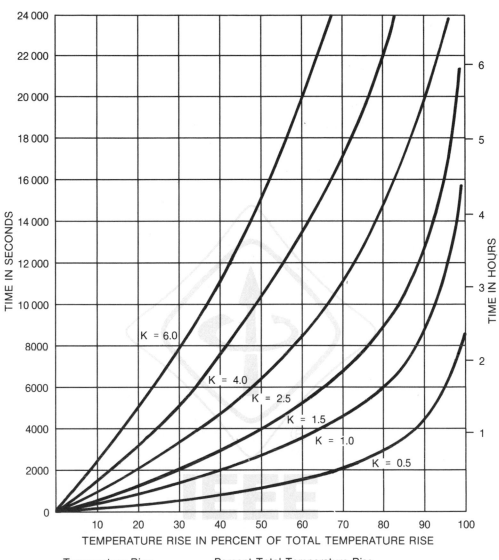

Temperature Rise	Percent Total Temperature Rise
Temperature at any Time	Initial Temperature and Temperature Rise
Final Temperature	Initial Temperature and Total Temperature Rise

K = 0.5	Small Cable in Air	
K = 1.0	Medium Size Cable in Air	
	Small Cable Underground	
K = 1.5	Large Cable in Air	
	Small Cable Direct Burial	
K = 2.5	Medium Cable Underground	(See Table 43)
	Medium Cable Direct Burial	
K = 4.5	Large Cable Underground	
K = 6.0	Large Cable Direct Burial	

**Fig 158
Rate of Temperature Rise Due to Current Increase**

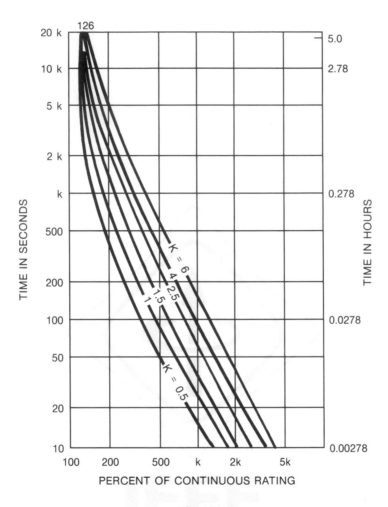

Fig 159
(a) Emergency Overload Current Percent of
Continuous Rating EPR — XLP Insulated 40 °C Ambient

away from the cable, raising the soil thermal resistivity and resulting in overtemperature of the cables. So for purposes of intermediate time emergency overload, a maximum conductor temperature of 80 °C has been selected as suitable to preserve this thermal resistivity condition for the times involved. Consequently, the tables and curves shown for air and duct use will not be applicable. Table 45 lists values applicable for direct buried installations. The short-time ratings for 250 °C are still applicable for this service since the times involved will not cause moisture migration.

8.5.2.6 Additional Observations. The absolute values of the short-time temperature and the emergency operating loading temperature are not precise.

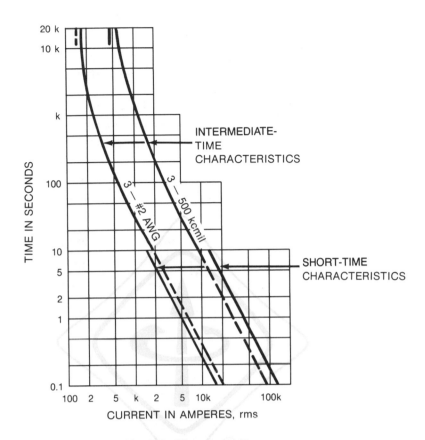

Fig 159 *(Continued)*
**(b) Ratings of Small and Large Cable in Conduit
in Air Intermediate and Short Time EPR and XLP**

They are values selected and proven to apply to the respective cable types without undue deterioration. For example, tests by Georgia Power Company of fault conditions imposed on medium voltage cable showed no appreciable degradation even where the nominal short-time temperature was exceeded by about 50 °C. Likewise, the 130 °C emergency operating temperature has an applicable time value of 36 h for no undue deterioration. It is only logical to deduce that this insulation can tolerate a somewhat higher temperature, say 150–175 °C, for a shorter time than 36 h. This condition is undoubtedly true, but its inclusion in calculations would complicate them unduly.

There is a compensating factor in the intermediate time range. An overcurrent of from 10–100 s range, for example, would not have sufficient time to cause heat to be dissipated by earth that was in contact with the cable. Times of over 100 s, and certainly 1000 s, would see this region of the heat dissipation chain contributing to the action. So, it is somewhat illogical to attribute the surrounding media's heat dissipation characteristics in the shorter portion of the interme-

Table 45
Emergency Overload Current
I_E—Percent of Continuous Rating
at 20 °C Ambient
Direct Buried, $T_N = 65$ °C, $T_E = 80$ °C

Time		Values of K		
s	h	1.5	3	6
10	0.00278	1313	1853	2616
100	0.0278	427	594	834
1000	0.278	168	213	282
10 000	2.78	115	121	134
18 000	5.0	113	116	123

diate time zone. Yet, a rigorous mathematical consideration would again substantially complicate the analysis.

So there is a tradeoff; the ability of insulation to withstand higher than nominal operating temperatures for shorter periods is considered adequate compensation for the lack of contribution of the surrounding media in absorbing heat during the shorter portion of the intermediate zone. Without this convention, it would be necessary to establish both varying allowable temperatures and K-factors over the whole range of the intermediate zone, an undue burden when it is apparent that the present method yields satisfactory results.

Even the 36 h nominal limit for 130 °C operation for MV cable does not mean that lower operating temperatures cannot be tolerated for longer periods. For example, 120 °C might be tolerated for 75 h, 110° for 150 h, and 100° for 500 h, just to illustrate the nature of the situation. Setting a continuous protective device to trip at precisely the 90 °C ampacity is almost certain to result in nuisance tripping on power surges, and so forth. So the setting of the device will, in all likelihood, be at something like 110% of rated cable ampacity, or an operating temperature of 100 °C. It will be left to visual or similar monitoring to keep the continuous loading of a cable from exceeding its rated ampacity for long periods of time.

8.5.3 Overload Protective Devices

8.5.3.1 Time–Current Characteristics. The time–current overload characteristics (Fig 159) of the cables differ from the short-circuit current characteristic (Figs 150 and 151). The overloads can be sustained for a much longer time than the short-circuit current, but the principle of protection is the same. A protective device will provide maximum protection if its time–current characteristic closely matches that of the cable overload characteristic. Thermal overcurrent relays generally offer better protection than do induction overcurrent relays because thermal relays operate on a long-time basis and their response time is proportional to the temperature of the cable or the square of its current.

8.5.3.2 Overcurrent Relays. Very inverse or extreme inverse relays of the induction disk type provide better protection than do the moderate inverse relays. However, all induction type overcurrent relays afford the cables sufficient protection. Figure 153 (b) shows the cable protection given by overcurrent relays (Device 51) and by thermal overcurrent relays (Device 49).

8.5.3.3 Thermal Overcurrent Relays or Bimetallic Devices. Thermal overload relays or bimetallic devices more closely resemble the cable's heating characteristic, but they are generally not as accurate as an overcurrent relay. Figure 153 (b) shows the cable protection given by thermal overcurrent relays (Device 49), and Fig 154 (b), protection given by bimetallic heaters.

8.5.3.4 Fuses. Where selected to match the ampacity of the cable, fuses provide excellent protection against high magnitude short circuits. Additionally, at 600 V and below, fuses provide protection for overloads or low current faults. Figures 153, 154, and 155 illustrate these applications. Figure 155 illustrates a combination of fast-acting 400 A fuse and motor overload relays. Had a 225 A dual-element fuse (selected for the ampacity) been used, the fuse alone would have provided overload protection.

Detailed treatment of fuses is given in Chapter 5 of this standard as well as in Chapter 3 of ANSI/IEEE Std 141-1986 [1], Chapter 5 of ANSI/IEEE Std 241-1983 [2], and IEEE Committee Report [15].

8.5.3.5 Magnetic Trip Coil or Static Sensor on 480 V Switchgear. The magnetic trip coils have a wide range of tripping tolerances. Their long-time characteristics match the cable overload curves for almost three quarters of an hour (Fig 154). Static trip devices provide better protection than magnetic direct-acting coils. However, for safe cable protection, the coils should be set below the heating curves of the cable by sizing the cable with normal loading current slightly greater than the coil pickup current.

8.5.3.6 Thermal Magnetic Coil on Molded-Case Circuit Breakers. The characteristics of the thermal magnetic coils resemble those of magnetic trip coils. They do not provide adequate thermal protection to cables during the long-term overloads [Fig 155 (a)]. The cable should be selected and protected in the same manner as described in the preceding paragraph.

8.5.4 Application of Overload Protective Devices

8.5.4.1 Feeder Circuits to Panels. A single or multiple cable feeder leading to a panel with or without an intermediate pull box should be protected from excessive overload by a thermal overcurrent device. If there are splice joints and a different type of installation, such as from an exposed conduit to an underground duct, the cable segment with the lowest current-carrying capacity should be used as the basis for protection [Fig 156 (b) and (c)].

A single cable feeder with tap to individual panels cannot be protected from excessive overload by a single protective device at the sending end, unless the cable is oversized. Therefore, overload protection should be provided at the receiving end. The protection should be based on the current-carrying capacity of the cable supplying power to the panel [Fig 156 (f)]. A multiple feeder with only a common protective device does not have overload protection for each cable feeder. In this case, overload protection should be provided at the receiving end [Fig 156 (g)]. See the NEC [3], 240-21.

8.5.4.2 Feeder Circuit to Transformers. A feeder circuit to one or more transformers should be protected in a manner similar to that used for feeder circuits to the panels. However, a protective device selected and sized for transformer protection also provides protection for the cable. This is due to the fact that the cables sized for a full transformer load have higher overload capability than the transformer. [See Fig 153 (a) for a comparison of the time–current curves between cable and transformer.]

8.5.4.3 Cable Circuit to Motors. A cable circuit to one or more motors should be protected in a manner similar to that for cable circuits to panels. Again, a protective device selected and sized for motor protection also provides cable protection because the overload capability of the cable is higher than that of the motor [Fig 155 (b)].

8.5.4.4 Protection and Coordination. Protective devices should be selected and cables sized for coordinated protection from short-time overload. The method of coordination is the same as for the short-circuit protection, that is, the time–current curve of the protective device should be below and to the left of the cable overload curve (Fig 159). Figures 153–155 illustrate the protective characteristics of relays and devices commonly used in cable circuits for overload protection.

8.6 Physical Protection of Cables. Cables require protection against physical damage as well as from electrical overload and short-circuit conditions. The physical conditions that should be considered are divided into three categories: mechanical hazards, adverse ambient conditions (excluding high temperatures), and attack by foreign elements. Cables can also be damaged (and frequently are) by improper handling during installation.

8.6.1 Mechanical Hazards. Electric cables can be damaged mechanically by vehicles, falling objects, misdirected excavation, or failure of adjacent circuits. Mechanical protection should serve the dual function of protecting cables and limiting the spread of damage in the event of an electrical failure.

Isolation is one of the most effective forms of mechanical protection. Conduit, tray, and duct systems are more effective if they are physically out of the way of probable accidents. A highly elevated cable is adequately protected against vehicles and falling objects. Where conduits or other enclosures must be run adjacent to roadways, large steel or concrete barriers provide adequate protection.

8.6.2 Exposed Raceways. The most popular form of mechanical cable protection is the use of conduits or metallic raceways. In addition to the electrical benefits of the grounded enclosure, the metallic conduit or raceway protects the cable against most types of mechanical damage. Cable trays are also becoming popular because they are economical and convenient for power and control cable systems. Cables may have increased protection from mechanical damage through the use of solid metal tray covers and metal barriers in the trays between different circuits. Covers incur derating, however.

8.6.3 Underground Systems. Underground ducts or embedded conduits provide similar mechanical protection. Ducts should be concrete encased for best results. Where they are subject to heavy traffic or poor soil conditions, reinforcement of the concrete envelope is desirable. Since excavation near underground cable runs is always a problem, it has been found advisable to color the concrete around electrical ducts. The addition of approximately three pounds of iron oxide per sack of cement provides a readily identifiable red color which is meant as a warning to anyone digging into the run. The color is effective even in mud or similar colored soil since it is conspicuous as soon as the concrete is chipped.

8.6.4 Direct-Buried Cables. The cables should be carefully routed to minimize damage from traffic and digging and to avoid areas where plant expansion is predicted. Cables should be covered with some type of special material, such as a brightly colored plastic strip or a wooden or concrete plank. Warning signs should also be placed above ground at frequent intervals along the cable route. Additionally, plant drawings accurately locating the buried cable run may also prevent accidental dig-ins.

8.6.5 Aerial Cable Systems. Insulated cables on a messenger require special care. These are especially susceptible to installation damage also. They must be located away from possible interference from portable cranes and support systems and protected from vehicle damage. Space or solid barriers provide reasonable protection for supports, whereas warning signs and nonelectric cables strung between electric cables and roadways offer protection against cranes and high vehicles.

8.6.6 Portable Cables. Exposed portable cables require extra consideration from a mechanical standpoint. Since they must remain portable, enclosures are not practical. The proper selection of a portable cable type provides one of the best methods of protection. It should be selected to match operating conditions. Moisture resistance, resistance to cutting or abrasion, and type of armor are all considerations that influence cable life. However, even the best cables require mechanical consideration in service. They should not be subjected to vehicular or steel-tired hand-pushed traffic. Means should be arranged to allow traffic to pass over or under cables without contacting the cable. Care should also be taken in moving portable cables to avoid snags or cuts. They should be located so that they are clear of welding, and placed where falling objects are not a serious hazard. A conspicuous color on the jacket is beneficial in warning personnel of the location of a portable cable.

8.6.7 Adverse Ambient Conditions. In the previous paragraphs, protection from overtemperature caused by short-circuit current or overload conditions has been discussed. There are other ambient conditions, however, that are not responsive to overcurrent devices or to compensation for elevated ambient temperatures.

In any type of cable enclosure, water or dampness should be considered, although underground installations are the most susceptible. Repeated cycles of high and low temperature, combined with humid air, can fill conduits or enclosures with water produced by "breathing" and condensation. It is almost impossible to stop the breathing, but suitable drains at low points will remove water as it collects. It is always desirable to prevent immersion, and duct systems and other raceways should be designed to slope so that the water can be removed.

Many of the available cable insulations are highly resistant to moisture, but where moisture is expected, extra care should be taken in selecting the insulation appropriate for that application.

The moisture problem may be amplified in industrial plants by the presence of various chemicals, and the possibility of chemical contamination should be considered for cables run through any process facility. Chemical seepage into ground water, or direct contact due to process misoperation, may result in chemicals coming into contact with a cable system. The enclosure, insulation, and conductor should all be tested to determine the effect of any particular chemical, and selected to be most resistant to the chemical involved. Where chemical contamination is severe, rerouting of the system should be considered. Acids and organic solvents are especially harmful.

Fires, which may result directly from cable failure or from unrelated external conditions, can cripple almost any cable system. It is easier to protect against damage to one cable caused by the failure of an adjacent cable than damage caused by an external fire. The enclosure of individual power circuits and fireproof coatings are the most effective means of limiting this type of damage, since it is unusual for a cable with proper electrical protection to burn through its individual conduit or raceway.

Pullboxes, pits, and manholes used as pulling points or sorting areas are the greatest fire hazards with respect to fault conditions, and elimination of the common enclosure for several circuits, where possible, offers the best protection.

The combustion of materials adjacent to a cable system is a difficult condition to protect against. The obvious solution is to remove all combustibles from the vicinity of the cables. As much as possible, this should be the practice. Critical circuits should be separated to lessen the extent of fire damage, and the use of multiple circuits following different routes can assure continuous service. Higher temperature-rated cable types might be considered for increased safe shut-down time.

In all cases the selection of proper enclosures or coverings can minimize fire damage; however, the method chosen should be based upon the possible hazards involved. For underground installation, heavier enclosures, higher racks, and overinsulation using materials with greater temperature resistance are all considerations, and under severe conditions, the use of MI cables may help.

8.6.8 Attack by Foreign Elements. In some environments, cable systems may be subject to attack by animals, insects, plants, and fungus, all of which may possibly cause cable failure. Small gnawing-type animals have been known to chew through cables, and insects and small animals such as lizards and snakes can cause difficulties at terminations where they or their nests may bridge the gap between terminals. The use of more resistant enclosures, armor, or indigestible cable materials are effective protection.

In tropical atmospheres, fungus may grow on cable and wire systems. The creation of a dry atmosphere is an effective deterrent, although fungus-resistant coatings and insulations are protection methods most often applied.

8.7 Code Requirements for Cable Protection. Codes and regulations are established to control the installation and operation of electric cable systems. Although there are many different codes and regulations that may be applied, depending on governmental, geographical, or company requirements, the NEC [3] is most often quoted and portions of it are mandatory by OSHA part 1910.302–1910.309. It is the responsibility of the engineer involved to determine which codes are applicable to the particular project. This discussion is limited to the NEC [3]. The NEC [3] is principally concerned with overtemperature (overcurrent), short-circuit, and mechanical protection in regard to cable applications.

Overcurrent protection is covered in Article 240 under the provision requiring all conductors to be protected in accordance with their current-carrying capacity. In general, the current-carrying capacity of cables is determined from the tables contained in Article 310, which concerns the installation of conductors.

In Article 240, short-circuit currents and overload currents are considered generally in the same manner, and rules are presented for the selection and setting of these devices. This article allows the use of fuses or circuit breakers for overcurrent protection. Short-time overcurrent characteristics of cables, and the coordination of protective devices with this, is recognized and permitted in critical industrial locations, by the NEC [3], 240-12. The tables in Article 310 also offer rules for derating cables for elevated ambient conditions.

Motor feeders receive particular attention in Article 430. In general, Article 430 governs the selection of the current-carrying capacity of cables used for motor circuits. After the cable size is selected in accordance with this article, the actual protection is applied in accordance with Article 240. It should be noted, however, that Article 430 provides rules for the overcurrent protection of the motors themselves, and although this discussion concerns only cables, it is possible that motor protective devices will also provide the required cable protection.

Article 310 insures that cables are adequate for their service applications by specifying currents which may be carried by particular conductors with specific insulation classifications and under specific governing conditions. It also requires the selection of cable materials that are suitable for application conditions, including moisture, chemicals, and nonstandard temperatures. This article permits the use of multiple cables, if means are provided to ensure the equal division of current, and that essentially identical conditions and materials are used for each of the parallel paths. Article 310 also covers installation methods designed to ensure the installation of cables without damage and with adequate working space.

Article 300 specifies wiring methods and protection required for cables subject to physical damage.

These articles pertain specifically to cable protection, but are not the only provisions of the NEC [3] that deal with the subject. Any specific cable or cable system will come under the provisions of one or more sections of the NEC [3], and responsible parties should ensure that the protective methods they have selected comply with both the relevant provisions and any special requirements that they may impose.

8.8 Busway Protection. Due to their economies, convenience, and excellent electrical characteristics, 600 V busway systems have gradually assumed a role of greater importance in today's industrial and commercial buildings. It is because of this consolidation of numerous cable runs into a single large bus duct run that the reliability of duct runs has become a critical factor in building design. Today's busways are very well designed for their intended use but, because of the critical nature of their purpose, they not only must suffer fewer outages, they must also be returned to service with a minimum of downtime. Thus, while the duct manufacturers can incorporate improved design concepts, better insulation, and so forth, it behooves the system designer to spend an extra amount of time on incorporating the best possible protection into the integrated system so that outages due to factors beyond his control are minimized in duration. This is an important concept in that it does not suggest that the number of outages can be controlled. It does, however, suggest that by minimizing the duration and extent of the outage, disruption of the normal activities can also be minimized. The duration of the outage is to a certain degree directly proportional to the amount of damage suffered by the busway during the fault, and this amount of damage is determined by the protective elements in the circuit.

8.8.1 Types of Busways. There are several different designs of busways available, and each offers certain features that are significant when considered in an integrated building plan. These can be identified as follows:

(1) Low-impedance busways
  (a) Feeder type
  (b) Plug-in type
  (c) High-frequency type
(2) High-impedance busways
  (a) Service entrance
  (b) Current limiting
(3) Simple plug-in busways

8.8.1.1 Low-Impedance Busways. Low-impedance busway designs (Fig 160) achieve their low-reactance characteristics by a careful positioning of each bus bar in close proximity to other bars of an opposite polarity. This close physical spacing (ranging from 0.050–0.250 in) demands that each bar be coated with some form of insulation to maintain satisfactory protection from accidental bridging. The losses in such designs are very low. Low-impedance busways are offered in feeder construction for the purpose of transmitting substantial blocks of power to a specific location or in plug-in construction. Plug-in designs feature door-like provisions at approximately 24 in increments along the length of the busway which, when opened, expose the bus bars so that plug-in taps may be made with minimum effort. Although most low-impedance designs are intended for use on 60 Hz applications, there are some that may be used at higher frequencies. Low-impedance designs are offered in voltage ratings of 600 V or less and current ratings up to 4000 A or more. Low-impedance busways may be ventilated or nonventilated and offered in indoor or outdoor construction, except that plug-in is available for indoor use only.

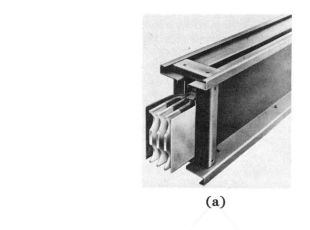

(a)

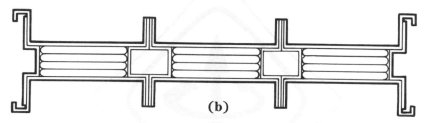

(b)

Fig 160
Low-Impedance Busway
(a) Sandwich-Type Construction for Heat Dissipation;
Close Proximity of Conductors Reduces Reactance
(b) Higher Current Ratings Often Employ Multiple Sets of Bars

8.8.1.2 High-Impedance Busways. High-impedance busways (Fig 161) are of two general types: those with deliberate impedance introduced to minimize fault current levels, and those which achieve high-impedance characteristics as an incidental by-product of their construction.

In the first type, high reactance is obtained just as low reactance was, by a careful placement of each bar relative to every other bar. In this case, however, the goal is to maximize the spacing between pairs of bars of opposite polarity. Since these high-impedance designs experience high losses and since these losses appear as heat, ventilated construction and insulated conductors are frequently employed. They are offered in generally the same ratings as low-impedance designs.

Under the provisions of some standards a special-purpose bus duct design may be built in total installed lengths of 30 ft or less for the purpose of connecting between an incoming service and a switchboard. This duct generally is constructed with a nonventilated enclosure and bars that may or may not be insulated. The bars are physically separated, and it is this large separation, introduced for

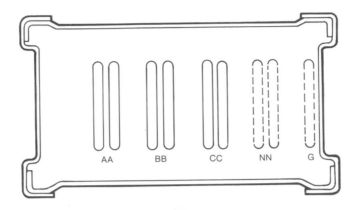

**Fig 161
High-Impedance Busway;
Wide Phase-to-Phase Spacings Result
in Increased Reactance**

safety, that results in high reactance characteristics. Short-run busways are limited to 2000 A and 600 V.

8.8.1.3 Simple Plug-in Busways. Among the first busway designs (Fig 162) introduced in the mid 1930's was a simple construction that supported bare conductors on insulators inside a nonventilated casing with periodic plug-in access doors. Like the short-run busway, this type of duct offered generous bar-to-bar spacings, but because it was not used to carry large quantities of current for lengthy runs, its losses were not objectionable. Short-circuit ratings are usually modest. It is still widely used today and is available in ratings of 100–1000 A and voltages of 600 V or less.

8.8.1.4 Bolted Faults. Faults associated with busways are either of the bolted type or the arcing type. Due to the prefabricated nature of busways, bolted faults are rare. Bolted fault, in this context, refers to the inadvertent fastening together of bus bars in a solid fashion resulting in an unintended connection between phases. Bolted faults can occur during the initial installation or at a later date when modifications are made to the system. The actual offending connection might be found in a bus duct cubicle, but it will more often be found in some pieces of equipment connected to the busway, such as a switchboard connection or a load served from a bus plug. Because a bolted connection implies a low-resistance connection, we may expect the maximum level of fault current to flow, and circuit protective elements should, therefore, be sized in accord with this maximum fault level. Bolted faults result in a distribution of energy through the entire length of the bus duct circuit. This energy flow results in an intense magnetic field around each conductor that opposes or attracts fields around adjacent conductors. The mechanical forces thus created are very high and are

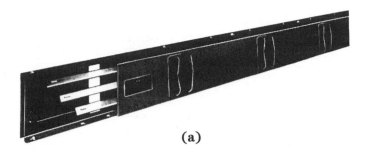

(a)

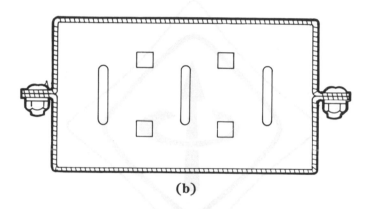

(b)

Fig 162
Simple Plug-In Busway
(a) Bare Bars Spaced Far Apart Offer
Generous Electrical Clearances and Low Cost
(b) Efficiency Is Low but Acceptable for Loads of 1000 A or Less

quite capable of bending bus bars (Fig 163), tearing duct casings apart, or shattering insulation. For this reason, the busway should have a short-circuit withstand rating that is greater than the maximum available fault current. Such ratings are published by the various busway manufacturers and are based on a 3 cycle duration. Table 46 reflects the NEMA standard ratings (see NEMA BU1-1983 [13] and ANSI/UL 857-1981 [4]).

8.8.1.5 Arcing Faults. In contrast to the bolted fault, arcing faults can occur at any time in the life of a system. Although there can be many individual agencies initiating an arcing fault, they generally involve one or more of the following: loose connections, foreign objects, insulation failure, voltage spikes, water entrance. Because of the resistance of the arc and the impedance of the return path, current values are substantially reduced from the bolted fault level.

The interaction of the magnetic fields around the conductors and around the arc results in an unbalanced force that causes the arc to try to move away from the power source. If the path is unobstructed, the arc will accelerate and move

Table 46
Busway Minimum
Short-Circuit Current Ratings

Continuous Current Rating of Busway (amperes)		Minimum Short-Circuit-Current Ratings (amperes)	
Plug-In	Feeder	Symmetrical	Asymmetrical
100		10 000	10 000
225		14 000	15 000
400		22 000	25 000
600		22 000	25 000
	600	42 000	50 000
800		22 000	25 000
	800	42 000	50 000
1000		42 000	50 000
	1000	75 000	85 000
1350		42 000	50 000
	1350	75 000	85 000
1600		65 000	75 000
	1600	100 000	110 000
2000		65 000	75 000
	2000	100 000	110 000
2500		65 000	75 000
	2500	150 000	165 000
3000		85 000	100 000
	3000	150 000	165 000
	4000	200 000	225 000
	5000	200 000	225 000

very quickly toward the remote end of the run. The only mark of its passage may be a scarcely noticeable pinhead size pit every several inches along the edge of the bus bar. These tiny marks, however, provide a clear trail for the investigator and can often lead him back to the origin of the fault. As the arc travels away from the power source, the length of the circuit becomes greater and the forces causing movement become smaller. Eventually, the arc reaches some obstruction that causes it to hesitate long enough to cause serious burning or even hang up until the bus duct is burned open. Busways employing insulated conductors, of course, do not permit traveling arcs. Arcing, therefore, remains at the point of initiation, or may burn slowly toward the source.

Although the magnitude of current present in an arcing fault is usually less than that present in a bolted fault, the entire thermal effect is concentrated at the arc location and results in major damage at that point. Figure 164 indicates the damage anticipated in terms of the quantity of conductor material vaporized by a phase-to-phase arc at 480 volts. A 15 000 amp arc persisting at one location for 9 cycles would remove about half of the $\frac{1}{4} \times 2$ in copper conductor. These charts are based on a simple plug-in busway design with bars on $2\frac{1}{4}$ in centers. Designs with bars closer together or designs employing aluminum conductors may be

**Fig 163
Simple Plug-In Busways Subjected to
Fault Currents above Their Ratings; Busway
on Left Was Protected by Current-Limiting
Fuses, while that on Right Was Protected
by a Circuit Breaker**

expected to show much more extensive damage. If the designer intends to minimize the duration of system outages, it is toward the arcing fault that he must direct his concern. Even in an insulated bus, arcing faults can persist, the arc extending from within the insulating tube to the burned-out spot in the insulation, several inches to a similar crater within the adjacent bus. Simple close spacing of an insulated bus does not guarantee against bolted faults.

8.8.2 Types of Protection. Like any other circuit, busways are subject to overloads, arcing faults, and bolted faults. Each of these is characterized by an entirely different set of parameters and, therefore, requires an entirely different set of protective concepts. No single protective element suits all requirements. An examination of the protection required will suggest the need for several protective devices.

8.8.2.1 Overload Protection. Overloads are, of course, those temporary conditions that cause the busway to carry currents greater than its continuous-current rating. Overloads such as stalled rotor currents or motor starting currents generally are not harmful to the busway because each motor served is usually small in comparison to the capacity of the busway. Overloads are more likely to occur as a result of adding more or larger pieces of equipment, over a period of years, to an existing busway circuit until its capacity is exceeded. Since busways tend to use large conductors, they exhibit considerable thermal inertia. For this reason, overloads of a temporary nature require a substantial time before

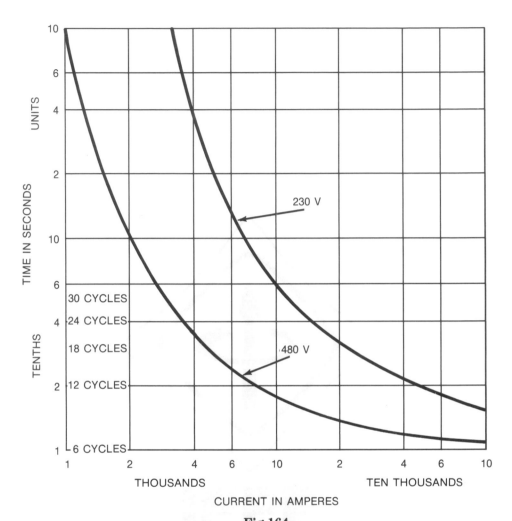

Fig 164
Time-Current Curves for a Power Arc
to Burn a ¼ × 2 in Copper Bar Halfway Through
(Bars Are Spaced on 2¼ in Centers in Standard Plug-in Bus Duct)

their effect is noticed. Figure 165 displays time versus temperature rise for three different loading conditions. Note that this particular busway required only 12 min to reach 55 °C at 200% current. The same busway took over an hour to reach 55 °C at 125% loading. When operating at 100% loading, it required 25 min to raise the temperature up to the 55 °C limit with a 25% overload. Naturally, this time will vary from size to size depending on the stable temperature produced by 100% loading. Most busway sizes are designed to operate at close to a 55 °C rise at full current. The value of 55 °C was selected because this is the maximum rise allowed by Underwriters Laboratories on plated bus bar joints. It was generally assumed that the busway will operate in a 30 °C (86 °F) ambient, and for this reason, busway manufacturers have employed 85 °C insulation for many years. (30 °C ambient + 55 °C = 85 °C operating temperature.) Newer designs of busways employ higher temperature insulation, although they are still limited to a 55 °C rise at the hottest spot. This suggests that busways operated in a 50 °C ambient could still carry full load (producing a 55 °C rise) without exceeding the 105 °C total temperature limit of the newer insulations. From the foregoing it is apparent that the older busways could easily suffer insulation damage should they be subject to a high ambient or a moderate overload, or both. For this reason, any protective device should be sensitive to overload conditions.

The new 105 or 130 °C insulations are intended to provide increased protection from the danger of high ambients or temporary overloads. The designer cannot apply long duration overloads even to these newer insulations without eliminating all the extended life factors that they provide.

8.8.2.2 Arcing-Fault Protection. Arcing-fault currents are found to be as low as 38% of the bolted-fault current calculated for the same circuit (arcing ground-fault currents can be much smaller). Such faults, because of their destructive nature, should be removed with no intentional time delay. Unfortunately, the magnitude of this current may be so low that the time-delay characteristics of the overload protective device confuses the low-magnitude arcing fault with the moderate overload or temporary inrush current and allows it to persist for lengthy periods. For example, a 200% overcurrent on a fuse might require 200 s or more before the fuse functions.

Since it is the arc resistance and circuit impedance that limit the current

Fig 165
Time-Temperature Curves of 100, 125, and 200 Percent Loads
on a 600 A Rated Low-Impedance Busway
(1¼ × 2¼ in Aluminum Bar per Phase)

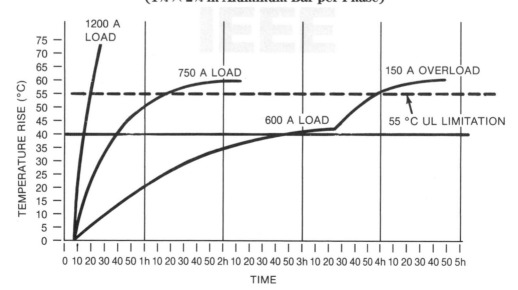

flowing in an arcing fault, most busway manufacturers offer an optional ground conductor, located inside the busway casing, to provide a low-reactance ground path for arcing current. Without this conductor, the arcing current to ground would be forced to travel on the high-impedance steel enclosure, including its many painted joints. This would have a tendency to reduce the already low current to an even lower level and further confuse the low-level protective device.

The best protection against this type of fault is found to be ground-fault protection and the second most suitable is a circuit breaker with a low-range instantaneous trip. Chapter 7 gives a more thorough discussion of this problem.

8.8.2.3 Bolted-Fault Protection. Bolted-fault currents can approach the maximum calculated available fault levels, and therefore protective elements must be capable of interrupting these maximum values. Circuit breakers and current-limiting fuses are suitable here.

Under certain conditions busways may be applied on circuits capable of delivering fault currents substantially above the busway's short-circuit rating. While it is true that electromagnetic forces increase as the square of the current, the use of very fast fuses permits busways to be applied on circuits having available fault currents higher than the busway short-circuit rating. The reason for this is that the busway rating is based on a 3 cycle duration while class J, R, T, or L current-limiting fuses function in much less time, generally less than 1 half-cycle, during high-level faults. The property of inertia exhibited by the heavy bus bars causes the bars to resist movement during the very short period of time that such fuses allow current to flow. Current-limiting fuses limit the magnitude of fault current to their let-through values (see Fig 166).

In general, a busway may be protected by a class J, R, T, or L fuse against the mechanical or thermal effects of the maximum energy the fuse will allow to flow, providing the fuse continuous current rating is equal to the bus-duct continuous current rating. Most manufacturers have conducted tests and will certify that their designs are satisfactory for use with fuses at least one rating larger than the busway. These higher fuse ratings are often needed for coordination with a circuit breaker in series with the fuse.

8.8.2.4 Typical Busway Protective Device. While no single element incorporates all the necessary characteristics, it is possible to assemble several elements into a single device. A particularly effective device is the fused circuit breaker equipped with ground-fault protection. In such a device, the circuit breaker elements provide operation in the overload or low-fault range, while the coordinated current-limiting fuse functions during high-level faults. The ground-fault sensor detects those arcing faults that go to ground and, regardless of their low magnitude, signal the circuit breaker to open.

8.8.3 Busway Testing and Maintenance. There are several well-known tests that should be performed before any busway is energized:

(1) Continuity check
(2) Voltage test
(3) High-potential test

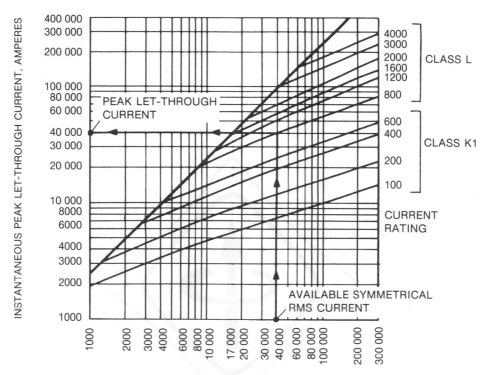

NOTE: Check specific manufacturer for exact data.

Fig 166
Typical 60 Hz Peak Let-Through Current as a Function of Available
RMS Symmetrical Current; Power Factor Below 20 Percent

These are conducted with the busway disconnected from the supply source and without bus plugs attached.

8.8.3.1 Continuity Check. By using a low-voltage source and a bell, the system should be checked to be sure that there is no accidental solid connection between phases or from phase to ground.

8.8.3.2 Voltage Test. Application of a 500 V megohmmeter test to the system will indicate the insulation resistance values between phases and from phase to ground. While it is not practical to assign specific acceptability limits to meter readings, any reading of less than 1 MΩ for a 100 ft run should be investigated.

8.8.3.3 High-Potential Test. Application of 2200 V (two times maximum design voltage plus 1000 V) for 600 V equipment between phases for 1 min while measuring the leakage current should disclose incipient insulation failures.

8.8.3.4 Visual Inspection and Joint Tightening. None of the preceding tests will disclose the presence of loose joints in the system. It is therefore essential that one individual be assigned the task of inspecting each joint to see that it is bolted properly to the manufacturer's recommended torque values. The importance of this step cannot be overemphasized. Periodic inspections to ensure that the joint integrity has not been affected by creep, vibration, heat cycling, or accumulation of dust or foreign matter should be performed as part of a normal preventive maintenance program. The frequency of such inspections will be determined by the nature of the installation, but ideally it might take place after three months, six months, and one year to build a history and provide a basis for scheduling future inspections.

8.9 References. The following publications shall be used in conjunction with this chapter.

[1] ANSI/IEEE Std 141-1986, IEEE Recommended Practice for Electric Power Distribution for Industrial Plants.

[2] ANSI/IEEE Std 241-1983, IEEE Recommended Practice for Electric Power Systems in Commercial Buildings.

[3] ANSI/NFPA 70-1984, National Electrical Code.

[4] ANSI/UL 857-1981, Busways and Fittings.

[5] ICEA P-32-382-1969, Short-Circuit Characteristics of Insulated Cable.

[6] ICEA P-45-482-1979, Short-Circuit Performance of Metallic Shielding and Sheaths of Insulated Cable.

[7] ICEA P-54-440/NEMA WC51-1979, Ampacities of Cables in Open-Top Cable Trays.

[8] ICEA S-19-81/NEMA WC3-1984, Rubber-Insulated Wire and Cable for the Transmission and Distribution of Electrical Energy.

[9] ICEA S-61-402/NEMA WC5-1984, Thermoplastic-Insulated Wire and Cable for the Transmission and Distribution of Electrical Energy.

[10] ICEA S-65-375/NEMA WC4-1983, Varnished-Cloth-Insulated Cables.

[11] ICEA S-66-524/NEMA WC7-1984, Cross-Linked Thermosetting—Poly-ethylene Insulated Wire and Cable for the Transmission and Distribution of Electrical Energy.

[12] IEEE S-135/ICEA P-46-426, Power Cable Ampacities for Copper and Aluminum Conductors (SH07096).

[13] NEMA BU1-1983, Busways.

[14] AIEE Committee Report. The Effect of Loss Factor on Temperature Rise of Pipe Cable and Buried Cables, Symposium on Temperature Rise of Cables. *AIEE Transactions on Power Apparatus and Systems,* pt III, vol 72, June 1953, pp 530–535.

[15] IEEE Committee Report. Protection Fundamentals for Low-Voltage Electrical Distribution Systems in Commercial Buildings, *IEEE JH 2112-1, 1974.*

[16] NEHER, J.H., and McGRATH, M.H. The Calculation of the Temperature Rise and Load Capability of Cable Systems. *AIEE Transactions on Power Apparatus and Systems,* pt III, vol 76, Oct 1957, pp 752–772.

[17] SHANKLIN, G.B. and BULLER, F.H. Cyclical Loading of Buried Cable and Pipe Cable. *AIEE Transactions on Power Apparatus and Systems,* pt III, vol 72, June 1953, pp 535–541.

[18] WISEMAN, R.J. An Empirical Method for Determining Transient Temperatures of Buried Cable Systems. *AIEE Transactions on Power Apparatus and Systems,* pt III, vol 72, June 1953, pp 545–562.

9. Motor Protection

9.1 General Discussion. This chapter is intended to apply specifically to three-phase integral horsepower motors. There are many variables involved in choosing motor protection: motor importance, motor rating (from one to several thousand horsepower), type of motor controller, etc. Therefore, it is recommended that protection for each specific motor installation be chosen to meet the requirements of the specific motor and its use. All items in 9.2 and 9.3 should be referred to as check lists when deciding upon protection for a given motor installation. After the types of protection have been selected, manufacturers' bulletins should be studied to ensure proper application of the specific protection chosen.

9.1.1 Low voltage systems (see voltage classes) are those nominally 1000 or less.

9.1.2 Medium voltages start above 1000 and may reach as high as 35 kV. NEMA ICS2-1983 [3][21], Section 324, limits equipment voltage to 7.2 kV on the basis that such a limit embraces upper voltages found in industrial and commercial systems that concern the plant electrical engineer. However, more recently since the revision of that standard, industrial and commercial power systems are being engineered as high as 15 kV, with isolated cases above this reaching into the 35 kV bracket. This is a trend that is the responsibility of the plant electrical engineer.

9.2 Items to Consider in Protection of Motors

9.2.1 Motor Characteristics. These include type, speed, voltage, horsepower rating, service factor, power factor rating, type of motor enclosure, lubrication arrangement, arrangement of windings and their temperature limits, thermal capabilities of rotor and stator during starting, running and stall conditions, etc.

9.2.2 Motor Starting Conditions. Included are full voltage or reduced voltage, voltage drop and degree of inrush during starting, repetitive starts, frequency and total number of starts, and others.

9.2.3 Ambient Conditions. Temperature maxima and minima, elevation, adjacent heat sources, ventilation arrangement, exposure to water and chemicals, exposure to rodents and various weather and flood conditions, and others should be considered.

9.2.4 Driven Equipment. Characteristics will influence chances of locked rotor, failure to reach normal speed, excessive heating during acceleration, overloading, stalling, etc.

9.2.5 Power System. Types of system grounding, exposure to lightning and switching surges, fault capacity, exposure to automatic reclosing or transfer, possibilities of single-phasing supply (broken conductor, open disconnect or circuit breaker pole, blown fuse), and other loads that can cause voltage

---

[21] The numbers in brackets correspond to those of the references listed at the end of this chapter.

unbalance should be considered.

9.2.6 Motor Importance. Motor cost, cost of unplanned down time, amount of maintenance and operating supervision to be provided to motor, ease and cost of repair, etc, have to be evaluated.

9.2.7 Load Side Faults for Motor Controllers. (Intro) Fault current available in a circuit is treated in Chapter 2; fuse and breaker protection for conductors in feeder and branches are treated in Chapters 5 and 6. Note that fuses and breakers are rated for connection to *available* fault current sources on the basis of protecting the conductors down stream from the breaker or fuse.

In a motor controller the above philosophy does not necessarily extend to protect the motor controller or its cubicle. For proper protection of the motor controller the controller manufacturer will indicate the correct breaker or fuse in the motor controller for safe levels of protection.

Such motor controllers for best results should bear UL listing for connection to available currents at least as high as that found in the power supply of the plant system under consideration. The UL-listed controller may still be substantially damaged in the event of the *load side* fault, but not to the extent of loss of life or property beyond the controller itself. If protection is desired to minimize damage to the controller itself, control manufacturer's recommendations must be sought out.

In any case, controllers connected to *available currents* above a 10 000 A symmetrical should be provided with breakers having ratings equally above 10 000 A or with fuses similarly rated and provided with rejection fuse clips to prevent substitution of lower and inadequately rated fuses. The steady increase of available currents in new and upgraded systems makes the above philosophy mandatory and required by ANSI/NFPA 70-1984 [2] (National Electrical Code [NEC]).

9.2.8 Arcing ground faults are treated in 7.4.3.

9.2.9 Any adjustment of overload protection must have adequate security to prevent inadvertent change due to accident, normal vibration, or ambient conditions.

9.3 Types of Protection
9.3.1 Undervoltage

9.3.1.1 Purpose. The usual reasons for using undervoltage protection are as follows:

(1) To prevent possible safety hazard of motor automatic restarting when voltage returns following an interruption

(2) To avoid excessive inrush to the total motor load on the power system, and the corresponding voltage drop, following a voltage dip, or when voltage returns following an interruption

9.3.1.2 Instantaneous or Time Delay. Undervoltage protection will be either instantaneous (no intentional delay) or of the time-delay type. Time-delay undervoltage protection should be used with motors important to production continuity of service, providing it is satisfactory in all respects, to avoid unnecessary tripping on voltage dips that accompany external short circuits. Examples follow of nonlatching starters where time-delay undervoltage protection is not satisfactory and instantaneous undervoltage must be used.

(1) Fused or circuit breaker combination motor starters having ac voltage held contactors, used on systems of low three-phase fault capacity. With the usual time-delay undervoltage scheme the contactor could drop out on the low voltage accompanying the fault before the fuse or circuit breaker opens. The contactor could then reclose into the fault. This problem does not exist if the fault capacity is high enough to open the fuse or circuit breaker before the contactor interrupts the fault current. (See note following the next example.)

(2) Synchronous motors used with starters having ac voltage held contactors. With the usual time-delay undervoltage scheme the contactor could drop out on an externally caused system voltage dip and then reclose reapplying the system voltage to an out-of-phase internal voltage in the motor. The high initial inrush could damage the motor winding, shaft, or foundation. This problem could also occur for large-horsepower high-speed squirrel-cage induction motors. It usually is not a problem with the 200 hp and smaller induction motors with which voltage held contactor starters are used because the internal voltages of these motors decay quite rapidly.

NOTE: The foregoing two limitations could be overcome by using a separate ac power source for control or dc battery control on the contactor to prevent its instantaneous dropout. In other words, the time-delay undervoltage feature can be applied directly to the main contactor.

(3) Motors used on systems having fast automatic transfer or reclosing where the motor must be tripped to protect it before the transfer or reclosure takes place.

(4) When the total motor load having time-delay undervoltage protection will result in more inrush and voltage drop after an interruption than the system can satisfactorily cope with. The least important of the motors should have instantaneous undervoltage protection. Time-delay undervoltage protection of selectively chosen delays could be used on the motors whose inrush the system can handle.

9.3.1.3 With Latching Contactor or Circuit Breaker. These motor switching devices inherently remain closed during periods of low or zero ac voltage. The following methods are used to trip (open) them:

(1) Energize shunt trip coil from dc battery.

(2) Energize shunt trip coil from a separately generated reliable source of ac. This ac source must be electrically isolated from the motor ac source in order to be reliable.

(3) Energize shunt trip coil from a capacitor charged through a rectifier from the ac system. This is commonly referred to as capacitor trip.

(4) Deenergize a solenoid and allow a spring to be released to trip the contactor or circuit breaker. This is commonly referred to as a dc trip scheme.

Items (1)–(3) are usually used in conjunction with voltage-sensing relays (see 9.3.1.6).

Item (4) could have the solenoid operating directly on the ac system voltage. Alternatively, the solenoid could operate on dc from a battery, in which case a relay would sense loss of ac voltage and deenergize the solenoid. The solenoid could be either instantaneous or time delayed using a dashpot arrangement.

9.3.1.4 With AC Voltage Held Main Contactor. Since the main contactor (which switches the motor) will drop out on loss of alternating current, it provides an instantaneous undervoltage function. There are two common approaches to achieve time-delay undervoltage protection:

(1) Permit the main contactor to drop out instantaneously but provide a timing scheme (which will time when ac voltage is low or zero) to reclose the main contactor providing normal ac voltage returns within the preset timing interval. Some of the timing schemes in use are as follows:

(a) Capacitor charged through a rectifier from the ac system. The charge keeps an instantaneous dropout auxiliary relay energized for an adjustable interval, which is commonly 2 or 4 s.

(b) Standard timer that times when deenergized (pneumatic or induction disk, etc).

(2) Note that two-wire control is sometimes used with an ac voltage held main contactor. This control utilizes a *maintained closed* start button, or operates from an external contact responsive to some condition such as process pressure, temperature, level, etc. The main contactor drops out with loss of ac but recloses when ac voltage returns. This arrangement does not provide undervoltage protection, and should not be used if automatic restarting could endanger personnel or equipment.

9.3.1.5 With DC Voltage Held Main Contactor. With this arrangement the contactor remains closed during low or zero ac voltage. Time-delay undervoltage protection is achieved using voltage-sensing relays (see 9.3.1.6).

9.3.1.6 Voltage-Sensing Relays. The most commonly used type is the single-phase induction disk undervoltage time-delay relay. Since a blown control fuse will cause tripping, it is sometimes desirable to use two or three of these relays connected to different phases and wire them so that all must operate before tripping will occur.

Three-phase undervoltage relays are available. Many operate in response to the area of the voltage triangle formed by the three-phase voltages.

In applications requiring a fixed time delay of a few cycles, an instantaneous undervoltage relay is applied in conjunction with a suitable timer (see 9.3.19).

When applying undervoltage protection with time delay, the time-delay setting should be chosen so that time-delay undervoltage tripping does not occur before all external fault-detecting relays have an opportunity to clear all faults from the system. This recognizes that the most frequent causes of low voltage are system faults, and when these are cleared most induction motors can continue normal operation. In the case of induction disk undervoltage relays it is recommended that their trip time versus system short-circuit current be plotted to ensure that they do not trip before the system overcurrent relays. This should be done for the most critical coordination condition, which exists when the system short-circuit capacity is minimum.

Typical time delay at zero voltage is 2 to 5 s.

For motors extremely important to continuity of service, such as some auxiliaries in electric generating plants, the undervoltage relays are used to alarm only.

9.3.2 Phase Unbalance

9.3.2.1 Purpose. The purpose is to prevent motor overheating damage. Motor overheating occurs when the phase voltages are unbalanced, for two reasons:

(1) Increased phase currents flow in order that the motor can continue to deliver the same horsepower as it did with balanced voltages.

(2) Negative-sequence voltage appears and causes abnormal currents to flow in the rotor. Since the motor negative-sequence impedance is approximately the same as the locked rotor impedance, a small negative-sequence voltage produces a much larger negative sequence current.

9.3.2.2 Single Phasing. Overcurrent protection in each phase is required by the NEC [2] for new installations. However, an understanding of the effect of having overload devices in only two phases is useful.

Whenever an older installation involving two overloads is encountered, any rework should include conversion to one properly sized overload relay or dual element fuse per phase.

When single phasing occurs at the same voltage level as the motor operates from, one of the phase conductors to the motor carries zero current. If overcurrent devices in only two phases are relied upon for single-phasing protection, and one of them is in the zero current phase, then there is only one overcurrent device available to sense the current. If an overcurrent device is used in each of the three phases, then failure of one to operate still leaves another to sense this single-phase condition.

Single phasing on the supply voltage side of a delta–delta transformer results in zero current in one of the phase conductors to all motors connected to the other side of the transformer.

When a motor is supplied from a delta–wye or wye–delta transformer, single phasing on the supply voltage side of the transformer results in currents to the motor in the ratio of 1:1:2. In two phases the current will be only slightly greater than prior to single phasing, while it will be approximately doubled in the third phase. It is this situation that requires a properly sized overload relay or dual element fuse in each phase, if the motor does not have suitable phase unbalance protection.

Many motors, especially in the higher horsepower ratings, can be seriously damaged by negative-sequence current heating, even though the stator currents are low enough to go undetected by overload (overcurrent) protection.

Therefore, phase unbalance protection is desirable for all motors where its cost can be justified relative to the cost and importance of the motor.

Phase unbalance protection should be provided in all applications where single phasing is a strong possibility due to the presence of fuses, overhead distribution lines subject to conductor breakage, or disconnect switches which may not close properly on all three phases, etc.

A general recommendation is to apply phase unbalance protection to all motors 1000 hp and above. For motors below 1000 hp, the specific requirements should be investigated.

9.3.2.3 Instantaneous or Time Delay. Unbalanced voltages accompany unbalanced system faults. Therefore it is desirable that phase unbalance protection has sufficient delay to permit the system overcurrent protection to clear external faults without unnecessarily tripping the motor or motors.

Delay is also desirable to avoid the possibility of tripping on motor starting inrush. Therefore unbalance protection having an inherent delay should be chosen, or a suitable additional timer used. If more than 2 or 3 s is used, the motor designer should be consulted.

9.3.2.4 Relays. There are several types of relays available to provide phase unbalance protection, including single phasing. Most of these are described in [6]. Further information about specific relays should be obtained from the various manufacturers. Most of the commonly used relays can be classified as follows:

(1) *Phase Current Balance.* These relays detect unbalance in the currents in the three phases. The induction disk type has an inherent time delay. Occasionally, a timer may be required to obtain additional delay.

(2) *Negative-Sequence Voltage.* This relay operates instantaneously using a negative-sequence voltage filter. A timer, either internal to the relay or external, is required for time delay.

(3) *Negative-Sequence Current.* An induction disk time-delay relay is available for application to generators, but is not intended for motors since it has a relatively high pickup. Instantaneous negative-sequence current relays are available with low pickup to provide good motor protection. A timer is required with these to delay tripping.

9.3.3 Instantaneous Phase Overcurrent

9.3.3.1 Purpose. The purpose is to detect phase short-circuit conditions with no intentional delay. Fast clearing of these faults results in the following:

(1) Limits damage at the fault

(2) Limits the duration of the voltage dip accompanying the fault

(3) Limits the possibility of fault spreading, fire, or explosion damage

9.3.3.2 Instantaneous Overcurrent Relays. These are normally used with phase current transformers. Relays are required in just two phases if a ground overcurrent relay is also provided (see 9.3.6 and 9.3.7); otherwise, one relay per phase is required. However, one relay per phase is often provided, as well as a ground relay. The third phase relay then provides backup protection to the other two phase relays. Requirements of the NEC [2] specify one overcurrent device per phase. These relays are used with the following equipment:

(1) Medium-voltage circuit breaker type motor starters

(2) Medium-voltage contactor-type starters which do not have power fuses. (The motor controller may have limited fault interrupting ability. Instant trip may be used to allow the contactor to open these faults.)

(a) Instant trip relays are set to open if the currents exceed normal starting (usually locked rotor) current.

(b) A second set of relays operate on rate of rise and a higher current intended to maintain the contactor closed during the fault so that the upstream protection fuse (or breaker) must interrupt this fault. Note that the upstream device should have been selected with the capability of interrupting the available fault current at that point in the circuit.

(c) The running overcurrent protection to the motor, either thermal or induction disk type, is usually served through current transformers that saturate on the fault currents, thereby controlling possible damage to the relays used for this protection.

(d) Relays for the above function are further described in Chapter 3.

(3) Low-voltage circuit-breaker-type motor started used with motors whose

importance or horsepower rating justifies the cost of this protection instead of, or in addition to, direct-acting instantaneous overcurrent protection (see 9.3.3).

These relays are available in several forms:

(a) In individual cases, one relay per case

(b) Grouped, two or three relays per case

(c) As additional element(s) in case with induction overcurrent or thermal overcurrent element(s).

9.3.3.3 Direct-Acting Instantaneous Overcurrent Trip Devices. The comments in the first paragraph of 9.3.3.2 apply here. These trip devices are commonly provided on low-voltage circuit-breaker-type motor starters.

9.3.3.4 Fuses. These are used to provide fast phase short-circuit protection on medium- and low-voltage fused-type motor starters. Refer to 5.7.4 for more details on the application of fuses.

9.3.3.5 Instantaneous Settings. Circuit breakers in motor branch circuits having instantaneous overcurrent relays or direct-acting trip should have their trip setting sufficiently high that they do not trip on current occurring:

(1) At initiation of the motor starting inrush

(2) When the motor contributes fault current to an external short-circuit condition

(3) Upon automatic transfer or fast reclosing

For many smaller squirrel-cage induction motors (that are installed on what may be considered a routine basis) it is usual to set the instantaneous trip at 10 or 11 times the motor full-load current.

For the larger squirrel-cage motors (say, above 200 hp) and synchronous motors it is recommended that the value of maximum symmetrical starting inrush be determined by the motor manufacturer and the instantaneous pickup be set 75% above this value. The settings should be even higher for automatic transfer or fast reclosing.

Wound-rotor induction motors usually have reduced inrush due to starting with external rotor resistance. It should be remembered that their contribution to an external fault will exceed their inrush if they are operated with their rotor sliprings short circuited. To avoid unnecessary tripping, their instantaneous pickup protection should be set on the basis of their contribution to an external fault.

Instantaneous settings are frequently determined by trial and error starting of the specific motor. This approach can result in unnecessary tripping at a later date if the maximum asymmetry possible never occurs during the trial and error starts. A minimum of three trial starts is recommended.

9.3.4 Time-Delay Phase Overcurrent

9.3.4.1 Purpose. The purpose is to detect:

(1) Failure to accelerate to rated speed in the normal starting interval

(2) Motor stalled condition

(3) Low-magnitude phase fault conditions.

In many motor protection schemes the overload protection (overcurrent type) is relied upon to provide all three protective functions. Actually, this overload protection is relatively slow, especially the thermal type, since it should not trip on normal motor-accelerating inrush. (Schmidt [7] provides data on the magnitude of currents for internal faults in motors.)

9.3.4.2 Overcurrent Relays and Settings. Thermal or induction disk may be chosen for this function. Thermal types have the advantage of memory of the heat condition of the motor while the induction disk allows relatively precise "tuning" of trip time delay. Both types provide the increase time function as well as sensing sustained overloads. The trip characteristic should be chosen to allow trip before motor damage.

Sometimes another characteristic is chosen, such as extremely inverse, to get faster operation at high currents or to facilitate overcurrent coordination with the supply feeder relays. However, it should be ensured that this characteristic will not trip on normal accelerating inrush.

These same induction disk relays can also be set to provide overload (overcurrent) protection (see 9.3.5), if desired. However, if used for overload protection, they will usually trip much sooner on overload than is necessary to protect the motor. Accordingly they prevent utilizing all the motor inherent thermal overload capability. (This limitation is also true of many thermal overcurrent relays commonly used for overload protection.) The NEC [2] requires one overcurrent device in each phase.

The instantaneous protection (9.3.3) can be provided in the same relay cases with the time-delay elements.

Section 9.3.3.2 applies with regard to the quantity of relays required for phase fault protection. For stall protection, or detection of failure to accelerate normally, only one relay is required on the assumption of balanced phase currents. For important motors two or three relays are recommended as backup to each other, and also to detect failure to accelerate due to single phasing.

The relay settings are normally chosen as follows:

(1) To also provide overload protection, set pickup at 5-25% above the motor continuous service factor rating.

(2) When it is not intended that overload protection be provided, the pickup would be set at 200–350% of motor ratings to avoid tripping for overload conditions.

NOTE: In (1) and (2) the time delay would be set to be a small margin longer than that required to prevent tripping on normal acceleration inrush.

(3) In some applications it might be desired to set the pickup slightly above the starting symmetrical inrush. In this case the relays would not "see" motor inrush, and only fault protection is provided. The time delay should then be very short, just sufficient to not trip on inrush asymmetry. The short-time induction disk relay should be chosen in this application.

9.3.4.3 Instantaneous Relay and Timer. These have been used to provide the protection of (3) in combination with (1) or (2) of 9.3.4.2. It is available as a standard relay in combination with an induction disk. It is also available as an inverse instantaneous curve in some new solid state overcurrent relays.

9.3.4.4 Direct-Acting and Solid-State Trip Devices. Trip devices integral with circuit breakers are often used at 600 V and below. Solid-state trip systems generally have a straight-line long-time characteristic with a negative slope 2 on a log–log plot. Refer to Chapter 6.

9.3.5 Overload (Phase Overcurrent)

9.3.5.1 Purpose. The purpose is to detect sustained stator current in excess of motor continuous rating and trip prior to motor damage.

On motors having winding temperature devices and close operator supervision, this protection is sometimes arranged just to alarm.

Sometimes two sets of overload protection are provided:

(1) One set to alarm only, at relatively low pickup and fast time setting. This would normally be overcurrent relay(s) as in 9.3.5.3

(2) Second set to trip at higher pickup or slower time than the overload alarm relays, or both. Use relays as in 9.3.5.3 or 9.3.5.4

9.3.5.2 Quantity of Relays. In the past it has been common to provide relays in only two phases. However, there is a definite trend to use one relay per phase, or a single relay responsive to individual currents in each of the three phases. This is now required by the NEC [2].

To the limited extent that overloads relays will detect a single-phasing condition, one overload element per phase is desirable in order to respond to single phasing on the supply side of a delta–wye or wye–delta transformer (see 9.3.2.2).

9.3.5.3 Induction Disk Relays and Settings. For larger motors, employing circuit-breaker-type motor starters, long time overcurrent relays may be used.

For this application, they have the following desirable and undesirable features.

Desirable:
(1) Continuously adjustable time delay
(2) Pickup tap settings cover wide range of currents
(3) Quite accurate
(4) Easy and fast to test
(5) Have operation indicator
Undesirable:
(6) Shape of time-delay curve usually results in tripping much faster than necessary, thus preventing all the motor inherent thermal overload capability from being utilized.
(7) Not being thermally operated, they reset quickly after an overload trip, and hence provide no protection against starting again too soon.
(8) Not being thermally operated, they do not "remember" overloads that may come in cycles and progressively overheat the motor.
(9) They are self-resetting and so the hand reset feature is not available without use of a suitable auxiliary relay.

(10) These relays are not significantly affected by change of ambient temperature. This is acceptable and may be considered an advantage for the frequent situation when the motor and relays are in different ambient temperatures. It may be an undesirable characteristic, however, when the motor and relays are in the same ambient and there are significant ambient temperature changes.

9.3.5.4 Thermal Overcurrent Relays. These relays occur in melting alloy and bimetallic types. They are used normally with contactor type motor starters.

(1) Selection of heater element is used rather than adjusting trip value to match motor current.

(2) After selection of heater, the unit is considered nontemperable in melting alloy type.

(3) Bimetallic types may have limited adjustment of trip setting intended to compensate for ambient temperature.

(4) Their thermal *memory* provides desirable protection for cyclic overloading and closely repeated motor starts.

(5) A hand reset feature is available. It is normally *trip free*. That is, manual *override* is not possible.

(6) Some relays are available either as ambient temperature compensated or as noncompensated. Noncompensated is an advantage when the relay and motor are in the same ambient since the relay time changes with temperature in a similar manner as the motor overload capability changes with temperature.

(7) Industry has now standardized on three classes denoting time delay to trip on locked rotor current: class 10 for fast trip, 10 s or less; class 20, for 20 s or less; class 30 for long time trip, 30 s or less; normal choice is class 20, considered compatible with present "T" frame motor design.

9.3.5.5 Dual Element Fuses. These fuses, available from 1/10 through 600 A, are sized at 100–115% of the running current of 1.0 S.F. motors and at 115–125% of the running current of 1.15 S.F. motors.

9.3.6 Instantaneous Ground Overcurrent

9.3.6.1 Purpose. The purpose is to detect ground-fault conditions with no intentional delay.

9.3.6.2 Zero-Sequence Current Transformer or Sensor and Ground Relay—Device 50G. It is recommended that a zero-sequence (window-type) current transformer that has been designed for this function be used to feed the ground relay (Fig 167).

The instantaneous relay is normally set to trip at a primary ground-fault current in the range of 5–20 A.

The following precautions should be observed in applying the relay and zero-sequence current transformer and in installing the cables through the current transformer.

(1) If the cable passes through the current transformer window and terminates in a pothead on the source side of the current transformer, the pothead should be

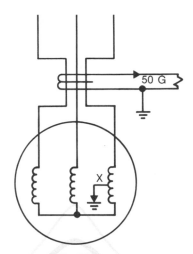

Fig 167
Ground Overcurrent Protection Using Window-Type Current Transformer

mounted on a bracket insulated from ground. Then the pothead should be grounded by passing a ground conductor through the current transformer window and connecting it to the pothead.

(2) If metallic covered cable passes through the current transformer window, the metal covering is kept on the source side of the current transformer insulated from ground. The terminator for the metal covering may be grounded by passing a ground conductor through the current transformer window and then connecting it to the terminator.

(3) Cable shield(s) should be grounded by passing a ground conductor through the current transformer window and then connecting it to the shield(s).

(4) It is important to test the overall current transformer and ground relay scheme by passing current in a test conductor through the current transformer window. Since normally there is no current in the relay, an open circuit in the current transformer secondary or wiring to the relay can only be discovered by this overall test.

9.3.6.3 Residually Connected Current Transformers and Ground Relay. Applications have been made using the residual connection from three current transformers (one per phase) to feed the relay. This arrangement is not ideal since high phase currents (due to motor starting inrush or phase faults) may cause unequal saturation of the current transformers and produce a false residual current resulting in undesired tripping of the ground relay.

Sometimes decreasing the relay pickup setting will overcome the problem because this has the effect of increasing the relay burden. In cases where this is effective, decreasing the relay pickup setting actually increases the relay pickup in terms of primary current since the higher relay burden requires more current transformer exciting current than is gained with the lower relay setting.

Some improvement is obtained by inserting an impedance in the residual connection in series with the relay. Usually, the best solution is to use device 5IN (9.3.7).

9.3.6.4 Combination-Type Starters. When an instantaneous ground relay is applied to this type of starter on a solidly grounded system, it should be remembered that the contactor might not have the capability to clear the maximum ground-fault capacity. Therefore, it is necessary to ensure that the fuses or circuit breaker instantaneous trip devices will clear a ground fault before the contactor is damaged in trying to clear it. In some cases it may be necessary to delay tripping from the ground relay or use a time-delay type ground relay (9.3.7).

9.3.6.5 When Surge Arresters Are Installed at Motor Terminals. There is a possibility that a surge discharge through an arrester would cause the ground relay to trip unnecessarily. To avoid this possibility it has been usual to recommend that an overcurrent time-delay relay be used in this situation. If the instantaneous relay has sufficient inertia, however, it may override a surge discharge without tripping.

9.3.7 Time-Delay Ground Overcurrent

9.3.7.1 Purpose. The purpose is to detect ground-fault conditions. Early applications of ground protection used current transformers and relays. However, both instantaneous and time-delay ground-fault protection is now available with solid-state tripping systems on low-voltage (up to 600 V) circuit breakers (see Chapter 6).

9.3.7.2 Zero-Sequence Current Transformer and Time-Delay Ground Relay. When the zero-sequence current transformer is used for motor ground protection, it is usual to use an instantaneous overcurrent ground relay. When a time-delay relay is used (9.3.6.5), it is usually a short time, or an extremely inverse, induction disk relay set at 0.5 A tap and 1.0 time dial. The comments in 9.3.6.2 apply here.

9.3.7.3 Residually Connected Current Transformers and Ground Relay. The relay is usually a short time, or an extremely inverse, induction disk type set at 0.5 A tap and 1.0 time dial. To get lower pickup, with high ratio current transformers, a 0.2 A relay is sometimes used.

If one of the current transformer secondary phase conductors becomes open circuited, the other two current transformers feed phase current through the residual ground relay causing it to trip.

9.3.7.4 Choice of Resistor for System Grounding. The object of resistance grounding is normally to limit the motor damage caused by a ground fault. (In mine distribution systems the object is to limit equipment frame to earth voltages for safety reasons.) However, the ground-fault current should not be limited to the extent that very much of the neutral end of motor wye windings goes unprotected. In the past, protection with within 5–10% of the neutral has often been considered adequate. Fawcett [4] recommends that the ground overcurrent relay should have at least 1.5 multiples of pickup for a ground fault one turn away from the neutral.

Noting the foregoing, it is recommended that the ground resistor rating and the ground protection be chosen together after having determined the winding arrangements of the various motors to be served. On this basis, the ground resistor chosen will normally limit the ground-fault current within the range of 100–2000 amps. A 10 second time rating is usually chosen for the resistor.

Note also that to avoid excessive transient overvoltages the resistor should be chosen so that the following zero-sequence impedance ratio is achieved:

$$\frac{R_0}{X_0}$$

and should be equal to or greater than 2.

9.3.8 Phase Current Differential

9.3.8.1 Purpose. The purpose is to quickly detect fault conditions.

9.3.8.2 Conventional Phase Differential. This scheme uses six identical current transformers (one pair for each phase) and three relays (one per phase). Since the current transformers carry load current, they must have primary current ratings chosen accordingly. (See Fig 168.)

The currents from each pair of current transformers are subtracted and their difference is fed to the relay of the associated phase. For normal (nonfault) conditions the two currents in each pair are equal and their difference is, therefore, zero. For a fault located between the two current transformers of any pair, the currents from the two current transformers will differ. This difference will operate the relay of the associated phase.

While sometimes applied to delta-connected motors, this scheme is usually used with wye-connected motors. (Wye-connected motors are much more common than delta-connected ones in the larger horsepower ratings.) With the wye-connected motor three of the current transformers are normally located at the starter (or motor switchgear) and the other three in the three phases at the motor

**Fig 168
Conventional Phase Differential Protection
Using Three Percentage Differential Relays (One Shown)**

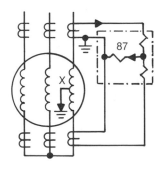

winding neutral. Occasionally, the three neutral phases are cabled back to the starter; for example, if there is a neutral starting reactor and it is located at the starter. In this case the neutral current transformers would also be at the starter.

There are three types of relays generally applicable for conventional phase differential protection; three identical current transformers, one recommended for use with these relays.

(1) High-speed (instantaneous) differential. While this protection is the most expensive, it is not much more expensive when the total cost of relays, wiring, current transformers, and current transformer mounting space and installation is considered. Therefore, this is the type now recommended. There are no settings to be chosen for these relays.

(2) Slow-speed induction disk percent differential. In the past this type has usually been used. There are no settings to be chosen for these relays.

(3) Standard induction disk overcurrent relays differentially connected. These relays have not been used very frequently. However, they are less expensive than those in (2) and quite satisfactory provided they are used with identical current transformers. They would normally be set between 0.5 and 2.5 A tap and 1.0 time dial. However, the high-speed type relays are the best choice.

9.3.8.3 Self-Balancing Differential Using Zero-Sequence Current Transformers. Three window-type current transformers are used. These are normally installed at the motor. One current transformer per phase is used with the motor line and neutral leads of one phase passed through it such that the two currents normally cancel each other. A winding phase-to-phase or ground fault will result in an output from the current transformer of the associated phase and operate the associated relay (see Fig 169).

The current transformers and relays would normally be the same as those used for zero-sequence instantaneous ground overcurrent protection (9.3.6.2) with the

Fig 169
Self-Balancing Differential Protection (One Relay Shown)

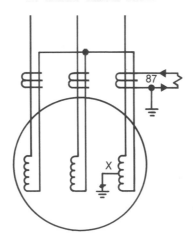

relay set between 0.25 and 1.0 A pickup. Therefore this differential scheme will usually have a lower primary pickup in amperes than the conventional differential scheme (since the current transformer ratio is usually greater with the conventional scheme). This differential scheme has a slight advantage over that of 9.3.6.2 in detecting ground fault. For motors installed on grounded systems this is quite significant, since most faults will be ground faults. If the fault does not involve ground, then the stator iron is not being damaged. The usual objective of motor-fault protection is to remove the fault before the stator iron is significantly damaged.

With the current transformers located at the motor, this scheme does not detect a fault in the cables to the motor. A fault in these cables would normally be detected by the overcurrent protection of 9.3.3–9.3.7. In the case of large motors it is often a problem to coordinate the supply phase overcurrent protection. The presence of motor differential protection is sometimes considered to make the above coordination less essential. In this regard the conventional differential is better than the zero-sequence differential since the motor cables are also included in the differential protection zone, and hence coordination between the motor differential and supply phase overcurrent relays is complete.

As with zero-sequence ground overcurrent protection, it is important during initial startup to test the overall current transformer and relay combinations by passing current in a test conductor through the window of each current transformer. Since normally the relays do not carry current, an open circuit in a current transformer secondary or wiring to a relay can only be discovered by this overall testing.

9.3.8.4 Contactor-Type Starters with Power Fuses. The comments of 9.3.6.4 also apply when using differential protection with these starters.

9.3.8.5 When to Apply Differential Protection. The following general recommendations are made:

(1) With all motors 1000–2000 hp and above used on ungrounded systems.

(2) With all motors 1000–2000 hp and above used on grounded systems where the ground protection applied is not considered sufficient without differential protection to protect against phase-to-phase faults.

(3) For large horsepower ratings (2500–5000 hp) the cost of differential protection compared to the cost of the motor would generally justify the use of this relay. However, differential protection is frequently justified for much smaller motors, especially at voltages above 2400 V.

9.3.9 Split Winding Current Unbalance

9.3.9.1 Purpose. The purpose is to quickly detect low-magnitude fault conditions. This protection also serves as backup to instantaneous phase overcurrent (9.3.3) and ground overcurrent protection (9.3.6 and 9.3.7).

This protection is normally only applied to motors having two (or three) winding paths in parallel per phase (see Fig 170).

9.3.9.2 Arrangement of Current Transformers and Relays. The usual application is with a motor having two winding paths in parallel per phase. The

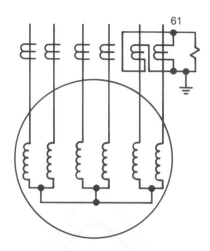

Fig 170
Split-Phase Motor Overcurrent Protection Can Be Used with
Two Paths per Phase (One Relay Shown)

six line leads (two per phase) of the motor are brought out and one current transformer is connected in each of the six leads. The primary current rating of the current transformers should be chosen to carry full-load current.

The current transformers may be installed at the motor. It is often convenient, however, to use six cables to connect the motor to its starter (or switchgear), and in this case the current transformers can be located in the starter.

The currents from each pair of current transformers, associated with the same phase, are subtracted and their difference is fed to a short-time induction disk overcurrent relay. Three of these relays are required (one per phase), and each is set at 1.0 time dial and between 0.5 and 2.5 A. The relay should be set above the maximum current unbalance that can occur between the two parallel windings for any motor loading condition.

9.3.9.3 Evaluation of This Protection

(1) Total cost would be somewhat less than conventional phase differential (9.3.8.2) and more than self-balancing differential (9.3.8.3).

(2) The primary pickup current for this protection would be about half that of conventional phase differential (since both schemes require the current transformer primaries to be rated to carry normal load currents). Self-balancing differential would usually have a lower primary pickup in amperes.

(3) This protection has a slight time delay compared to the phase differential schemes.

(4) When the current transformers are located in the motor starter, split winding protection has the same advantage over self-balancing differential as does conventional phase differential; namely, it detects a fault in the motor cables and may facilitate coordination with the supply feeder overcurrent relays. (This

is discussed in Section 9.3.8.3.)

(5) The salient feature that this protection provides, and no other motor protection has, is the ability to sense short-circuited winding turns. The number of turns that must be short-circuited in order that detection will occur depends upon the motor winding arrangement as well as the relay pickup and current transformer ratio. An analysis of the specific motor winding would be required to determine how worthwhile this feature is. It may be that the short-circuited turns would cause a ground fault and be detected by the self-balancing differential scheme before this split winding protection would sense the short-circuited turns condition.

(6) This protection could be applied to a motor with four winding paths in parallel per phase by grouping them as two pairs and then treating them as if there were only two paths in parallel (that is, six current transformers and three relays are used).

(7) A split differential scheme is often effectively used where one current transformer is in one of the parallel paths and the other current transformer sees the total phase current.

9.3.9.4 When to Apply Split Winding Protection. This protection has not been used in the past with very many motors. It is probably desirable on important motors that have two or four winding paths in parallel per phase and are rated above 5000–10 000 hp.

9.3.10 Stator Winding Overtemperature

9.3.10.1 Purpose. The purpose is to detect excessive stator winding temperature prior to the occurrence of motor damage. This protection is often arranged just to alarm on motors operated with competent supervision. Sometimes two temperature settings are used, the lower setting for alarm, the higher setting to trip.

9.3.10.2 Resistance Temperature Detectors. Six of these detectors are commonly provided (when specified) in motors rated 1500 hp and above. They are installed in the winding slots when the motor is being wound. The six are spaced around the circumference of the motor core to monitor all phases.

Commonly used types are rated either 10 or 120 Ω, at a specified temperature. Their resistance increases with temperature and wheatstone bridge devices are used to provide temperature indication or contact operation, or both.

For safety reasons the detectors should be grounded. This places a ground on the wheatstone bridge control and, therefore, makes it undesirable to directly operate the wheatstone bridge control from a switchgear dc battery (since these dc control schemes should normally operate ungrounded in order to achieve maximum reliability). However, use of ac control has the following limitations:

(1) Loss of ac due to a fuse blowing, etc, removes the protection, unless (2) applies.

(2) If the null point of the wheatstone bridge, for loss of ac power, is near the trip contact setting, then loss of ac or a voltage dip may cause a trip of the motor.

The foregoing should be evaluated in applying this protection.

An open circuit in the detector leads appears as an infinite resistance and will cause a false trip since the output is the same as a very high-temperature condition.

The following arrangements are frequently used:

(1) Determine which detector normally runs hottest and permanently connect a trip relay to it. Use one temperature indicator and a manual switch to monitor the other five detectors.

(2) Use manual switch and combination indicator and alarm relay. *Precaution:* An open circuit in the switch contact will cause a false trip.

(3) Use manual switch and indicator only.

(4) Use one, two, or three (one per phase) alarm relays and one, two, or three (one per phase) trip relays set at a higher temperature.

9.3.10.3 Thermocouples. Thermocouples are used to indicate temperatures and for alarm and trip functions, in a similar manner as resistance temperature detectors.

An open circuit in the thermocouple leads will not cause a trip since the output is the same as a low-temperature condition.

The output from thermocouples is compatible with central control room temperature monitoring and data logging schemes.

9.3.10.4 Thermistors. These are used to operate relays for alarm or trip functions, or both. They are not used to provide temperature indication. However, they are available combined with thermocouples—the thermocouples to provide indication and the thermistor to operate a relay. Relays to operate from thermistors are relatively inexpensive.

There are two types of thermistors:

(1) *Positive-Temperature-Coefficient Type:* With this type its resistance increases with temperature. An open circuit in the thermistor would appear as a high-temperature condition and would operate the relay. This is the failsafe type of arrangement.

(2) *Negative-Temperature-Coefficient Type:* Resistance decreases as temperature increases. An open circuit in the thermistor appears as a low-temperature condition and does not cause relay operation.

9.3.10.5 Thermostats and Temperature Bulbs. These devices are used on some motors. They have not been commonly used on the larger motors and will not be discussed here.

9.3.10.6 When to Apply Stator Winding Temperature Protection. It is commonly applied to all motors rated 1500 hp and above. In the following situations it is particularly desirable:

(1) Motors in high ambient temperatures or at high altitudes

(2) Motors whose ventilation systems tend to become dirty, losing cooling effectiveness

(3) Motors subject to periodic overloading due to load characteristics of the drive or process

(4) Motors likely to be subjected to steady-state overloading in order to increase production

(5) Motors for which continuity of service is very important

9.3.11 Rotor Overtemperature

9.3.11.1 Synchronous Motors. Rotor winding overtemperature protection is available for brush-type synchronous motors, but is not normally used. One well-known approach is to use a Kelvin bridge type strip chart recorder with contacts adjustable to the temperature settings desired. The Kelvin bridge measures the field resistance in order to determine the field winding temperature, using field voltage and field current (from a shunt) as inputs.

9.3.11.2 Wound Rotor Induction-Motor Starting Resistors. Some form of temperature protection should be applied for these resistors on drives having severe starting requirements such as long acceleration intervals or frequent starting. Resistance temperature detectors as well as other types of temperature sensors have been used in proximity to the resistors.

9.3.12 Synchronous Motor Protection

9.3.12.1 Damper Winding Protection. When a synchronous motor is starting, high currents are induced in its rotor damper winding. If the motor takes longer to accelerate than it has been designed to be suitable for, the damper winding may overheat and be damaged.

Several different relay-type and static-type protective schemes are available. None of these schemes directly senses damper winding temperature. Instead, they try to simulate what it must be by evaulating two or more of the following quantities:

(1) Magnitude of induced field current that flows through the field discharge resistor. This is a measure of the relative magnitude of induced damper winding current.

(2) Frequency of induced field current that flows through the discharge resistor. This is a measure of rotor speed and provides an indicator, therefore, of the increase in damper winding thermal capability resulting from the ventilation effect and the decrease of induced current.

(3) Time interval after starting.

9.3.12.2 Field Current Failure Protection. Field current may drop to zero or a low value when a synchronous motor is operating for several reasons.

(1) Tripping of the remote exciter, either motor-generator set-type or static-type. (Note that control for these should be arranged so that it will not drop out on an ac voltage dip.)

(2) Burnout of the field contactor coil. (Note that the control should be arranged such that the field contactor will not drop out on an ac voltage dip.)

(3) Accidental tripping of the field. (Note that field circuit overcurrent protec-

tion is usually omitted from field breakers and contactors in order to avoid unnecessary tripping. The field circuit is usually ungrounded and should have ground detection lights or relay applied to it to detect the first ground fault before a short circuit occurs; (see 9.3.15.1.)

(4) High resistance contact or open circuit between slip ring and brushes due to excessive wear or misalignment.

Reduced field current conditions should be detected because:

(1) Heavily loaded motors will pull out of step and stall
(2) Lightly loaded motors are not capable of accepting load when required to
(3) Intermediate loaded motors, which do not pull out of step, are likely to do so on an ac voltage dip that they otherwise might ride through
(4) The excitation drawn from the power system by large motors may cause a serious voltage drop and endanger continuity of service of other motors

A common approach is to use an instantaneous dc undercurrent relay to monitor field current. This application should be investigated to ensure that there are no transient conditions that would reduce the field current and cause unnecessary tripping of the instantaneous relay. A timer could be used to obtain a delay of one or more seconds, or the relay could be connected to alarm only where competent supervising personnel are available.

Field current failure protection is also obtained by the generator-type loss of excitation relay that operates from potential transformers and current transformers monitoring motor stator voltages and current. This has been done on some large motors (4000 hp and above). This relay may also provide pullout protection (9.3.12.4)

9.3.12.3 Excitation Voltage Availability. A simple voltage relay is used as a permissive start to ensure that voltage is available from the remote exciter. This avoids starting and then having to trip because excitation was not available. Loss of excitation voltage is not normally used as a trip; the field current failure protection is used for this function.

9.3.12.4 Pullout Protection. Pulling out of step is usually detected by one of the following relay schemes:

(1) An instantaneous relay connected in the secondary of a transformer whose primary carries the direct field current. The normal direct field current is not transformed. When the motor pulls out of step, alternating currents are induced in the field circuit and these are transformed and operate the pullout relay. This relay, while inexpensive, is sometimes subject to false tripping on ac transients accompanying external system fault conditions and also ac transients caused by pulsations in reciprocating compressor drive applications. Device 95 has sometimes been used to designate this relay.

(2) A power factor type relay responding to motor stator voltage and current obtained from potential and current transformers.

(3) The generator-type loss of excitation relay has been used (see 9.3.12.2)

9.3.12.5 Incomplete Starting Sequence. This protection is normally a timer that blocks tripping of the field current failure protection (9.3.12.2) and the pullout protection (9.3.12.4) during the normal starting interval. The timer is started by an auxiliary contact on the motor starter and times for a preset interval that has been determined during test starting to be slightly greater than the normal interval from start to reaching full field current. The timer puts the field current failure and pullout protection in service at the end of its timing interval.

This timer is often the de-energize-to-time type so that it is failsafe with regard to applying the field current failure and pullout protection.

9.3.12.6 Operation Indicator for Foregoing Protections. Many types of the foregoing protective devices do not have operation indicators. It is suggested that separate operation indicators be used with these protective devices.

9.3.13 Induction Motor Incomplete Starting Sequence Protection. It is recommended that wound-rotor induction motors and reduced voltage start motors have a timer applied to protect against failure to reach normal running conditions within the normal starting time. The timer would be started by an auxiliary contact on the motor starter and would time for a preset interval that has been determined during test starting to be slightly greater than the normal starting interval. The timer trip contact would be blocked by an auxiliary contact of the final device that operates to complete the starting sequence. This final device would be the final secondary contactor in the case of a wound-rotor motor, or it would be the device which applies full voltage to the motor stator.

This timer is often the de-energized-to-time type so that it is failsafe from a protection point of view.

Incomplete sequence protection should also be applied to part winding and wye–delta motor starting control, as well as pony motor and other sequential start arrangements.

9.3.14 Protection Against Too Frequent Starting. The following protections are available:

9.3.14.1 A timer, started by an auxiliary contact on the motor starter, with contact arranged to block a second start until the preset timing interval has elapsed.

9.3.14.2 Stator thermal overcurrent relays (9.3.5.4) provide some protection. The degree of protection depends upon:

(1) The normal duration and magnitude of motor inrush
(2) The relay operating time at motor inrush and the cool-down time of the relay
(3) The thermal type damper winding protection on synchronous motors (9.3.12.1)
(4) Rotor overtemperature protection (9.3.11)

Note that large motors are often provided with nameplates giving their permissible frequency of starting.

9.3.15 Rotor Winding Protection

9.3.15.1 Synchronous Motors. The field and field supply should not be grounded. While the first ground does not cause a fault, a second ground probably will. Therefore, it is important to detect the first ground. The following methods are used.

(1) Connect two lights in series between field positive and negative with the midpoint between the lamps connected to ground. A ground condition will show by unequal brilliancy of the two lamps.

(2) Connect two resistors in series between field positive and negative with the midpoint between the resistors connected through a suitable instantaneous relay to ground. The maximum resistance to ground which can be detected depends upon the relay sensitivity and the resistance in the two resistors.

This scheme will not detect a ground fault at midpoint in the field winding. If a varistor is used instead of one of the resistors, then the point in the field winding at which a ground fault cannot be detected changes with the magnitude of the excitation voltage. This approach is used to overcome the limitation of not being able to detect a field midpoint ground fault.

(3) Some schemes apply to low ac voltage between the field circuit and ground and monitor the ac flow to determine when a field circuit ground fault occurs. Before using one of these schemes, it should be determined that a damaging ac will not flow through the field capacitance to the rotor iron and then through the bearings to ground, thus causing damage to the bearings.

(4) If a portion of the field becomes faulted, damaging vibration may result. Vibration detectors should be considered (9.3.21.3)

9.3.15.2 Wound Rotor Induction Motors. Section 9.3.15.1 applies here, except that the rotor winding is three-phase ac instead of a dc field (see Fig 171). Yuen *et al.* [8] give experience confirming the desirability and effectiveness of this protection.

Wound-rotor motor damage can occur due to high resonant torques resulting from operation with unbalanced impedances in the external rotor circuit on speed-controlled motors. Protection to detect this is available although it has seldom been used.

9.3.16 Lightning and Surge Protection

9.3.16.1 Types of Protection. Surge arresters are often used, one per phase connected between phase and ground, to limit the voltage to ground impressed upon the motor stator winding due to lightning surges and switching surges.

Surge capacitors are used, connected between each phase and ground, to decrease the slope of the wavefront of lightning surge voltages and switching surge voltages. As the surge voltage wavefront travels through the motor winding, the surge voltage between adjacent turns and adjacent coils of the same phase will be lower for a wavefront having a decreased slope. (A less steep wavefront is another way of designating a wavefront having a decreased slope.)

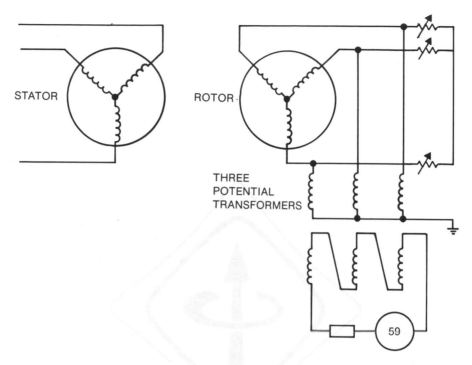

Fig 171
Rotor Ground Protection of Wound-Rotor Motor

9.3.16.2 Locations of Surge Protection. The surge arresters and surge capacitors should be connected within 3 circuit ft of the terminals of each motor. The supply circuit should connect to the surge equipment first and then go to the motor terminals. The two important points are:

(1) The surge protection should be as close to the motor terminals (in circuit feet) as feasible

(2) The supply circuit should connect directly to the surge equipment first and then go to the motor

It is becoming more common to specify the surge protection to be supplied in a terminal box on the motor or in a terminal box adjacent to the motor.

If the motors are within 100 feet of their starters or the supply bus, it is sometimes compromised for cost reasons to locate the surge protection in the starters or supply bus switchgear. In the latter case, one set of surge protection is used for all the motors within 100 ft of the bus. Alternatively, this approach may be used for the smaller motors, and individual surge protection installed at each larger motor.

When surge protection is supplied in a motor terminal box, it is necessary to disconnect it before high-voltage testing the motor. This is a recognized inconvenience of this arrangement.

9.3.16.3 When to Apply Surge Protection. The following general guides are given:

(1) Apply to each medium-voltage motor rated above about 500 hp.
(2) Apply to each motor rated above 200 hp that is connected to open overhead lines at the same voltage level as the motor.
(3) When there is a transformer connecting the motor(s) to open overhead lines, surge protection is still required sometimes to protect against lightning surges. Techniques are available to analyze this situation. If in doubt it is best to provide surge protection. Refer to 9.3.16.2 for surge protection on the supply bus for motors located within 100 ft of the bus.

9.3.17 Protection Against Overexcitation from Shunt Capacitance

9.3.17.1 Nature of Problem. When the supply voltage is switched off, an induction motor initially continues to rotate and retain its internal voltage. If a capacitor bank is left connected to the motor, or if a long distribution line having significant shunt capacitance is left connected to the motor, the possiblity of overexcitation exists. Overexcitation results when the voltage versus current curves of the shunt capacitance and the motor no-load excitation characteristic intersect at a voltage above the rated motor voltage.

The maximum voltage that can occur is the maximum voltage on the motor no-load excitation characteristic (sometimes called magnetization or saturation characteristic). This voltage, which decays with motor speed, can be damaging to a motor (see Fig 172 as an example).

Damaging inrush can occur if automatic reclosing or transfer takes place on a motor which has a significant internal voltage due to overexcitation.

9.3.17.2 Protection. It is assumed here that sufficient knowledge of the system and motor exists to determine whether protection against overexcitation is required. The simplest protection would be instantaneous overvoltage relays.

An alternative is to use a high-speed underfrequency relay. However, this may not be fast enough on drives having high inertia or light loading.

The underfrequency relay is not suitable for drives whose frequency may not decrease following loss of the supply overcurrent protective disconnecting device. Examples of these are:

(1) Mine hoist with overhauling load characteristic at time of loss of supply overcurrent device
(2) Motor operating as induction generator on shaft with process gas expander

With these applictions a loss of power relay could be used [see 9.3.18.2, item (6)].

9.3.18 Protection Against Automatic Reclosing or Automatic Transfer

9.3.18.1 Nature of Problem. When the supply voltage is switched off, motors initially continue to rotate and retain an internal voltage. This voltage decays with motor speed and internal flux. If system voltage is restored out of phase with a significant motor internal voltage, high inrush can occur and damage the motor windings or produce torques damaging to the shaft, foundation

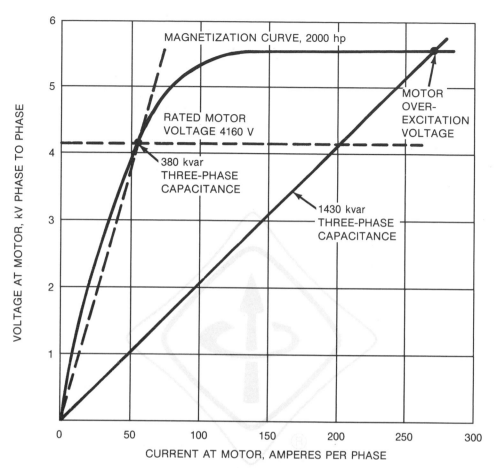

Fig 172
Excess Shunt Capacitance from Utility Line Is Likely
to Overexcite a Large High-Speed Motor
(Voltage May Shoot up to 170 Percent)

or drive coupling, or gears.

ANSI/IEEE C37.96-1976 [1] discusses considerations as to the probability of damage occurring for various motor and system parameters.

9.3.18.2 Protection. The following protection is used.

(1) Delay restoration of system voltage, using timer, for a preset interval known from actual tests to be sufficient for adequate decay of the motor internal voltage.

(2) Delay restoration of system voltage until the internal voltage fed back from the motor(s) has dropped to a low enough value. This value is commonly considered to be 25% of rated voltage. Note that the frequency also decreases as the voltage decays (due to the motor(s) slowing down). The undervoltage relay and its setting should be chosen accordingly. Sometimes a full-wave rectifier and dc relay are chosen so that the relay dropout will be independent of frequency. If an ac frequency-sensitive relay is used, it should be set, based on motor and

system tests, to actually drop out at 25% voltage (and the frequency that will exist when 25% voltage is reached).

(3) Use of a high-speed underfrequency relay to detect the supply outage and trip the motor before supply voltage is restored. The limitations of this relay given in 9.3.17.2 also apply here. A further limitation exists if the motor operates at the same voltage level as the supply lines on which faults may occur followed by an automatic reclosing or transfer operation. The problem is that the underfrequency relay requires some voltage in order to have operating torque. If there is no impedance (such as a transformer) between the motor and the system fault location, then there may not be sufficient voltage to permit the underfrequency relay to operate.

(4) Single-phase or three-phase undervoltage relays are used as follows:

(a) One relay with a sufficiently fast time setting can be connected to the same potential transformer as the underfrequency relay [see (3)] and take care of the fault condition that results in insufficient voltage to operate the underfrequency relay.

(b) One, two, or three relays (each relay connected to a different phase) can be used to detect the supply outage and trip the motor(s) when sufficient time delay exists before the supply will be restored.

(5) Loss of power relay. This relay should be sufficiently fast and sensitive. A high-speed three-phase relay has been used frequently. Being three-phase, it is more difficult to apply and connect properly than the other schemes listed. Since this is a loss of power relay, it should be blocked at start-up until sufficient load is obtained on the circuit or motor with which it is applied.

(6) Reference is sometimes made to using a reverse power relay to detect a separation between motors and their source. While this approach is suitable in some circumstances, its limitations should be recognized:

(a) During the time the fault is one and the source still connected to the motors, net power flow will continue into the motors for low-level faults. This is not the case for three-phase bolted faults, however, for these faults there is very low impedance into which reverse power flows.

(b) In view of the foregoing, tripping by reverse power can usually only be relied upon if there is a definite load remaining to absorb power from high-inertia motor drives after the source fault detecting relays isolate the source from the motors.

(c) Reverse power relays responsive to reactive power (vars) instead of real power (watts) usually do not provide a suitable means of isolating motors prior to automatic reclosing or automatic transfer operations.

Therefore, the loss of power relay application is generally more suitable than the reverse power relay application.

9.3.19 Protection Against Excessive Shaft Torques Developed During Phase-to-Phase or Three-Phase Short Circuits Near Synchronous Motors. A phase-to-phase or three-phase short circuit at or near the motor terminals produces very high shaft torques that may be damaging to the motor or driven machine. Computer programs have been developed for calculation of these torques. Refer to ANSI/IEEE C37.96-1976 [1] for information on this potential problem.

To minimize exposure to damaging torques, a three-phase high-speed voltage relay can be applied to detect severe phase-to-phase or three-phase short-circuit conditions for which the motor(s) should be tripped. This relay is often the type whose torque is proportional to the area of the three-phase voltage triangle. When a severe reduction in phase-to-phase or three-phase voltage occurs, a spring in the relay will overcome the torque produced by the voltage and cause tripping. An additional delay before tripping of 1–8 cycles may be satisfactory from a protection point of view, and desirable to avoid unnecessary shutdowns. This can be achieved using a suitable form of timer. Selection of protection and settings for this application should be done in conjunction with the motor and driven machine suppliers as well as the protection supplier.

9.3.20 Protection Against Excessive Shaft Torques Developed During Transfer of Motors Between Out-of-Phase Sources. A rapid transfer of large motors from one energized power system to another energized power system could cause very high motor inrush currents and severe mechanical shock to the motor. The abnormal inrush currents may be high enough to trip circuit breakers or blow fuses, and could damage motor system components. The mechanical jolt could physically damage the motor, shaft, and couplings.

These effects can occur in emergency or standby power systems when a motor is deenergized and then rapidly reconnected to another source of power that is out of phase with the motor's regenerated voltage. Motors above 50 hp driving high-inertia loads (for example, centrifugal pumps and fans) may require special consideration.

The problem can be eliminated if the motor circuits can be deenergized long enough to permit the residual voltage to decay before power is again applied to the motor. This can be done in two ways. In one method, auxiliary contacts or a relay on the transfer switch can open the motor holding coil circuits, while the transfer is delayed about 3 s. Such contacts are available. This is completely effective, but requires interwiring between the transfer switch and the motor starters. Another method utilizes a transfer switch with a timed center-off position. The switch opens, goes to the neutral or off position, is timed to stay there about 3–10 s, and then completes the transfer. This eliminates any interwiring to the motors. The required time delay should be carefully set and may vary as the system conditions change. A third position (neutral) creates the possibility of the transfer switch remaining indefinitely in a neutral position in the event of a control circuit or contactor malfunction.

Another solution is to momentarily parallel the two power sources on transfer, connecting both sources together and then dropping one out. This is completely effective, since power to the motors is never interrupted. However, it is very costly, and if one source is utility power, there may be a problem in that some utilities will not permit paralleling another source with their system.

Another solution to the problem is known as *in-phase* transfer. An accessory on the transfer switch, known as an *in-phase monitor,* measures the phase angle difference between the two power sources. An on-site engine generator set is not locked in with the utility source, and the two sources continually go in and out of phase. At the proper *window* or acceptable phase angle difference between the sources, the monitor initiates transfer. The design allows for the operating time of the transfer switch so that the oncoming source is connected to the motors in phase or at a phase difference small enough to eliminate excessive inrush currents and mechanical shock. No special field adjustments or interwiring to the motors are required. For typical transfer switches with transfer times of 10 cycles (166 ms) or less and for frequency differences between the sources of up to 2 Hz, the in-phase monitor will provide a safe transfer of motors.

9.3.21 Protection Against Failure to Rotate or Reverse Rotation

9.3.21.1 Failure to Rotate. This condition will occur if the supply is single phased, or if the motor or driven machine is jammed in some way. The following protection is available:

(1) Section 9.3.2.4 discusses types or relays to detect single phasing.

(2) The direct means to detect failure to rotate is to use a shaft speed sensor and timer to check whether a preset speed has been reached by the end of a short preset time interval after energizing the motor. This protection is desirable for induction and brushless synchronous motors that have a permissible locked rotor time less than normal acceleration time.

(3) For induction and brushless synchronous motors having a permissible locked rotor time greater than normal acceleration time, it is normal to rely upon the time-delay phase overcurrent relays (9.3.4 or 9.3.5).

(4) For brush-type synchronous motors having a permissible locked rotor time less than normal acceleration time, it is possible to use a frequency-sensitive relay connected to the field discharge resistor and a timer to achieve this protection. This is because the frequency of the induced field current flowing through the discharge resistor indicates the motor speed. An induction disk frequency-sensitive adjustable time-delay voltage relay is available to provide this protection.

(5) For brush-type synchronous motors having a permissible locked rotor time greater than normal acceleration time, it is normal to rely upon the damper winding protection (9.3.12.1) and incomplete starting sequence protection (9.3.12.5).

9.3.21.2 Reverse Rotation. A directional speed switch mounted on the shaft and a timer can be used to detect starting with reverse rotation.

Some motor drives are equipped with a ratchet arrangement to prevent reverse rotation.

A reversal in the phase rotation can be detected by a reverse phase voltage relay (9.3.2.4) if the reversal occurs in the system on the supply side of the relay. This relay cannot detect a reversal that occurs between the motor and the point at which the relay is connected to the system.

9.3.22 Mechanical and Other Protection

9.3.22.1 Bearing and Lubricating Systems. Various types of temperature sensors are used on sleeve bearings to detect overheating: resistance temperature detectors, thermocouples, thermistors, thermostats, temperature bulbs. Excessive bearing temperature may not be detected soon enough to prevent bearing damage. However, if the motor is tripped before complete bearing failure occurs, more serious mechanical damage to the rotor, and hence to the stator, may be prevented. Accordingly, for maximum effectiveness it is recommended that:

(1) A fast-responding type of temperature sensor be used
(2) The temperature sensor be located in the bearing metal where it is close to the source of overheating
(3) The temperature sensor be used for tripping instead of alarm; there may be situations where both alarm and trip sensors can be used, the former having a lower temperature setting

Alarm and trip devices should be provided with bearing lubricating systems to monitor

(1) Lubricating oil temperature, preferably from each bearing, and also to the bearings
(2) Cooling water temperature, both temperature in and out
(3) Lubricating oil flow and cooling water flow
(4) In lieu of the flow monitoring recommended in (3), a suitable arrangement of pressure switches is often used. However, flow monitoring is strongly recommended for important or high-speed machines.

It is generally considered that temperature sensors cannot detect impending failure of ball or roller bearings soon enough to be effective. Vibration detectors should be considered (9.3.22.3).

Protection to detect currents that may cause bearing damage should be considered for motors having insulated bearing pedestals.

9.3.22.2 Ventilation and Cooling Systems. Alarm and trip devices should be considered

(1) To detect high-pressure drop across filters in motor ventilation systems
(2) To detect loss of air flow from external blower(s) in motor ventilation systems
(3) In lieu of the air flow monitoring in (2), a suitable arrangement of pressure switches is often used; flow monitoring is preferable, however
(4) With water-cooled motors, water temperature, flow, or pressure monitoring
(5) With inert gas-cooled motors, suitable pressure and temperature sensors

9.3.22.3 Vibration Detectors. Experience indicates, especially on the higher speed drives, that serious damage can be prevented using vibration detectors for tripping, or alarm and tripping.

9.3.22.4 Liquid Detectors. On large machines liquid detectors are sometimes provided to detect liquid (usually water) inside the stator frame. This can occur because of one of the following.

(1) Excessive condensation. Use of motor space heaters, or low-voltage winding heating with automatic control should normally avoid this problem.

(2) Exposure to hosing down cleaning operations, flooding, or outdoor weather conditions. A suitable choice of motor enclosure should normally avoid this problem.

(3) Leak from motor cooler. Double-tube coolers are sometimes used to help avoid this problem.

9.3.22.5 Fire Detection and Protection. The following items should be considered:

(1) Installation of suitable fire detectors to alert operators to use suitable hand-type fire extinguishers.

(2) Installation of suitable fire detectors and automatic system to apply carbon dioxide into motor.

(3) Some old large motors have internal piping to apply water for fire extinguishing. Possible false release of the water is a serious disadvantage.

(4) Use of synthetic lubricating oil that does not burn is worth considering for drives having large lubricating systems and reservoirs and also for use in hazardous atmospheres.

9.4 References. The following publications shall be used in conjunction with this chapter.

[1] ANSI/IEEE C37.96-1976, IEEE Guide for AC Motor Protection.

[2] ANSI/NFPA 70-1984, National Electrical Code.

[3] NEMA ICS2-1983, Industrial Control Devices, Controllers Assemblies.

[4] FAWCETT, D.V. Protection of Large Three-Phase Motors. *IEEE Transactions on Industry and General Applications,* vol IGA-3, Jan/Feb 1967, pp 52–55.

[5] GILL, J.D. Transfer of Motor Loads Between Out-of-Phase Sources. *IEEE Transactions on Industry and General Applications,* vol IGA-15, July/Aug 1979, pp 376–381.

[6] GLEASON, L.L. and ELMORE, W.A. Protection of Three-Phase Motors against Single-Phase Operation. *AIEE Transactions on Power Apparatus and Systems,* pt III, vol IGA-77, Dec 1958, pp 1112–1120.

[7] SCHMIDT, R.A. Calculation of Fault Currents for Internal Faults in AC Motors. *AIEE Transactions on Power Apparatus and Systems,* pt III, vol IGA-75, Oct 1956, pp 818–824.

[8] YUEN, M.H., RITTENHOUSE, J.D., and FOX F.K. Large Wound-Rotor Motor with Liquid Rheostat for Refinery Compressor Drive. *IEEE Transactions on Industry and General Applications,* vol IGA-1, Mar/Apr 1965, pp 140–149.

10. Transformer Protection

10.1 General Discussion. Increased use of electric power in industrial plants has led to the use of larger and more expensive primary and secondary substation transformers. This chapter is directed towards the proper protection of these transformers.

Primary substation transformers normally range in size between 1000 and 12 000 kVA, with a secondary voltage between 2400 and 13 800 V. Secondary substation transformers normally range in size between 300 and 2500 kVA, with secondary voltages of 208, 240, or 480 V. Larger and smaller transformers may also be protected by the devices discussed in this chapter.

10.2 Need for Protection. Transformer failure may result in loss of service. However, prompt fault clearing from the system, in addition to minimizing the damage and cost of repairs, usually minimizes system disturbance, the magnitude of the service outage, and the duration of the outage. Prompt fault clearing will usually prevent catastrophic damage. Proper protection is, therefore, important for transformers of all sizes, even though they are among the simplest and most reliable components in the plant's electrical system.

Previous studies have indicated that transformers above 500 kVA had a failure rate that was lower than that of most other system components. In that study, transformers averaged only 76 failures per 10 000 transformer-years. This might incorrectly be taken to imply that little or no transformer protection is required.

The need for transformer protection is strongly indicated when the average forced hours of downtime per transformer-year is considered. In this category, transformers ranked just below the utility power supply in most cases, as shown in Tables 47A and 47B. The large value of 342 h average out-of-service time per transformer failure is a challenge to the system engineer to properly protect the transformer and minimize any damage that could occur.

The failure of a transformer can be caused by any of a number of internal or external conditions that make the unit incapable of performing its proper function electrically or mechanically. Transformer failures may be grouped as follows:

Table 47A
Reliability Data from IEEE
Reliability Survey of Industrial Plants***

Equipment Category	λ (Failures per Year)	τ (Hours of Downtime per Failure)	$\lambda.\tau$ (Forced Hours of Downtime per Year)
Protective relays	0.0002	5.0	0.0010
Metal-clad drawout circuit breakers			
0–600 V	0.0027	4.0	0.0108
Above 600 V	0.0036	83.1*	0.2992
Above 600 V	0.0036	2.1**	0.0076
Power cables (1000 circuit ft)			
0–600 V, above ground	0.00141	10.5	0.0148
601–15 000 V, conduit below ground	0.00613	26.5*	0.1624
601–15 000 V, conduit below ground	0.00613	19.0**	0.1165
Cable terminations			
0–600 V, above ground	0.0001	3.8	0.0004
601–15 000 V, conduit below ground	0.0003	25.0	0.0075
Disconnect switches enclosed	0.0061	3.6	0.0220
Transformers			
601–15 000 V	0.0030	342.0*	1.0260
601–15 000 V	0.0030	130.0**	0.3900
Switchgear bus–bare			
0–600 V (connected to 7 breakers)	0.0024	24.0	0.0576
0–600 V (connected to 5 breakers)	0.0017	24.0	0.0408
Switchgear bus–insulated			
601–15 000 V (connected to 1 breaker)	0.0034	26.8	0.0911
601–15 000 V (connected to 2 breakers)	0.0068	26.8	0.1822

*Repair failed unit.
**Replace with spare.
***IEEE Committee Report. Report on Reliability Survey of Industrial Plants. *IEEE Transactions on Industry Applications,* Mar/Apr, July/Aug, Sept/Oct 1974, pp 213–252, p 455–476, p 681.

(1) Winding failures are the most frequent cause of transformer failure. Reasons for this type of failure include insulation deterioration or defects in manufacturing, overheating, mechanical stress, vibration, and voltage surges.

(2) Terminal boards and no-load tap changers. Failures are attributed to improper assembly, damage during transportation, excessive vibration, or inadequate design.

(3) Bushing failures can be caused by vandalism, contamination, aging, cracking, or animals.

Table 47B
IEEE Survey of the Reliability
of Electric Utility Power Supplies
to Industrial Plants[***]

Number of Circuits (All Voltages)	λ (Failures per Year)	τ (Hours of Downtime per Failure)	$\lambda \cdot \tau$ (Forced Hours of Downtime per Year
Single circuit	1.956	1.32	2.582
Double circuit Loss of both circuits[*]	0.312	0.52	0.1622
Calculated value for loss of Source 1 (while Source 2 is functional)	1.644	0.15[**]	0.2466

[*]Data for double circuits had all circuit breakers closed.
[**]Manual switchover time of 9 min to source 2.
[***]Power Systems Reliability Subcommittee. Reliability of Electric Utility Supplies to Industrial Plants. *IEEE Industrial and Commercial Power Systems Technical Conference,* Toronto, Ontario, May 5–8, 1975.

(4) Load tap changer failures can be caused by mechanism malfunction, contact problems, insulating liquid contamination, vibration, improper assembly, or excessive stresses within the unit. Load tap changing units are normally applied on utility systems rather than on industrial systems.

(5) Miscellaneous failures would include core insulation breakdown, bushing current transformer failure, liquid leakage due to poor welds or tank damage, shipping damage, and foreign materials left within the tank.

Failure of other equipment within the transformer protective device's zone of protection could cause the loss of the transformer to the system. This would include any equipment between the next upstream protective device and the next downstream device. Included may be such components as cables, bus ducts, switches, instrument transformers, surge arresters, and neutral grounding devices.

10.3 Objectives in Transformer Protection. Protection is achieved by the proper combination of system design, physical layout, and protective devices as required to economically meet the requirements of the application and to:

(1) Protect the electrical system from the effects of transformer failure

(2) Protect the transformer from disturbances occurring on the electrical system to which it is connected

(3) Protect the transformer as much as possible from incipient malfunction within the transformer itself

(4) Protect the transformer from physical conditions in the environment that may affect reliable performance

10.4 Types of Transformers. Under the broad category of transformers there are two types that are widely used in industrial and commercial power systems, namely, liquid-type and dry-type. The former are constructed to have the essential element, the core and coils of the transformer, contained in the liquid-filled enclosure, said liquid serving both as an insulating medium and as a heat-transfer medium. The dry-type transformers are constructed to have the core and coils surrounded by an atmosphere, which may be the surrounding air, free to circulate from the outside to the inside of the transformer enclosure. The dry-type coils may be the conventional type with exposed, insulated conductors, or the encapsulated type, wherein the coils are completely vacuum cast in an epoxy resin.

An alternative to the free circulation of outside air through the dry-type transformer is the sealed enclosure in which a gas or vapor is contained. In either case, this surrounding medium acts both as a heat-transfer medium and as an insulating medium. It is important, with both liquid and dry transformers, that the quality and function of the surrounding media be monitored to avoid damage to the core and coil structures. Systems to preserve or protect the medium within the transformer enclosure are presented briefly in the following paragraphs.

10.5 Preservation Systems

10.5.1 Dry Type Preservation Systems. Dry type preservation systems are used to assure an adequate supply of clean ventilating air at an acceptable ambient temperature. Contamination of the insulating ducts within the transformer can lead to reduced insulation strength and severe overheating. The protection method most commonly used in commercial applications consists of a temperature indicating device with probes installed in the transformer winding ducts and contacts to signal dangerously high temperature by visual and audible alarm. Figure 173 illustrates this feature.

The following types of dry type systems are commonly used:
(1) Open ventilated
(2) Filtered ventilated
(3) Totally enclosed, nonventilated
(4) Sealed air or gas filled

10.5.2 Liquid Preservation Systems. Liquid preservation systems are used to preserve the amount of liquid and to prevent its contamination by the surrounding atmosphere that may introduce moisture and oxygen leading to reduced insulation strength and to sludge formation in cooling ducts.

The importance of maintaining the purity of insulating oil becomes increasingly critical at higher voltages because of increased electrical stress on the insulating oil.

The sealed tank system is now used almost to the total exclusion of other types in the US in industrial and commercial applications. The following types of systems are commonly used:

**Fig 173
Tamper-Proof Fan-Cooled Dry Type
Ventilated Outdoor Transformer with
Microprocessor Temperature-Control System**

(1) Sealed tank
(2) Positive pressure inert gas
(3) Gas-oil seal
(4) Conservator tank

Liquid preservation systems have historically been called oil-cooled systems, even though the medium was askarel or a substitute for askarel.

10.5.2.1 Sealed Tank. The sealed-tank design is the one most commonly used and is standard on most substation transformers. As the name implies, the transformer tank is sealed, isolating it from the outside atmosphere.

A gas space equal to about one-tenth of the liquid volume is maintained above the liquid to allow for thermal expansion. This space may be purged of air and nitrogen filled.

A pressure-vacuum gauge and bleeder device may be furnished on the tank to allow monitoring of the internal pressure or vacuum and to allow relieving any excessive static pressure buildup that could damage the enclosure or cause the pressure-relief device to operate. This system is the simplest and most maintenance-free of all of the preservation systems.

10.5.2.2 Positive-Pressure Inert Gas. The design shown in Fig 174 is similar to the sealed-tank design with the addition of a gas (usually nitrogen) pressurizing the assembly. This assembly provides a slight positive pressure in the gas supply line to prevent air from entering the transformer during operating mode or temperature changes. Transformers with primary windings rated 69 kV and above, and rated 7500 kVA and above, can be equipped with this device.

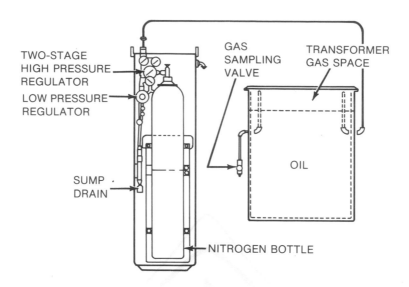

Fig 174
Positive-Pressure Inert-Gas Assembly Is Often Used on
Sealed-Tank Transformers above 7500 kVA and 69 kV Primary Voltage

10.5.2.3 Gas-Oil Seal. This design incorporates a captive gas space that isolates a second auxiliary oil tank from the main transformer oil as shown in Fig 175. The auxiliary oil tank is open to the atmosphere and provides room for thermal expansion of the main transformer oil volume.

The main tank oil expands or contracts due to changes in its temperature causing the level of the oil in the auxiliary tank to raise or lower as the captive volume of gas is forced out of or allowed to reenter the main tank. The pressure of the auxiliary tank oil on the contained gas maintains a positive pressure in the gas space, preventing atmospheric vapors from entering the main tank.

10.5.2.4 Conservator Tank. The conservator-tank design shown in Fig 176 does not have a gas space above the oil in the main tank. It includes a second oil tank above the main tank cover with a gas space adequate to absorb the thermal expansion of the main tank oil volume. The second tank is connected to the main tank by an oil-filled tube or pipe.

A large diameter stand pipe extends at an angle from the cover and is closed above the liquid level by a frangible diaphragm that ruptures in the event of rapid gas evolution and thereby releases pressure to prevent damage to the enclosure.

Because the conservator construction allows gradual liquid contamination, it has become obsolete in the US.

10.6 Protective Devices for Liquid Preservation Systems.

10.6.1 Liquid-Level Gauge. The liquid-level gauge, shown in Figs 177 and 178, is used to measure the level of insulating liquid within the tank with respect to predetermined level, usually indicated at the 25 °C (77 °F) level. An excessive-

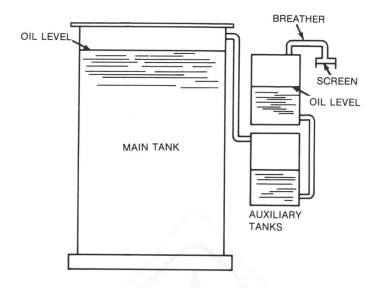

**Fig 175
Gas-Oil Seal System of Oil Preservation**

ly low level could indicate the loss of insulating liquid, which could lead to internal flashovers and overheating if not corrected. Periodic observation is normally performed to check that the liquid level is within acceptable limits. Alarm contacts for low liquid level are normally available as a standard option when specified. It is recommended that alarm contacts be specified for unattended stations to save transformers from loss-of-insulation failure. The alarm contact is set to close before an unsafe condition actually occurs. The alarm contacts should be connected through a communication link to an attended station.

10.6.2 Pressure-Vacuum Gauge. The pressure-vacuum gauge in Fig 179 indicates the difference between the transformer internal gas pressure and atmospheric pressure. It is used on transformers with sealed-tank oil preservation systems. Both the pressure-vacuum gauge and the sealed-tank oil preservation system are standard on most small and medium power transformers.

The pressure in the gas space is normally related to the thermal expansion of the insulating liquid and will vary with load and ambient temperature changes. Large positive or negative pressures could indicate an abnormal condition, such as a gas leak, particularly if the transformer has been observed to remain within normal pressure limits for some time, or if the pressure-vacuum gauge has remained at the zero mark for a long period of time. The pressure-vacuum gauge equipped with limit alarms may be used to detect excessive vacuum or positive pressure that could cause tank rupture or deformation. The need for pressure limit alarms is less urgent when the transformer is equipped with a pressure-relief device.

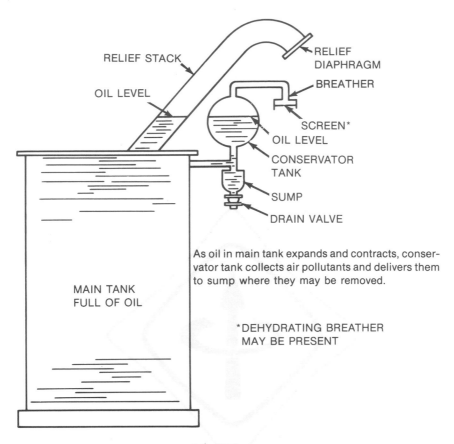

As oil in main tank expands and contracts, conservator tank collects air pollutants and delivers them to sump where they may be removed.

*DEHYDRATING BREATHER
 MAY BE PRESENT

Fig 176
Conservator-Tank System of Oil Preservation

10.6.3 Pressure-Vacuum Bleeder Valve. A transformer is designed to operate over a range of 100 °C, generally from -30 °C to +70 °C. Should the temperature exceed these limits, the pressure-vacuum bleeder valve automatically adjusts to prevent any pressure or vacuum in excess of 5 psig. This valve also prevents operation of the pressure-relief device in response to slowly increasing pressure caused by severe overload heating or extreme ambient temperatures. Also incorporated in the pressure-vacuum bleeder valve is a hose burr and a manually operated valve to allow purging or checking for leaks by attaching the transformer to an external source of gas pressure. The pressure-vacuum bleeder valve is usually mounted with the pressure-vacuum gauge as shown in Fig 179.

10.6.4 Pressure-Relief Device. A pressure-relief device is a standard accessory on all liquid insulated substation transformers, except on small secondary substation oil-insulated units, where it may be optional. This device shown in Fig 180 can relieve both minor and serious internal pressures. When the internal

Fig 177
Liquid-Level Indicator Depicts Level of Liquid
with Respect to a Predetermined Level, Usually 25 °C

pressure exceeds the tripping pressure (10 lbs/in², ± 1 lb/in² gauge), the device snaps open allowing the excess gas or fluid to be released. Upon operation, a pin (standard), alarm contact (optional), or semaphore signal (optional) is actuated to indicate operation. The device is normally automatic-resetting and self-sealing and requires little or no maintenance or adjustment.

This device is mounted on top of the transformer cover and usually has a visual type indicator. The indicator should be reset manually in order to indicate subsequent operation. This device, when equipped with an alarm contact in conjunction with a self-sealing relay, can provide remote warning. Any operation of the pressure-relief device that was not preceded by high temperature loading is indicative of possible trouble in the windings.

The major function of the pressure-relief device is to prevent rupture or damage to the transformer tank due to excessive pressure in the tank. Excessive pressure is developed due to high peak loading, long-time overloads or internal arc-producing faults.

10.6.5 Rapid Pressure Rise Relay. This pressure sensitive relay is normally used to initiate isolation of the transformer from the electrical system and to limit damage to the unit when there is an abrupt rise in the transformer internal pressure. The abrupt pressure rise is due to the vaporization of the insulating liquid by an internal fault, such as internal shorted turns, ground faults, or winding-to-winding faults. The bubble of gas formed in the insulating liquid creates a pressure wave that promptly activates the relay.

Since operation of this device is closely associated with actual faults in the windings, it is risky to re-energize a transformer that has been tripped off the line by the rapid pressure rise relay. The transformer must be taken out of service for thorough visual and diagnostic checks to determine the extent of damage.

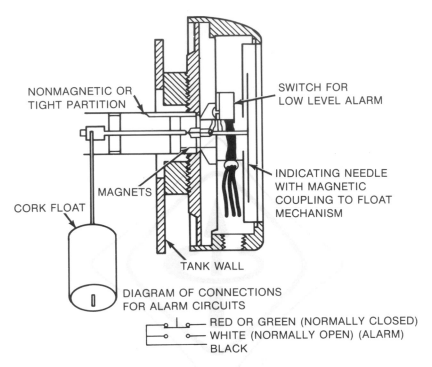

NONMAGNETIC OR
TIGHT PARTITION

SWITCH FOR
LOW LEVEL ALARM

MAGNETS

INDICATING NEEDLE
WITH MAGNETIC
COUPLING TO FLOAT
MECHANISM

CORK FLOAT

TANK WALL

DIAGRAM OF CONNECTIONS
FOR ALARM CIRCUITS

RED OR GREEN (NORMALLY CLOSED)
WHITE (NORMALLY OPEN) (ALARM)
BLACK

Fig 178
Liquid-Level Indicating
Needle Is Driven by a Magnetic
Coupling to the Float Mechanism

One type of relay, shown in Fig 181, uses the insulating liquid to transmit the pressure wave to the relay bellows. Inside the bellows a special oil transmits the pressure wave to a piston that will actuate a set of switch contacts. This type of relay is mounted on the transformer tank below oil level.

Another type of relay, shown in Fig 182, uses the inert gas above the insulating liquid to transmit the pressure wave to the relay bellows. Expansion of the bellows actuates a set of switch contacts. This type of relay is mounted on the transformer tank above oil level.

Both types of relays have a pressure-equalizing opening to prevent operation of

Fig 179
Gas-Pressure Gauge Indicates Internal Gas Pressure Relative to
Atmospheric Pressure; Bleeder Allows Pressure to Be Equalized Manually

the relay on gradual rises in internal pressure due to changes in loading or ambient conditions.

Both types of rapid pressure rise relays are very sensitive to the rate of rise in the internal pressure. The time for the relay switch to operate is on the order of 4 cycles for high rates of pressure rise [25 (lbf/in^2)/s of oil pressure rise; 5 (lbf/in^2)/s of air pressure rise]. These relays are designed to be insensitive to mechanical shock and vibration, through faults and magnetizing inrush current.

The use of rapid pressure rise relays increases as the size and value of the transformer increases. Most transformers 5000 kVA and above are equipped with this type of device. This relay provides valuable protection at low cost.

10.6.6 Gas-Detection Relay. The gas-detection relay shown in Fig 183 is a special device used to detect and indicate an accumulation of gas from a transformer with a conservator tank. Incipient winding faults or hot spots in the core normally generate small amounts of gas that are channeled to the top of the special domed cover. From there the bubbles enter the accumulation chamber of the relay through a pipe. Gas accumulation is indicated on the gauge in cubic centimeters. An accumulation of gas of 100–200 cm^3, for example, will lower a float and operate an alarm switch, indicating that an investigation is necessary. This gas can then be withdrawn for analysis and recording.

The rate of gas accumulation is a clue to the magnitude of the fault. If the chamber continues to fill quickly, with resultant operation of the relay, potential danger may justify removing the transformer from service.

Fig 180
Pressure-Relief Device Limits
Internal Pressure to Prevent Tank
Rupture Under Internal Fault Conditions

10.6.7 Combustible-Gas Relay. The combustible-gas relay as shown in Fig 184 is a special device used to detect and indicate the presence of combustible gas coming from the transformer. The combustible gas is formed by the decomposition of insulating materials within the transformer by a low-level fault or by discontinuous discharges (corona). These faults are normally not detected until they develop into larger and more damaging ones.

The combustible-gas relay can be used on transformers with positive-pressure inert gas–oil preservation systems. The relay periodically takes a sample of the gas in the transformer and tests it with a heated wire. If combustible gases are in the sample, they will ignite, further heating the wire, which in turn changes its resistance. The change is detected by a bridge network and activates a signal relay. The combustible-gas relay is expensive and is not normally applied on substation transformers.

Portable gas-analysis equipment can be used to test the composition of gases in the transformers. By analyzing the percentage of unusual or decomposed gases in the transformer, it can be determined if the transformer has a low-level fault, and if so what type of fault has occurred. This type of device is normally used on utility systems having large numbers of large-capacity transformers.

10.7 Transformer Primary Protective Device. A fault on the electrical system at the point of connection to the transformer can arise from failure (internal fault) of the transformer or from an abnormal condition on the circuit connected to the transformer secondary, such as a short circuit (through fault). The predominant means of clearing such faults is a current interrupting device on the primary side of the transformer, such as fuses or a circuit breaker or circuit switcher. Whatever the choice, the primary side protective device should

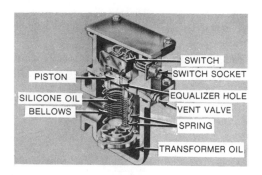

**Fig 181
Pressure Rise Type Relay Mounted on
Transformer Tank Below Normal Oil Level**

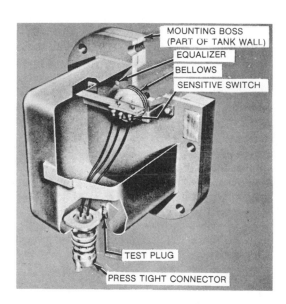

**Fig 182
Pressure Type Relay Mounted on Transformer Tank Above Normal Oil Level**

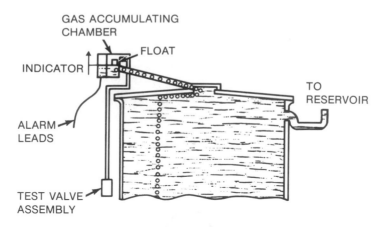

Fig 183
Gas-Detection Relay Accumulates
Gases from Top Air Space of Transformer
(Used Only on Conservator Tank Units)

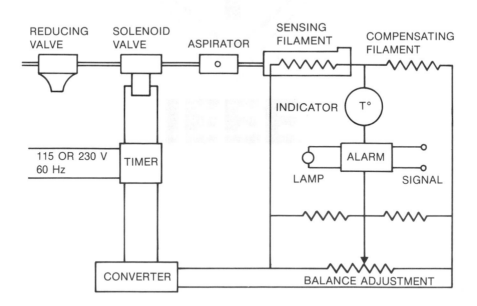

Fig 184
Combustible Gas Relay Periodically Samples Gas in
Transformer to Detect Any Minor Internal Fault
Before it Can Develop into a Serious Fault

have an interrupting rating adequate for the maximum short-circuit current that can occur on the primary side of the transformer or, alternately, it should be relayed so that it will only be called upon to clear lower current internal or secondary faults. Instantaneous relay elements used to protect transformer feeders and high-voltage windings are set greater than the maximum through fault on the transformer secondary. The operating current of the primary protective device should be less than the short-circuit current of the transformer as limited by the combination of system and transformer impedance. The above is true for a fuse or a time-overcurrent relay element. The point of operation should not be so low, however, as to cause circuit interruption due to inrush excitation current of the transformer or normal current transients in the secondary circuits. Of course, any devices operating to protect the transformer to remove it from the system by detecting abnormal conditions within the transformer will also operate to protect the system; but these are subordinate to the primary protection as discussed above.

10.8 Protecting the Transformer from Electrical Disturbances. Transformer failures arising from abusive operating conditions are caused by:
  (1) Continuous overloading
  (2) Short circuits
  (3) Ground faults
  (4) Transient overvoltages

10.8.1 Overload Protection. An overload will cause a rise in the temperature of the various transformer components. If the final temperature is above the design temperature limit, deterioration of the insulation system will occur causing a reduction in the useful life of the transformer. The insulation may be weakened such that a moderate overvoltage may cause insulation breakdown before expiration of expected service life. Transformers have certain overload capabilities, varying with ambient temperature, preloading, and overload duration. These capabilities are defined in ANSI/IEEE C57.92-1981 [7][22] and ANSI C57.96-1959 [1]. It should be recognized that whenever the temperature rise of a winding is increased, the insulation deteriorates more rapidly and the life expectancy of the transformer is shortened.

Protection against overloads consists of both load limitation and overload detection. Loading on the transformers may be limited by designing a system where the transformer capacity is greater than the total connected load, assuming a diversity in load usage. This is an expensive method to provide overload protection, since load growth and changes in operating procedures would quite often eliminate the extra capacity needed to achieve this protection. It is a good engineering practice to size the transformer at about 125% of the present load to allow for system growth and change in the diversity of loads. The specification of a lower than ANSI temperature rise also permits an overload capability.

---

[22] The numbers in brackets correspond to those of the references listed at the end of this chapter.

Fig 185
Liquid Temperature Indicator Is the Most Common
Transformer Temperature Sensing Device

Load limitation by disconnecting part of the load can be done automatically or manually. Automatic load shedding schemes, because of their cost, are restricted to larger units. However, manual operation is often preferred because it gives greater flexibility in selecting the expendable loads.

In some instances, load growth can be accommodated by specifying cooling fans or providing for future fan cooling.

The major method of load limitation that can be properly applied to a transformer is one that responds to transformer temperature. By monitoring the temperature of the transformer, overload conditions can be detected. A number of monitoring devices that mount on the transformer are available as standard or optional accessories.

These devices are normally used for alarm or to initiate secondary protective device operation. They include the following.

10.8.1.1 Liquid Temperature Indicator. The liquid temperature indicator shown in Fig 185 measures the temperature of the insulating liquid at the top of the transformer. Since the hottest liquid is less dense and rises to the top of the tank, the temperature of the liquid at the top partially reflects the temperature of the transformer windings and is related to the loading of the transformer.

The thermometer reading is related to transformer loading only insofar as that loading affects the liquid temperature rise above ambient. Transformer liquid has a much longer time constant than the winding itself, and responds slowly to changes in loading losses that directly affect winding temperature. Thus, the thermometer's temperature warning will vary between too conservative or too pessimistic, depending on the rate and direction of the change in loading. A high reading could indicate an overload condition.

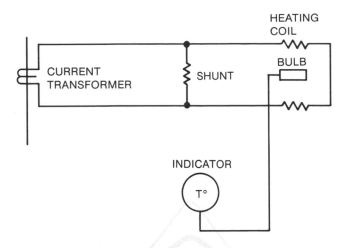

Fig 186
Thermal (or Winding Temperature) Relay Uses a Heating Element to
Duplicate Effects of Current in Transformer

The liquid temperature indicator is normally furnished as a standard accessory on power transformers. It is equipped with a temperature-indicating pointer and a drag pointer that shows the highest temperature reached since it was last reset.

This device can be equipped with one to three adjustable contacts that operate at preset temperatures. The single contact can be used for alarm purposes. When forced air cooling is employed, the first contact initiates the first stage of fans. The second contact either initiates a second stage of fans, if furnished, or an alarm. The third contact, if furnished, is used for the final alarm or to initiate load reduction on the transformer. The indicated temperatures would change for different temperature insulation system designs.

Similar devices as described earlier in this chapter are available for responding to air or gas temperatures in dry-type transformers. For unattended substations, these devices may be connected to central annunciators.

10.8.1.2 Thermal Relays. Thermal relays, diagramatically shown in Fig 186, are used to give a more direct indication of winding temperatures of either liquid or dry-type transformers. A current transformer is mounted on one of the three phases of the transformer bushing. It supplies current to the thermometer bulb heater coil, which contributes the proper heat to closely simulate the transformer hot-spot temperature.

Monitoring of more than one phase is desirable if there is a reason to expect an unbalance in the three-phase loading.

The temperature indicator is a bourdon gauge connected through a capillary tube to the thermometer bulb. The fluid in the bulb will expand or contract proportionally to the temperature changes and is transmitted through the tube to the gauge. Coupled to the shaft of the gauge indicator are three cams that operate individual switches at preset levels of indicated transformer temperature.

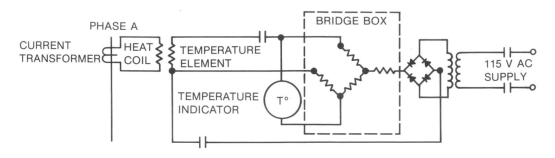

Fig 187
Hot-Spot Temperature Indicator Utilizes Wheatstone Bridge Method to
Determine Transformer Temperature

Thermal relays are used more often on transformers rated 10 000 kVA and above than on smaller transformers. They can be used on all sizes of substation transformers.

10.8.1.3 Hot-Spot Temperature Thermometers. Hot-spot temperature equipment shown in Fig 187 is similar to the thermal relay equipment on a transformer, since it indicates the hottest-spot temperature of the transformer. While the thermal relay does it with fluid expansion and a bourdon gauge, the hot-spot temperature equipment does it electrically using a Wheatstone bridge method, measuring the resistance of a resistance-type temperature detector (RTD) that is responsive to transformer temperature changes—increasing with higher temperature. Since this can be used with more than one detector coil location, temperatures of several locations within the transformer can be checked.

10.8.1.4 Forced Air Cooling. Another means of protecting against over-loads is to increase the transformer capacity by auxiliary cooling as shown in Fig 188. Forced air-cooling equipment is used to increase the capacity of a transform-er by 15–33% of base rating, depending upon transformer size and design. Refer to ANSI/IEEE Std 141–1986 [2], Chapter 8. Dual cooling by a second stage of forced-air fans or a forced-oil system will give a second increase in capacity applicable to three-phase transformers rated 12 000 kVA and above.

Forced air cooling can be added at a later date to increase the transformer capacity to take care of increased loads, provided that the transformer was ordered to have provisions for future fan cooling.

Auxiliary cooling of the insulating liquid helps keep the temperature of the windings and other components below the design temperature limits. Usually, operation of the cooling equipment is automatically initiated by the top liquid temperature indicator or the thermal relay, after a predetermined temperature is reached.

10.8.1.5 Overcurrent Relays. Transformer overcurrent protection may be provided by relays. Chapter 3 describes overcurrent relay construction character-istics and ranges. These relays are applied in conjunction with current trans-formers (CT's) and a circuit breaker or circuit switcher, sized for the maximum

Fig 188
Forced-Air Fans Are Normally Controlled Automatically from a
Top Liquid Temperature or Winding Temperature Relay

continuous and interrupting duty requirements of the application. A typical application is shown in Fig 189.

Overcurrent relays are selected to provide a range of overcurrent settings above the permitted overloads and instantaneous settings when possible within the transformer through-fault current withstand rating. The characteristics should be selected to coordinate with upstream and downstream protective devices.

Ground faults occurring in the substation transformer secondary or between the transformer secondary and main secondary protective device cannot be isolated by the main secondary protective device, which is located on the load side of the ground fault. These ground faults, when limited by a neutral grounding resistor, may not be seen by either the transformer primary fuses or transformer differential relays. They can be isolated only by a primary circuit breaker or other protective device tripped by either a ground relay in the grounding resistor circuit or a ground differential relay. A ground differential relay may consist of a simple overcurrent relay, connected to a neutral ground CT and the residual circuit of the transformer line CT's fed through a ratio matching auxiliary CT. Since this scheme is subject to error on through faults due to unequal CT saturation, a relay with phase restraint coils may be used instead of a simple overcurrent relay.

Overcurrent relays applied on the primary side of the transformer provide protection for transformer faults in the winding, as well as backup protection for the transformer for secondary-side faults. When overcurrent relays are also

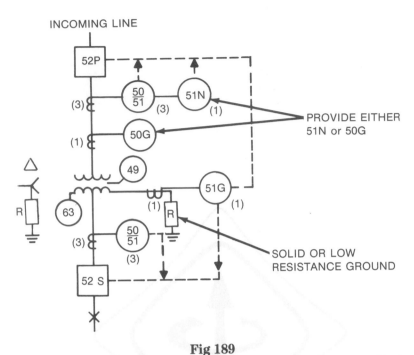

INCOMING LINE

PROVIDE EITHER
51N or 50G

SOLID OR LOW
RESISTANCE GROUND

Fig 189
Overcurrent Relays Are Frequently Used to Provide Transformer Protection in Combination with Primary Circuit Breaker or Circuit Switcher

applied on the secondary side of the transformer, these relays are the principal protection for transformer secondary-side faults. However, overcurrent relays applied on the secondary side of the transformer do not provide protection for the transformer winding faults.

The settings of the overcurrent relays should meet the requirements of applicable standards and codes as well as the needs of the power system. ANSI/NFPA 70-1984 [12] (National Electrical Code [NEC]) requirements represent upper limits that should be met when selecting overcurrent devices. These requirements, however, are not guidelines for the design of a system providing maximum protection for transformers. For example, setting a transformer primary or secondary overcurrent protective device at 2.5 times rated current could allow that transformer to be damaged without the protective device operating.

The best protection for the transformer would be provided by circuit breakers or fuses on both the primary side and secondary side of the transformer, set or selected to operate at minimum values. Common practice is for the secondary-side circuit breaker or fuses to protect the transformer for loading in excess of 125% of maximum rating.

Using a circuit breaker on the primary of each transformer is expensive, however, especially for small-capacity and small-value transformers. An economical compromise may be considered whereby one circuit breaker may be installed to feed two to six relatively small transformers. Each transformer has its own secondary circuit breaker and, in most cases, a primary disconnect. Overcurrent protection should satisfy the requirements prescribed by the NEC [12].

The major disadvantage of this system is that all of the transformers will be de-energized by the opening of the primary circuit breaker. Moreover, the rating or setting of a primary circuit breaker selected to accommodate the total loading requirements of all of the transformers would typically be so large that only a small degree of secondary-fault protection, and almost no backup protection, would be provided for each individual transformer.

By using fused switches on the primary of each transformer, short-circuit protection can be provided for the transformer and additional selectivity provided for the system. Using fused switches and time-delay dual-element fuses for the secondary of each transformer will allow close sizing (typically 125% of secondary full load current), thus giving excellent overload and short-circuit protection for 600 V or less applications.

10.8.2 Protection Against Overvoltages. The most common cause of transformer failure is transient overvoltages due to lightning, switching surges, and other system disturbances. High-voltage disturbances can be generated by certain types of loads as well as from the incoming line. There is a common misconception that underground services are isolated from these disturbances. System insulation coordination in the use and location of primary and secondary surge arresters is very important. Normally, liquid insulated transformers have higher basic impulse insulation level ratings than do standard ventilated dry-type and sealed dry-type transformers. Solid dielectric cast coil transformers have basic impulse levels equal to liquid insulated transformers. Ventilated dry-type transformers and sealed dry-type transformers can be specified to have basic impulse insulation levels equal to that of liquid transformers.

10.8.2.1 Surge Arresters. Ordinarily, if the liquid insulated transformer is supplied by enclosed conductors from the secondaries of transformers with adequate primary surge protection, additional protection may not be required, depending on the system design. However, if the transformer primary or secondary is connected to conductors that are exposed to lightning, the installation of surge arresters is necessary. For best protection, the surge arrester should be mounted as close as possible to the transformer terminals, preferably within 3 ft and on the load side of the incoming switch.

The degree of surge protection obtained is determined by the amount of exposure, the size and importance of the transformer to the system, and the type and cost of the arresters. In descending order of cost and degree of protection, the arresters available are station-type, intermediate-type, and distribution-type. Lower BIL transformers are less costly and are equally reliable if protected by compatible surge arresters.

Ventilated dry-type and sealed dry-type transformers are normally used indoors and surge protection is necessary. Since all systems have the potential for transmitting and reflecting primary and secondary surges caused by lightning and system disturbances, special low spark-over distribution-type arresters and low voltage arresters have been developed for the protection of dry-type transformers and rotating machinery.

The surge arrester selection (kV class) should be based on the system voltage and system conditions (grounded or ungrounded). The arrester kV class is not determined by the kV class of the primary winding of the transformer.

10.8.2.2 Surge Capacitors. Additional protection in the form of surge capacitors located as closely as possible to the transformer terminals may also be appropriate for all types of transformers. Transformer windings can experience a very nonuniform distribution of a fast-front surge in the winding, thus overstressing the turn insulation locally in parts of the windings. Surge capacitors serve a dual function of sloping-off fast-rising transients that might impinge on the transformer winding, as well as reducing the effective surge impedance presented by the transformer to an incoming surge. This type of additional protection is appropriate against voltage transients generated within the system due to circuit conditions such as prestriking, restriking, high-frequency current interruption, multiple reignitions, voltage escalation, and current suppression (chopping) as the result of switching, current-limiting fuse operation, thyristor-switching, or ferroresonance conditions.

10.8.2.3 Ferroresonance. Ferroresonance is a phenomenon resulting in the development of a higher than normal voltage in the windings of a transformer. These overvoltages may result in surge arrester operation, damage to the transformer, and an electrical shock hazard. The following conditions combine to produce ferroresonance:

(1) No load on the transformer

(2) An open circuit on one of the primary terminals of the transformer and, at the same time, an energized terminal; in the case of three-phase transformers, either one or two of the three primary terminals may be disconnected

(3) The point of disconnection is not close to the transformer

(4) There is a voltage potential between the disconnected terminal conductor and ground

The resonant circuit may be traced from the energized terminal through the transformer primary to one of the disconnected terminals, then through the capacitance of the isolated terminal conductor insulation to ground, and then back through the supply system to the energized terminal. See Fig 190. Although more common with underground distribution systems, ferroresonance can occur with overhead lines when the single-phase open point is far enough from the transformer. The typical scenarios for ferroresonance involve single-phase remote switching of an unloaded transformer, or remote primary fuse operation on one phase or failure of all three poles of a three-pole device to properly open accompanied by disconnection of the secondary load.

Ferroresonance may be minimized or eliminated by having load connected to the secondary when single-phase switching on the primary, by using gang-operated switches, circuit breakers, or circuit switchers on the primary, or by providing that current-interrupting devices are located next to or on the transformer.

The subject of ferroresonance is quite complicated; and the literature on this subject should be reviewed by concerned persons to avoid ferroresonance in transformer operation or system design.

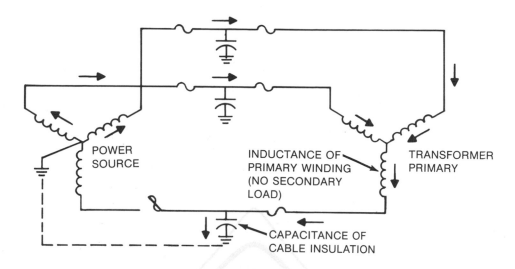

Fig 190
One-Line Diagram Showing Current Flow That May Result in Ferroresonance

10.8.3 Short-Circuit Current Protection. In addition to thermal damage from prolonged overloads, transformers are also adversely affected by internal or external short-circuit conditions, which can result in internal electromagnetic forces, temperature rise, and arc-energy release.

Secondary-side short circuits can subject the transformer to short-circuit current magnitudes limited only by the sum of transformer and supply-system impedance. Hence, transformers with unusually low impedance may experience extremely high short-circuit currents, resulting in mechanical damage. Prolonged flow of a short-circuit current of lesser magnitude can also inflict thermal damage.

Protection of the transformer for both internal and external faults should be as rapid as possible to reduce damage to a minimum. This protection, however, may be reduced by selective-coordination system design and operating procedure limitations.

There are several sensing devices available that provide varying degrees of short-circuit protection. These devices sense two different aspects of a short circuit. The first group of devices senses the formation of gases consequent to a fault and are used to detect internal faults. The second group senses the magnitude of the short-circuit current directly.

The gas-sensing devices include pressure-relief devices, rapid pressure rise relays, gas-detector relays, and combustible-gas relays. The current-sensing devices include fuses, overcurrent relays, and differential relays.

10.8.3.1 Gas-Sensing Device. Low-magnitude faults in the transformer cause gases to be formed by the decomposition of insulation exposed to high temperature at the fault. Detection of the presence of these gases can allow the transformer to be taken out of service before extensive damage occurs. In some cases, gas may be detected a long time before the unit fails.

High-magnitude fault currents will usually be first sensed by other detectors, but the gas-sensing device will respond with modest time delay. These devices are described in detail in 10.6.

10.8.3.2 Current-Sensing Devices. Fuses, overcurrent relays, and differential relays should be selected to provide the maximum degree of protection to the transformer. These protective devices should operate in response to a fault before the magnitude and duration of the overcurrent exceed the short-time loading limits recommended by the transformer manufacturer. In the absence of specific information applicable to an individual transformer, protective devices should be selected in accordance with applicable guidelines for the maximum permissible transformer short-time loading limits. Curves illustrating these limits for liquid-immersed transformers are discussed in the following section. In addition, ratings or settings of the protective devices should be selected in accordance with pertinent provisions of the NEC [12], Chapter 4, Article 450.

The following discussion is excerpted and paraphrased from Appendix A of ANSI/IEEE C37.91-1985 [3]. This Appendix is entitled "Application of Transformer Through-Fault Current Duration Guide to the Protection of Power Transformers." Similar information and through-fault protection curves can be found in ANSI/IEEE C57.109-1985 [11]. The following discussion is based on these two standards. Similar through-fault protection curves for dry-type transformers are currently being developed.

Overcurrent protective devices such as fuses and relays have well defined operating characteristics that relate fault-current magnitude to operating time. It is desirable that the characteristic curves for these devices be coordinated with comparable curves, applicable to transformers, which reflect their through-fault withstand capability. Such curves for Category I, II, III, and IV liquid-immersed transformers (as described in ANSI/IEEE C57.12.00-1980 [4]) are presented in this section as through-fault protection curves.

It is widely recognized that damage to transformers from through faults is the result of thermal as well as mechanical effects. The latter have, however, recently gained increased recognition as a major cause of transformer failure. Though the temperature rise associated with high magnitude through faults is typically quite acceptable, the mechanical effects are intolerable if such faults are permitted to occur with any regularity. This results from the cumulative nature of some of the mechanical effects, particularly insulation compression, insulation wear, and friction-induced displacement. The damage that occurs as a result of these cumulative effects is, therefore, a function of not only the magnitude and duration of through faults, but also the total number of such faults.

The through-fault protection curves presented in this section take into consideration the fact that transformer damage, as discussed above, is cumulative, and the fact that the number of through faults to which a transformer can be exposed is inherently different for different transformer applications. For example, transformers with secondary-side conductors enclosed in conduit or isolated in some other fashion, such as those typically found in industrial, commercial, and institutional power systems, experience extremely low incidence of through faults. In contrast, transformers with secondary-side overhead lines, such as those found in utility distribution substations, have a relatively high incidence of through faults, and the use of reclosers or automatic reclosing circuit breakers may subject the transformer to repeated current surges from each fault. Thus, for a given transformer in these two different applications, a different through-fault protection curve should apply, depending on the type of application. For applications in which faults occur infrequently, the through-fault protection curve should reflect primarily thermal damage considerations, since cumulative mechanical-damage effects of through faults will not be a problem. For applications in which faults occur frequently, the through-fault protection curve should reflect the fact that the transformer will be subjected to both thermal and cumulative-mechanical damage effects of through faults.

In using the through-fault protection curves to select the time-current characteristics of protective devices, the protection engineer should take into account not only the inherent level of through-fault incidence, as described above, but also the location of each protective device and its role in providing transformer protection. As noted, substation transformers with secondary-side overhead lines have a relatively high incidence of through faults. The secondary-side *feeder* protective equipment is the first line of defense against through faults and its time-current characteristics should, therefore, be selected by reference to the frequent-fault-incidence protection curve. More specifically, the time-current characteristics of feeder protective devices should be below and to the left of the appropriate frequent-fault-incidence protection curve. *Main secondary-side* protective devices (if applicable) and *primary-side* protective devices typically operate to protect against through faults in the rare event of a fault between the transformer and the feeder protective devices, or in the equally rare event that a feeder protective device fails to operate or operates too slowly due to an incorrect (higher) rating or setting. The time-current characteristics of these devices, therefore, should be selected by reference to the infrequent-fault-incidence protection curve. In addition, these time-current characteristics should be selected to achieve the desired coordination among the various protective devices.

In contrast, transformers with protected secondary conductors (for example, cable, bus duct, or switchgear) experience an extremely low incidence of through faults. Hence the feeder protective devices may be selected by reference to the infrequent-fault-incidence protection curve. The main secondary-side protective device (if applicable) and the primary-side protective device should also be selected by reference to the infrequent-fault-incidence protection curve. Again, these time-current characteristics should also be selected to achieve the desired coordination among the various protective devices.

For Category I transformers (5–500 kVA single-phase, 15–500 kVA three-phase), a single through-fault protection curve applies. Refer to Fig 191. This curve may be used for selecting protective device time-current characteristics for all applications regardless of the anticipated level of fault incidence.

For Category II transformers (501–1667 kVA single-phase, 501–5000 kVA three-phase), two through-fault protection curves apply. Refer to Fig 192. The left-hand curve in Fig 192 reflects both thermal and mechanical damage

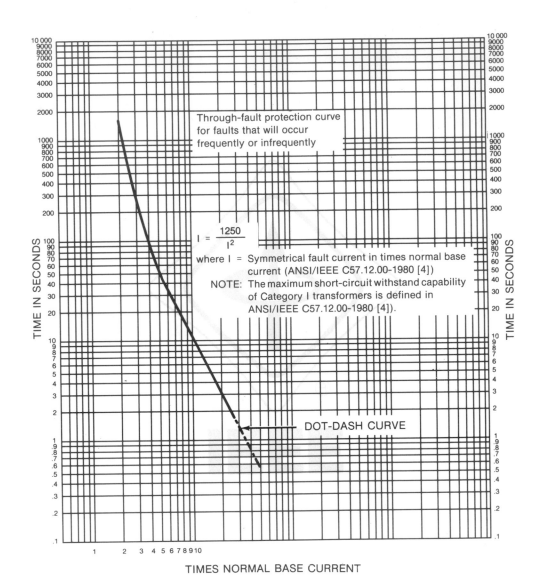

TIMES NORMAL BASE CURRENT

Fig 191
Through-Fault Protection Curve for Liquid-Immersed Category I Transformers
(5 kVA to 500 kVA Single-Phase, 15 kVA to 500 kVA Three-Phase)

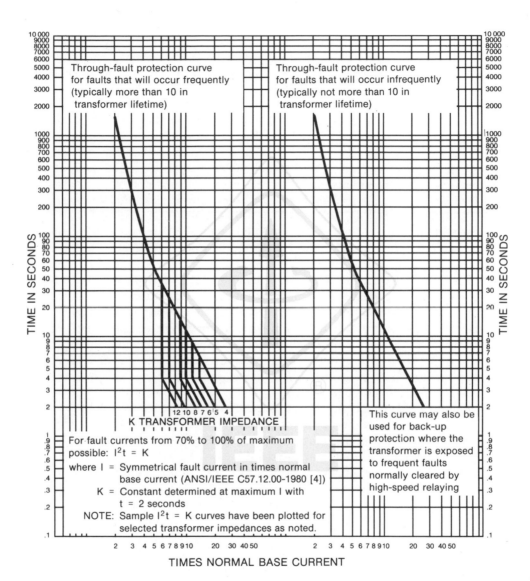

**Fig 192
Through-Fault Protection Curves for Liquid-Immersed Category II Transformers
(501 kVA to 1667 kVA Single-Phase, 501 kVA to 5000 kVA Three-Phase)**

considerations and may be used for selecting feeder protective device time-current characteristics for frequent-fault-incidence applications. The right-hand curve in Fig 192 reflects primarily thermal damage considerations and may be used for selecting feeder protective device time-current characteristics for infrequent-fault-incidence applications. This curve may also be used for selecting main secondary-side protective device (if applicable) and primary-side protective device time-current characteristics for all applications—regardless of the anticipated level of fault incidence.

For Category III transformers (1668–10 000 kVA single-phase, 500–30 000 kVA three-phase), two through-fault protection curves apply. Refer to Fig 193. The left-hand curve in Fig 193 reflects both thermal and mechanical damage considerations and may be used for selecting feeder protective device time-current characteristics for frequent-fault-incidence applications. The right-hand curve in Fig 193 reflects primarily thermal damage considerations and may be used for selecting feeder protective device time-current characteristics for infrequent fault-incidence applications. This curve may also be used for selecting main secondary-side protective device (if applicable) and primary-side protective device time-current characteristics for all applications—regardless of the anticipated level of fault incidence.

For Category IV transformers (above 10 000 kVA single-phase, above 30 000 kVA three-phase), a single through-fault protection curve applies. Refer to Fig 194. This curve reflects both thermal and mechanical damage considerations and may be used for selecting protective device time-current characteristics for all applications—regardless of the anticipated level of fault incidence.

The aforementioned delineation of infrequent versus frequent fault-incidence applications for Category II and III transformers can be related to the zone or location of the fault. Refer to Fig 195.

The primary-side protective device characteristic curve may cross the through-fault protection curve at lower current levels since low-current overload protection is a function of the secondary-side protective device or devices. (Refer to appropriate transformer loading guides, ANSI C57.91-1981 [6] and C57.92-1981 [7].) Efforts should be made, however, to have the primary-side protective device characteristic curve intersect the through-fault protection curve at as low a current as possible in order to maximize the degree of backup protection for the secondary-side devices.

The through-fault protection curve values are based on winding-current relationships for a three-phase secondary fault, and may be used directly for delta–delta and wye–wye connected transformers. For delta-wye connected transformers, the through-fault protection curve values must be reduced to 58% of the values shown, to provide appropriate protection for a secondary-side single-phase-to-neutral fault.

10.8.3.3 Fuses. Fuses utilized on the transformer primary are relatively simple and inexpensive one-time devices that provide short-circuit protection for the transformer. Fuses are normally applied in combination with interrupter switches capable of interrupting full-load current. By using fused switches on the

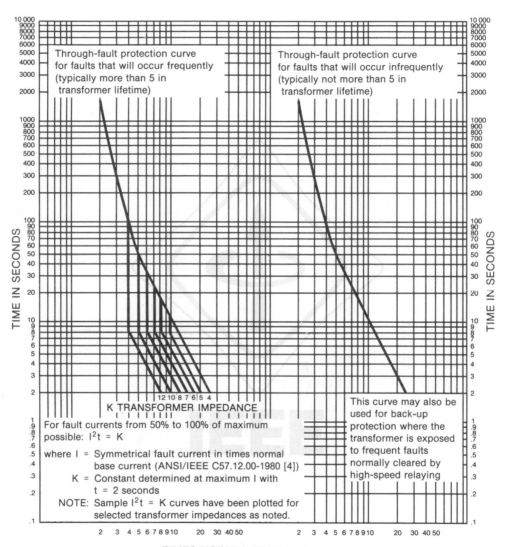

Fig 193
Through-Fault Protection Curves for Liquid-Immersed Category III Transformers
(1668 kVA to 10 000 kVA Single-Phase, 5001 kVA to 30 000 kVA Three-Phase)

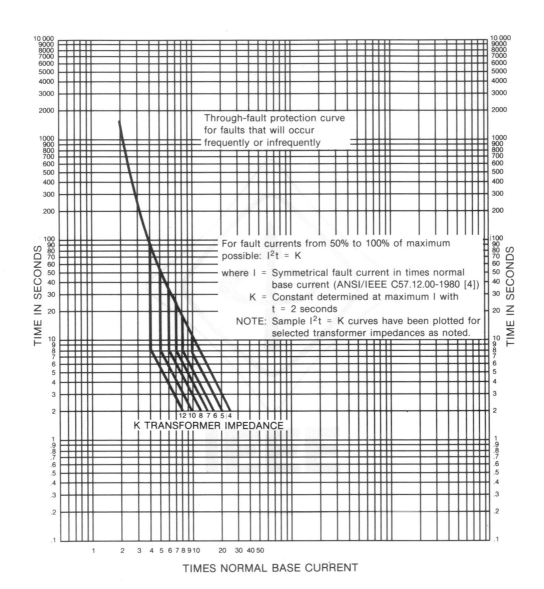

For fault currents from 50% to 100% of maximum
possible: $I^2 t = K$

where I = Symmetrical fault current in times normal
base current (ANSI/IEEE C57.12.00-1980 [4])

K = Constant determined at maximum I with
t = 2 seconds

NOTE: Sample $I^2 t = K$ curves have been plotted for
selected transformer impedances as noted.

Fig 194
Through-Fault Protection Curve for Liquid-Immersed Category IV Transformers
(Above 10 000 kVA Single-Phase, Above 30 000 kVA Three-Phase)

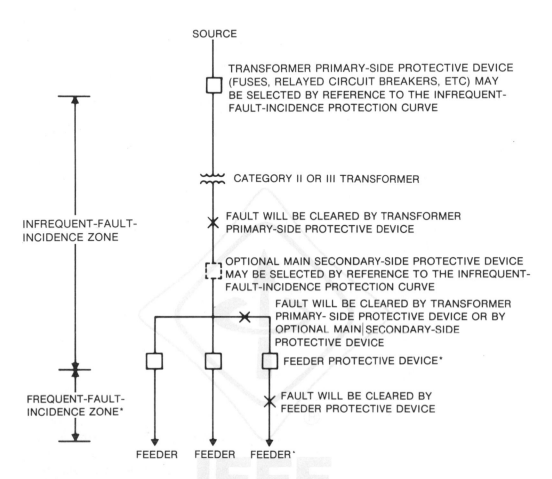

*Should be selected by reference to the frequent-fault-incidence protection curve. For transformers serving systems with secondary-side conductors enclosed in conduit, bus duct, etc, the feeder protective device may be selected by reference to the infrequent-fault-incidence protection curve.

Fig 195
Infrequent- and Frequent-Fault-Incidence Zones for Liquid-Immersed
Category II and III Transformers

primary where possible, short-circuit protection can be provided for the transformer, as well as a high degree of system selectivity.

Fuse selection considerations include having an interrupting capacity equal to or higher than the system fault capacity at the point of application, having a continuous current capability above the maximum continuous load under various operating modes, and having time-current characteristics that will pass the magnetizing- and load-inrush currents that occur simultaneously following a momenetary interruption without fuse operation, and will interrupt before the transformer withstand point is reached. Fuses so selected can provide protection for secondary faults between the transformer and the secondary-side overcurrent protective device, as well as backup protection for the latter.

The magnitude and duration of magnetizing inrush currents vary between different designs of transformers. Inrush currents of 8 or 12 times normal full-load current for 0.1 s are commonly used for coordination purposes.

Overload protection can be provided when fuses are used by utilizing a contact of the transformer temperature indicator to shed nonessential load or trip the transformer secondary-side overcurrent protective device.

When the possibility of backfeed exists, it is recommended that the switch, the fuse access door, and the transformer main secondary overcurrent protective device be interlocked to insure the fuse is de-energized before being serviced.

Relay protected systems can provide low-level overcurrent protection. Relay protection systems, as well as fused interrupter switches, can provide protection against single-phase operation when an appropriate open-phase detector is used to initiate opening of a circuit breaker or interrupter switch if an open-phase condition should occur.

10.8.3.4 Instantaneous Relays. Phase overcurrent relays with instantaneous elements provide short-circuit protection to the transformers in addition to overload protection. When used on the primary side, they usually coordinate with secondary protective devices. The setting of instantaneous relays is selected on its application with respect to secondary protective devices and circuit arrangements. The setting of instantaneous devices for short-circuit protection is described in [B10] for three-circuit arrangements.

10.8.3.5 Phase and Ground Differential Relays. Differential relaying compares the sum of currents entering the protected zone to the sum of currents leaving the protected zone; these sums should be equal. If more than a certain amount or percentage of current enters than leaves the protected zone, a fault is indicated in the protected zone and the relay operates to isolate the faulted zone.

Transformer differential relays operate on a percentage ratio of input current to through current; this percentage is called the slope of the relay. A relay with 25% slope will operate if the difference between the incoming and outgoing currents is greater than 25% of the through current and higher than the relay minimum pickup.

The fault-detection sensitivity of differential relays is determined by a combination of relay setting and circuit parameters. For most high-speed transformer differential relays, the relay pickup is about 30% of the tap setting. Depending on the setting, sensitivity will be about 25–50% of full-load current. For delta–wye connected transformers that are rated over about 10 000 kVA and that supply resistance-grounded systems, phase differential relays should be supplemented with secondary ground differential relays (87TG), as shown in Fig 196 to provide additional sensitivity to secondary ground faults. For more details on application of device 87TG, refer to Chapter 7 on ground fault protection.

Several considerations are involved in the application of differential relays:

(1) The system must be designed so that the relays can operate a transformer primary circuit breaker. If a remote circuit breaker is to be operated, a remote trip system must be used, by using either a pilot wire or a high-speed grounding

switch. Often the utility controls the remote circuit breaker and may not allow it to be tripped. Operation of a user-owned local primary circuit breaker presents no problem.

(2) Current transformers associated with each winding have different ratios, ratings, and characteristics when subjected to heavy loads and short circuits. Multiratio current transformers and relay taps may be selected to compensate for ratio differences. A less preferable but acceptable method is to use auxiliary transformers.

(3) Transformer taps can be operated changing the effective turns ratio. By selecting the ratio and taps for midrange, the maximum unbalance will be equivalent to half the transformer tap range.

(4) Current transformers of the same make and type are recommended to be used in the different windings to minimize error current due to the current transformer's different characteristics.

(5) Magnetizing current inrush appears as an internal fault to the differential relays. The relays should be desensitized to the current inrush, but they should be sensitive to short circuits within the zone during the same period. This can be accomplished using relays with harmonic restraint. The magnetizing current inrush has a large harmonic component, which is not present in short-circuit currents. This permits harmonic-restraint relays to distinguish between faults and inrush.

(6) Transformer connections often introduce a phase shift between high- and low-voltage currents. This is compensated for by proper current transformer connections. For a delta primary, wye secondary transformer, current transformers are normally wye connected in the primary and delta connected in the secondary.

(7) Heavy currents for faults outside the zone of protection can cause an unbalance between the current transformers. Percentage differential relays shown in Fig 197, which operate when the difference is greater than a definite percentage of the phase current, are designed to overcome this problem. Percentage differential relays also help in solving the tap-changing problem and the current transformer ratio balance problem. Percentage slopes available are 15% for standard transformers, 25% for load tap-changing transformers and 40% for special applications.

Harmonic-restraint percentage differential relays are recommended for transformers rated 5000 kVA and above.

Unlike the differential relays applied to protect high-voltage buses or large motors, the transformer differential relay application has both harmonics and phase shift to consider. Although all transformer differential relays do not include harmonic filters, the experience with harmonic filters has been beneficial and faster acting, and permits more sensitive pickups.

(8) A delta–wye, or wye–delta, transformer with the neutral grounded is a source (generator) of zero-sequence (ground) fault current. A ground fault on the wye side of the transformer, external to the differential protective zone, will cause zero-sequence currents to flow in the current transformers on the wye side of the transformer without corresponding current flow in the line CT's on the

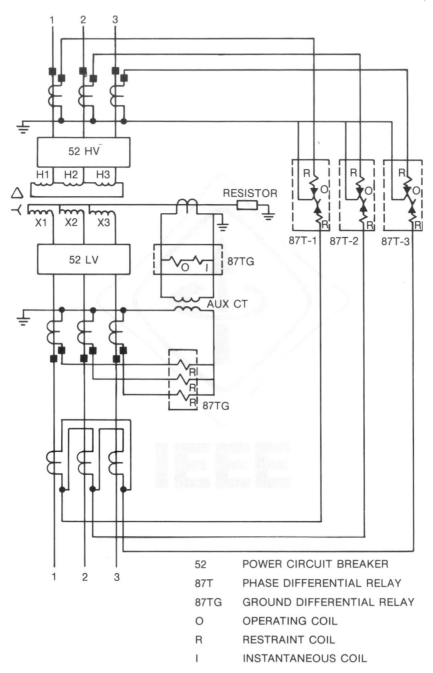

52	POWER CIRCUIT BREAKER
87T	PHASE DIFFERENTIAL RELAY
87TG	GROUND DIFFERENTIAL RELAY
O	OPERATING COIL
R	RESTRAINT COIL
I	INSTANTANEOUS COIL

Fig 196
Transformer Phase and Ground Differential Relay CT
and Current Coil Connections

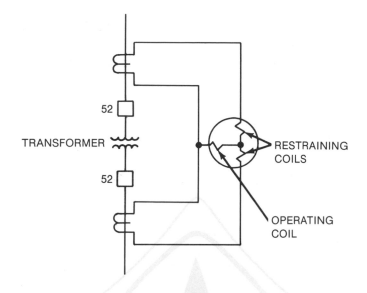

Fig 197
Percentage Differential Relays Provide Increased Sensitivity While Minimizing
False Operation As a Result of Current Transformer Mismatch Errors
for Heavy Through Faults

delta side of the transformer. If these zero-sequence currents are allowed to flow through the differential relays, they will cause immediate undesired tripping. To prevent such undesired tripping, the current transformer connections should be such as to cause the zero-sequence currents to flow in a closed-delta CT secondary connection of low impedance instead of in the differential relay operating coil. This is readily accomplished by connecting the CT's secondary in delta on the wye side of the transformer.

The protection for a single-phase transformer is shown in Fig 197, although most transformer differential relay applications would apply to three-phase transformers of 5 MVA and larger.

In Fig 197, there are shown two restraining windings and one operating coil. The CT ratios are selected so as to produce essentially equal secondary currents, such that under a no-fault condition the CT secondary current entering one restraining winding will continue through the other restraining winding, with no differential current to pass through the operating coil. Because of ratio mismatches in current transformers and relay tap settings, there may always exist some current in the operating coil circuit under a no-fault condition.

When a fault is internal to the differential relay zone, definite quantities of current will flow into the operating coil circuit. The relay will then respond to this differential current, and determine the ratio of the operating currents to the restraining winding currents. The relay will operate to trip when this ratio exceeds the slope setting (15, 25, 30, or 40% are usual available ratio settings) and is above the relay minimum sensitivity. The three-phase connection shown

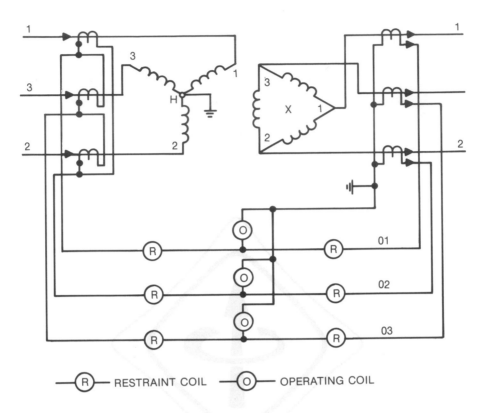

Fig 198
Typical Schematic Connections for Percentage Differential Protection
of a Wye-Delta Transformer

in Fig 198 illustrates a typical application for protection of a three-phase transformer. The transformer is connnected wye–delta, selected generally to provide an ungrounded secondary connection while permitting the primary wye neutral to be grounded solidly. Other configurations would be reversed, and the grounded wye would be the secondary connection. The basic delta–wye or wye–delta connection produces a phase shift between current entering the primary and current leaving the secondary. For this reason, the current transformers on the wye side have their secondaries connected in delta, and the current transformers on the delta side have their secondaries connected in wye.

In addition to the phase shift which is easily corrected, the magnitudes of the secondary currents rarely match each other because standard CT ratios are employed. To compensate for this, most percentage differential relays have selectable auto transformer taps (in a 3:1 range) at the input of each restraining winding. By following the relay instructions, the best match can be made so that the no-fault operating coil current is minimized. In some cases where high voltage switchyards are involved, the available relay adjustments are inadequate, and auxiliary current transformers or auto transformers are needed. This should be attempted only after a thorough examination of the effects of through faults and secondary burdens upon the primary current transformers.

Assuming that CT ratio and phase shift problems can be resolved, it should also be noted that often a transformer secondary may be connected to more than one bus. In that event, a separate restraining winding is required for each such bus. Paralleling CT secondaries in place of multiple restraining windings can lead to misoperation on through faults if the secondary buses are strong fault-current sources. If they are only weak sources, then paralleled CT secondaries are acceptable.

Harmonics in the primary circuit can develop during transformer energization, during overvoltage periods, and through faults, in other words, when a fault occurs outside of the transformer differential relay zone. The harmonics could cause differential relay misoperation if not recognized. For the most part, zero-sequence harmonics (3rd, 9th, etc) are excluded from the relays by the CT secondary connection.

Except for filtering the second harmonics for restraining purposes, experts disagree on the merits of filtering these other harmonics (5, 7, 11, 13, etc) for restraining. Present practice has been to filter the second harmonic and to apply it to the restraining winding when the magnitude of the second harmonic exceeds 20% of the fundamental current. Due to problems of misoperation, one manufacturer now begins the second harmonic restraint when the secondary harmonic current exceeds 7.5% of the fundamental current. During normal no-fault conditions, this earlier restraining action is beneficial, but this 7.5% setting makes the relay less sensitive on an internal fault.

Protection of the transformer by percentage differential relays improves the overall effectiveness in detecting phase-to-phase internal faults. However, ground faults in a wye winding cannot be discreetly detected if the transformer is resistance grounded and ground-fault current is limited to a low value below the differential relay pickup level. Such ground faults may lead to a destructive phase-to-phase fault. Where the transformer is solidly grounded, the transformer differential relay will operate for ground faults within the differential protective zone.

Two methods can be easily adapted for protecting the wye winding more effectively. Figure 199 illustrates one approach that employs an overcurrent relay in a differential connection. The zero-sequence currents are shown for an external fault. Properly connected, the secondary current will circulate for this external fault, but would be additive for an internal fault and cause the 51G device to operate.

The Fig 199 circuit is susceptible to through faults that may saturate the phase CT's and cause the 51G to operate. For this reason CT selections are more demanding and 51G settings are less sensitive than would originally appear.

Overcoming the CT and through-fault problems is the directional relay shown in Fig 200. The currents shown are for an external fault, and the secondary currents circulate as shown. However, upon an internal fault, the secondary currents are additive in the operating coil as shown in Fig 201. This directional relay has the additional element that prevents misoperation, and in fact permits

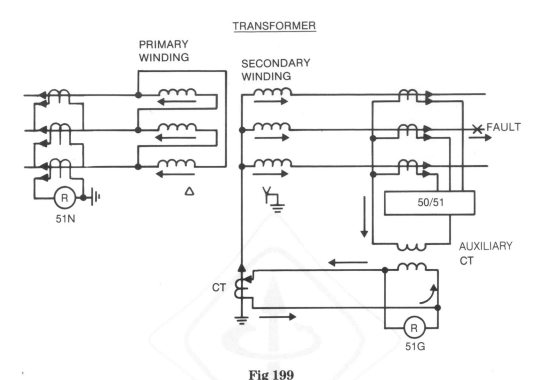

Fig 199
Complete Ground-Fault Protection for Delta–Wye Bank,
Using Residual Overcurrent and Differentially Connected Ground Relay;
Zero-Sequence Current Arrows Are for an External Ground Fault
for Which the Relay Will Not Operate

a faster acting relay, a product-type relay that can operate in less than a cycle. Comparing this operating time to the seconds taken by a 51G relay makes the choice more definitive.

In both ground-fault differential relay applications, selection of CT ratios is important. Should the neutral CT ratio be smaller than the phase CT (generally the case), the auxiliary CT in the residual secondary can correct this mismatch. Some users select the auxiliary CT ratio such that slightly more restraining current flows during an external fault, as shown in Fig 202. In effect, this excess secondary current will flow in the opposite direction in the operating winding, thus precluding false operation.

10.9 Protection from the Environment. In addition to electrical protection, protection is necessary against physical conditions in the environment that may affect reliable performance. Although most of these are obvious, they are important enough to be listed. Undesirable conditions are:

(1) Average ambient temperatures above 30 °C when the transformer is loaded at rated kVA or more

TRANSFORMER

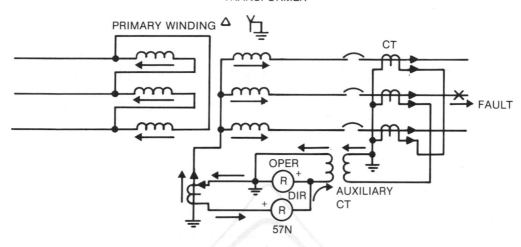

Fig 200
Directional Relay for Detection of Ground Faults in Grounded Wye
Connected Bank; Zero-Sequence Current Arrows Are for an
External Ground Fault for Which the Relay Will Not Operate

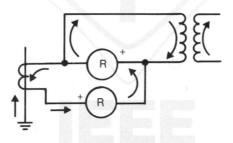

Fig 201
Relay Current During Transformer Internal Faults

(2) Corrosive agents, abrasive particulate matter, and surface contaminants derived from the surrounding atmosphere

(3) Conditions that can lead to moisture penetration or to condensation on windings and other internal electrical components

(4) Submersion in water or mud

(5) Obstruction to proper ventilation of liquid transformer radiators or, in the case of dry-type transformers, ventilating openings

(6) Access to damage from collision by vehicles

(7) Excessive vibration

(8) Access to vandalism

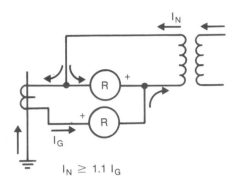

Fig 202
Relay Current During External Fault
When Auxiliary CT Ratio is Selected to Restrain

10.10 Conclusion. Protection of today's larger and more expensive transformers can be achieved by the proper selection and application of protective devices. Published application guides covering transformers are few in number; see, for example, ANSI/IEEE C37.91-1985 [3]. The system design engineer must rely heavily on his sound engineering judgment to achieve an adequate protection system.

10.11 References. The following publications shall be used in conjunction with this chapter.

[1] ANSI C57.96-1959, American National Standard Guide for Loading Dry-Type Distribution and Power Transformers.

[2] ANSI/IEEE Std 141-1986, IEEE Recommended Practice for Electric Power Distribution for Industrial Plants.

[3] ANSI/IEEE C37.91-1985, IEEE Guide for Protective Relay Applications to Power Transformers.

[4] ANSI/IEEE C57.12.00-1980, IEEE Standard General Requirements for Liquid-Immersed Distribution, Power, and Regulating Transformers.

[5] ANSI/IEEE C57.12.01-1979, IEEE Standard General Requirements for Dry-Type Distribution and Power Transformers.

[6] ANSI/IEEE C57.91-1981, IEEE Guide for Loading Mineral-Oil-Immersed Overhead and Pad-Mounted Distribution Transformers Rated 500 kVA and Less with 65 °C or 55 °C Average Winding Rise.

[7] ANSI/IEEE C57.92-1981, IEEE Guide for Loading Mineral-Oil-Immersed Power Transformers Up to and Including 100 MVA with 55 °C or 65 °C Winding Rise.

[8] ANSI/IEEE C57.94-1982, IEEE Recommended Practice for Installation, Application, Operation, and Maintenance of Dry-Type General Purpose Distribution and Power Transformers.

[9] ANSI/IEEE C57.104-1978, IEEE Guide for the Detection and Determination of Generated Gases in Oil-Immersed Transformers and Their Relation to the Serviceability of the Equipment.

[10] ANSI/IEEE C57.106-1977, IEEE Guide for Acceptance and Maintenance of Insulating Oil in Equipment.

[11] ANSI/IEEE C57.109-1985, IEEE Guide for Transformer Through-Fault-Current Duration.

[12] ANSI/NFPA 70-1984, National Electrical Code.

10.12 Bibliography

[B1] AIEE Committee Report. Bibliography of Industrial System Coordination and Protection Literature. *IEEE Transactions on Industry Applications,* vol IA-82, Mar 1963, pp 1–2.

[B2] *Applied Protective Relaying.* Newark, NJ: Westinghouse Electric Corporation.

[B3] *The Art of Protective Relaying.* Philadelphia, PA: General Electric Company, Bulletin 1768.

[B4] BEEMAN, D. L. Ed. *Industrial Power Systems Handbook.* New York: McGraw-Hill, 1955.

[B5] BOYARIS, E. and BUYOT, W. S. Experience with Fault Pressure Relaying and Combustible Gas Detection in Power Transformers. *Proceedings of the American Power Conference,* vol 33, Apr 1971, pp 1116–1126.

[B6] BRUBAKER, J. F. Fault Protection and Indication on Substation Transformers. *IEEE Transactions on Industry Applications,* vol IA–14, May/June 1977.

[B7] BURGIN, E. R. A Comparison of Protective Methods and Devices for Industrial Power Transformers. *Proceedings of the American Power Conference,* vol 26, Apr 1964, pp 931–938.

[B8] DICKINSON, W. H. Report on Reliability of Electric Equipment in Industrial Plants. *AIEE Transactions on Industry Applications,* pt II, vol IA-81, July 1962, pp 132–151.

[B9] MASON, C. R. *The Art and Science of Protective Relaying.* New York: Wiley, 1956.

[B10] MATHUR, B. K. *A Closer Look at the Application and Setting of Instantaneous Devices.* Presented at Industrial and Commercial Power Systems Technical Conference, 1979, Seattle, WA, Paper #CH1460-5/79/0000-0107.

[B11] ZOLAR, D. A. A Guide to the Application of Surge Arresters for Transformer Protection. *IEEE Transactions on Industry Applications,* vol IA–15, Nov/Dec 1979.

11. Generator Protection

11.1 Introduction. Industrial and commercial power systems may include generators as a local source of energy. These generators supply all or part of the total energy required, or they provide emergency power in the event of a failure of the normal source of energy. The application of generators can be classified as single isolated generators, multiple isolated generators, and large industrial generators.

Generator protection requires the consideration of many abnormal conditions that are not present with other system elements. Where the equipment is unattended, it should be provided with automatic protection against all harmful conditions. In those installations where an attendant is present it may be preferable to alarm on some abnormal condition rather than remove the generator from service. Generator protective schemes will vary depending on the objectives to be achieved.

11.2 Classification of Generator Applications

11.2.1 Single Isolated Generators. Single isolated generators are used to supply emergency power or for standby service and are normally shutdown. They are operated for brief periods of time when the normal source fails or during maintenance, testing, and inspection. They are connected to the system load through an automatic transfer switch or through interlocked circuit breakers and are not operated in parallel with other system power sources. They are diesel engine or gas turbine driven with ratings from less than 100 kW up to a few thousand kilowatts. Generation is usually at utilization voltage level, typically 480 or 480Y/277 V, but with larger machines the voltage may be 2.4 or 4.16 kV. These generators are designed to start, operate during a power failure, and to shutdown when normal power is restored through automatic controls.

11.2.2 Multiple Isolated Generators. This classification consists of several units operating in parallel without connection to any electric utility supply system. Examples of these installations are *total energy* systems for commercial and industrial projects, offshore platforms for drilling and production of energy sources, and other remote sites requiring continuous electric energy. The size of the individual generators may range from a few hundred kilowatts up to several thousand kilowatts depending on the system demand. The prime movers are typically gas turbines and oil or gas fueled diesel engines. These systems are normally operated manually but load sensing controls and automatic synchronizing relays can be used. The rated voltage of these generators is usually at the utilization voltage or the highest distribution voltage level, or both, such as 4.16 or 13.8 kV.

11.2.3 Large Industrial Generators. These are bulk power producing units that operate in parallel with an electric utility supply system. All power generated is normally utilized by the industrial user. These units are used where there is a demand for low pressure process steam, such as in petrochemical installations and pulp and paper plants. The generator size may range from 10 000–50 000 kVA. Operation is on a continuous basis at or near rated load, but may vary seasonally. The prime movers are usually steam or gas turbines depending on the process requirements, fuel availability, and system economics. Generation is usually at the industrial plant systems' highest voltage level, typically 12.47 or 13.8 kV. The majority of these machines are attendant-operated.

11.3 Short-Circuit Performance

11.3.1 General Considerations. The proper application of several generator protective device functions requires the knowledge of the short-circuit performance of the generator. The magnitude of generator fault current is a function of the armature and field characteristics, time, and the loading conditions immediately preceeding the fault. The ability of the generator to sustain an output current during a fault is determined by the characteristics of the excitation system.

11:3.2 Excitation Systems. There are excitation systems that do not have the ability to sustain the short-circuit current. The magnitude of fault current is determined only by the subtransient and transient reactances and will decay as determined by their respective time constants, and can be essentially zero in 1.0–1.5 s. An example of a system with these characteristics is the static exciter generator, which obtains all of its excitation energy from the generator terminal voltage. The excitation systems of round rotor generators, typically those two-pole machines above 10 MVA, have the capability to support a sustained three-phase fault current corresponding to that limited by transient reactance for a period of several seconds. These machines may have a brushless excitation system, although some units may be equipped with a static exciter using slip rings to obtain its excitation energy from both generator potential and current transformers.

Salient pole machines that range in size from a few kilowatts up to the round rotor machine sizes will typically be capable of supporting a fault current at 300% of generator full-load current. Such units are typically of the brushless exciter design where the exciter delivers three-phase ac to rotating rectifiers that are connected directly to the field. Excitation energy to the exciter field is supplied to maintain the maximum fault-current magnitude.

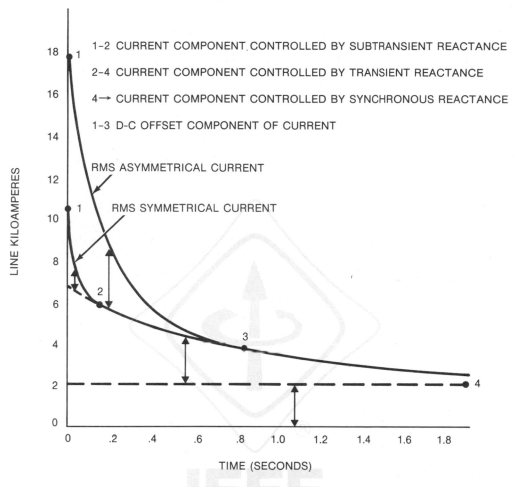

18 1-2 CURRENT COMPONENT CONTROLLED BY SUBTRANSIENT REACTANCE

2-4 CURRENT COMPONENT CONTROLLED BY TRANSIENT REACTANCE

16 4→ CURRENT COMPONENT CONTROLLED BY SYNCHRONOUS REACTANCE

1-3 D-C OFFSET COMPONENT OF CURRENT

14

RMS ASYMMETRICAL CURRENT

12

RMS SYMMETRICAL CURRENT

Fig 203
Typical Generator Decrement Curve

11.3.3 Generator Decrement Characteristics. The current output of a generator with a fault at or near its terminals consists of two components, both of which have a time variable rate of decay, depending on machine constants. The two components are the symmetrical ac current, i_{ac}, and the unidirectional offset current, i_{dc}. The ac component will decay with time according to the pattern shown in Fig 203. The sudden drop, 1–2, is a function of the subtransient values of machine internal voltage, reactance, and short-circuit time constant. The more gradual drop toward steady state, 2–4, is a function of the transient values of machine internal voltage reactance, and short-circuit time constant. The dc offset current, 1–3, is a function of the subtransient reactance and the armature short-circuit time constant. The steady-state component, 4, and beyond is a function of the generator synchronous reactance and field current. The maximum symmetrical current that a generator can deliver on a bolted three-phase fault is determined by the subtransient reactance x_d''. This reactance will range from a minimum of 9% for a two-pole round rotor machine to 32% for a low-speed salient-pole hydrogenerator. Thus, the initial symmetrical fault current (1–5 cycles) can be as great as 11 times the generator full-load current. In the intermediate time span 5–200 cycles) the transient reactance, x_d', will determine the magnitude of the ac component. The synchronous reactance, which may vary from a value of 120–240%, determines the sustained value of the ac component of the fault current. The sustained fault current, assuming no initial load and no change in the voltage regulator setting, can be as low as 42% of the generator full-load current. Since the no-load fixed field current results in the longest relay operating times, the *stuck regulator* condition should be considered as the criteria for the relay settings. Regulator response will normally occur, however, producing sustained fault currents at much higher levels. The generator decrement curves can be calculated from the following procedure. The initial loading condition, initial terminal voltage, and the field forcing capability can all be included in these calculations.

11.3.3.1 The total ac component of armature current consists of the steady-state (i_d) value and two components that decay at a rate according to their respective time constants.

$$i_{ac} = (i_d'' - i_d')\epsilon^{-t/T_d''} + (i_d' - i_d)\epsilon^{-t/T_d'} + i_d$$

(1) Subtransient Component, i_d'':

$$i_d'' = \frac{e''}{X_d''} \text{ pu}$$

$$e'' = e_t + X_d'' \sin \theta \text{ pu}$$

When machine is at no-load, $e'' = e_t$.

(2) Transient Component, i_d':

$$i_d' = \frac{e'}{X_d'}$$

$$e' = e_t + X_d' \sin \theta \text{ pu}$$

Again, at no-load, $e' = e_t$.

(3) Steady-State Component, i_d. The steady-state component is the current finally attained, and is a function of the field current:

$$i_d = \frac{e_t}{X_d} \left(\frac{I_F}{I_{Fg}}\right)$$

I_F is the actual prefault amperes at the initial loading conditions, either no-load or full-load; or, when regulator action is taken into consideration, it is the field amperes with maximum excitation voltage applied.

11.3.3.2 The dc component of armature current is controlled by the subtransient reactance and the armature time constant:

$$i_{dc} = \sqrt{2} \, i_d'' \epsilon^{-t/T_A}$$

11.3.3.3 The total rms current is the sum of the two components as follows:

$$i_{tot} = \sqrt{(i_{ac})^2 + (i_{dc})^2}$$

11.3.3.4 The terms used in the above expressions are defined as follows and are normally obtained from the generator manufacturer.

X_d'' = subtransient reactance, saturated value
X_d' = transient reactance, saturated value
X_d = synchronous reactance
e_t = machine terminal voltage, pu
e'' = machine internal voltage behind X_d''
e' = machine internal voltage behind X_d'
T_d'' = subtransient short-circuit time constant, in s
T_d' = transient short-circuit time constant, in s
T_A = armature short-circuit time constant, in s
I_{Fg} = field current at no-load rated volts
I_F = field current at given load condition
θ = load power factor angle

Example Calculation. The following data was obtained from the generator manufacturer for a round rotor machine. Decrement curves, shown in Fig 204, are plotted for (1) constant excitation at no initial load, (2) field current at 3 per unit of no-load value, and (3) total current trace of (2), which includes the dc component. The machine characteristics are:

kVA = 19 500	$x_d'' = 10.7\%$	$T_d'' = 0.015$ s
PF = 0.8	$x_d' = 15.4\%$	$T_d' = 0.417$ s
voltage = 12.47 kV	$x_d = 154\%$	$T_A = 0.189$ s
rated amperes = 903	$I_{Fg} = 1$ pu	
	$I_F = 3$ pu	

11.4 Protective Devices

11.4.1 Backup Overcurrent Protection—Device 51V*. The function of generator backup protection is to disconnect the generator if a system fault has not been cleared by other protective devices after a sufficient time delay has elapsed.

This function serves to protect the distribution system components against excessive damage and to prevent the generator and its auxiliaries from exceeding their thermal limitations. In industrial and commercial application, where the generator is connected to a bus that serves distribution and utilization equipment using overcurrent devices, the overcurrent relay, ANSI Device Function 51V, is used. Where the output of the generator is stepped up to a transmission voltage, an impedance relay, ANSI Device Function 21, is normally used.

Users and system designers are reluctant to use any relay that operates solely on overcurrent for fear that it might trip off the generator when the load demand on it is the greatest. The use of ordinary time overcurrent relays present a dilemma in attempting to determine the proper current and time settings. If the

---

* ANSI Device Function Numbers are defined in ANSI/IEEE C37.2-1979 [4].

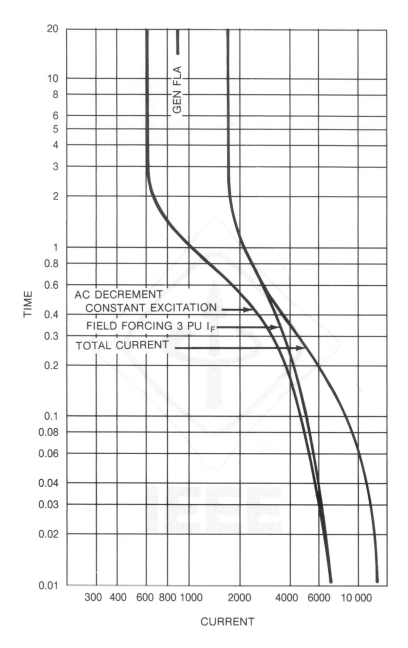

Fig 204
Generator Decrement Curve for a 19 000 kVA Generator

current and time settings are too low, the relay may trip the generator unnecessarily on normal overloads. If the settings are too high to allow for the proper coordinating time interval for selectivity with downstream devices, the relay may not respond at all due to the decaying characteristic of the generator fault current. Thus ordinary overcurrent relays cannot be used without the likelihood of false operation occuring. In the successful application of these relays it is, therefore, necessary to know not only the fault characteristics of the generator, but also the system emergency conditions, which will call for backup relay operation and the characteristics of the relays available for that purpose.

11.4.1.1 Relay Description. The overcurrent relays that are used are specially constructed to make their operating characteristics a function of voltage as well as current. As the magnitude of the voltage applied to the coil decreases from rated value, the time-current characteristic is modified so that the relay becomes more sensitive. There are two types of these relays that are customarily used: the *voltage restraint* and the *voltage controlled* overcurrent relay.

(1) The voltage restraint overcurrent relay consists of a conventional induction disk overcurrent unit with a voltage element so constructed that it applys a torque that opposes the operating torque produced by the current coil. This restraining torque is proportional to voltage and effectively controls the pickup current of the relay over a 4:1 range. The relay is calibrated and rated for a range of tap settings, for example, 4–16 A, with 100% V applied to the restraint coil. As the voltage is reduced the current required to operate the relay at a given tap setting drops, giving an infinite series of characteristic curves. The performance at selected values of voltage is given as follows:

% Rated Volts	Pickup, % Tap Setting
100	100
78	78
48	52
0	25

(2) The voltage-controlled overcurrent relay is a low-burden induction-disk time-overcurrent relay. It is torque controlled by a high-speed voltage relay continuously adjustable over a range of 65–83% of rated voltage. When the applied voltage is above the pickup setting, it's contacts connected in the shading coil circuit of the overcurrent element are open and no operating torque is produced regardless of current magnitude. When the applied voltage falls below the dropout value, the contacts in the shading coil circuit close enabling the relay to produce torque and operate as a conventional overcurrent relay. If the current is above the tap setting, then it will operate in accordance with its inherent time-current characteristics. A clear distinction can thus be made between normal *no fault* conditions and abnormal *fault* conditions. Several manufacturers offer a solid-state version of the voltage controlled overcurrent relay. It consists of time overcurrent logic circuits to give the performance just described.

11.4.1.2 Settings. An example of the considerations for selecting the tap and time dial settings for both the voltage restraint and voltage-controlled overcurrent relays will show the basic differences between the two types. Each relay is to be applied on the 19 500 kVA generator whose decrement curves used are those calculated and plotted in Fig 204. The criteria for selecting the relay tap setting are:

(1) The relay should pickup on synchronous current for 0 V to the relay (constant excitation).

(2) The relay should *not* pickup for moderate overloads up to 150% of generator full-load current with 100% V on the potential coil.

11.4.1.2.1 Voltage-Controlled Relay. The tap values of this relay are based on the current flowing when the voltage is below the dropout setting on the potential coil. With the voltage above the coil pickup the relay is effectively removed from the circuit. Calculate the tap setting so that it is less than the sustained fault current of 586 A.

$$I_{\text{fault}} = \frac{I_{FLA}}{X_d} = \frac{903}{1.54} = 586 \text{ A}$$

$$\text{tap value} = \frac{I_{\text{fault}}}{\text{CT ratio}} = \frac{586}{1200/5} = 2.44 \text{ A}$$

Select a 2 A tap.

The relay will operate since the sustained fault current is 1.22 multiples of relay pickup, as illustrated in Fig 205. This relay should have either a 2–6 or a 0.5–2.5 A tap range. (Using a relay with the lower tap range a 1.5 A tap could be selected for even greater sensitivity, that is, 1.63 multiples.)

11.4.1.2.2 Voltage Restraint Relay. Since the current tap values are based on normal operating conditions, that is, 100% restraint voltage applied, calculate the tap setting to be equal to or greater than 150% of full-load current.

$$\text{tap value} = \frac{I_{FL} \cdot 1.5}{\text{CT ratio}} = \frac{903 \cdot 1.5}{1200/5} = 5.6 \text{ A}$$

Select a 6 A tap (159% of full-load).

Verify that the relay will operate at 0 V. At 0 V the relay sensitivity is reduced to 25%, so

$$I_{\text{pickup}} = 6 \cdot 0.25 \cdot 1200/5 = 360 \text{ A}$$

From the decrement curve in Fig 206, it is determined that the relay will pickup since the sustained fault current is 586 A, or 1.63 multiples of pickup. The relay selected should have either a 2–8 or 4–16 A tap range.

Two observations can be made from the above selection process. First, the type of relay selected will make a significant difference in the proper choice of tap range. Second, with the tap set as described, it should be observed that from the inception of the fault the current in the relay will always be above its pickup value, and thus, the relay contacts will ultimately close.

11.4.2 Ground Overcurrent Protection–Device 51G. The ground-fault overcurrent relay provides backup protection for all ground relays in the system at the generator voltage level. It also affords protection against internal generator ground faults, but this protection is limited by the amount of time delay that the relay must have to coordinate with other ground relays. On small isolated machines this device and Device 51V (when the CT's for it are installed on the neutral side) provide the only protection for internal generator faults. In

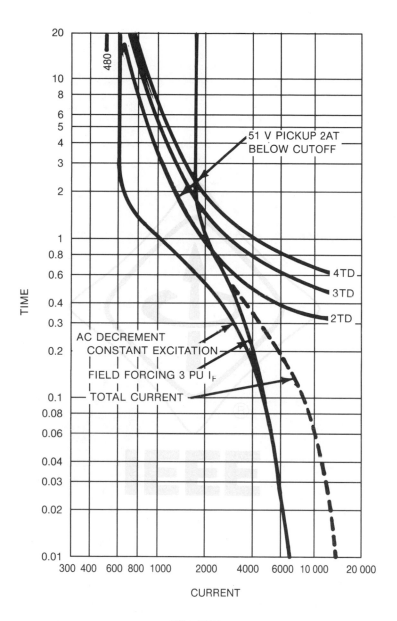

Fig 205
Voltage-Controlled Relay
19 500 kVA Generator Fault Current Decrement Curves
with Backup Overcurrent Relays

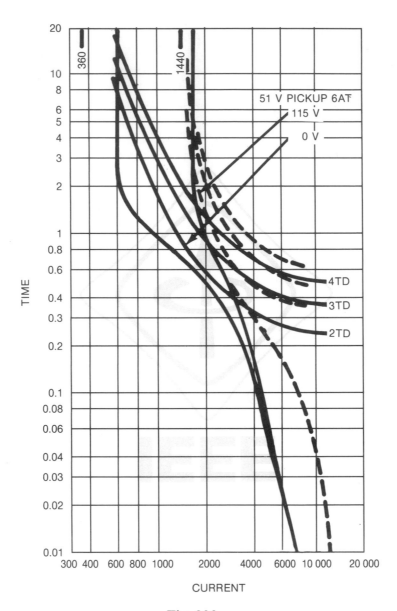

Fig 206
Voltage Restraint Relay
19 500 kVA Generator Fault Current Decrement Curves
with Backup Overcurrent Relays

applications where the output of the generator is utilized directly at the distribution or utilization voltage the method of grounding is determined by the system to which it is to be connected. The solid grounding of generator neutrals is normally restricted to systems rated 600 V and below, whereas low- or high-resistance grounding of generators is employed on voltage levels above 600 V.

When low-resistance method of grounding of a generator is selected, the neutral CT turns ratio and tap setting of the overcurrent relay should be selected to provide an operating current of 5–10 times its pickup setting for a bolted line to ground fault. The pickup setting should be at least equal to and preferably greater than the pickup setting of downstream ground overcurrent devices for selectivity between those relays. The pickup setting should also be set above the anticipated level of any harmonic current flowing in the neutral during normal conditions. Many European generator designs have a winding configuration that produces a higher magnitude of harmonic currents than domestic machines, sufficient to preclude the use of machine neutral grounding. In these cases the use of a zig-zag grounding transformer should be considered to establish the source grounding point.

When the high-resistance method of grounding is encountered, normally used on generators having its winding isolated by a delta–wye connected stepup transformer, a voltage relay, Device 59N, may be connected across the resistor to sense the voltage that will appear across the resistor during a line to ground fault. A common improvement of this method involves the use of a distribution transformer connected in the neutral. Since the secondary is rated at either 120 or 240 V the resistor connected in this winding is physically smaller and less fragile than one connected directly in the generator neutral.

This ground-fault protection requires a very sensitive overvoltage relay that is capable of discriminating between voltage produced by 60 cycle fault current and that produced by third-harmonic load current. Special relays are available that use tuned circuits to render them immune to the third-harmonic voltage or its multiples.

11.4.3 Differential Relay—Device 87. Differential relays provide a method for rapidly detecting internal generator phase-to-phase or phase-to-ground faults. After the detection of these faults, the generator is quickly removed from service to limit the extent of damage. For a further discussion on the basic principles of differential protection the reader is referred to Chapter 4 on protective relays. The self-balancing generator differential scheme, as shown in Fig 207, may be used for generator protection when suitable window-type current transformers can be applied. Since both ends of the generator windings must be passed through the opening in the current transformer, this scheme is normally employed only on generators where the flexibility of the conductors will permit this connection. Current transformers are usually mounted in the generator terminal box, thus excluding generator cables from the protective zone of the scheme. The ratios of the window-type current transformer may be quite small since their value is independent of the generator load current. In addition, since the relays are also insensitive to the generator load current, instantaneous

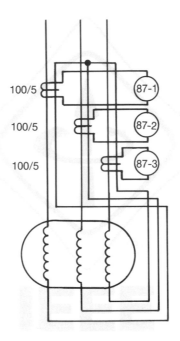

Fig 207
Self-Balancing Generator Differential Relay Scheme

overcurrent relays may be used, set at pickup levels as sensitive as 5–10 primary amperes.

Percentage differential relays connected, as shown in Fig 208, are normally used for protection of larger generators. Current transformers should have a primary current rating equal to or larger than the generator rated current, preferably at least 150%, and they should also have closely matched performance characteristics to prevent false operation on faults outside the differential protection zone. Variable percentage differential relays will make the scheme more tolerant of current transformer mismatch, thus reducing the relay's sensitivity to external faults. Typical characteristic slope values for variable percentage differential relays are 5% for very low values of fault current and increase to 50% at high fault current levels. Fixed percentage differential relays have constant slope characteristics of 10% or 25% for all fault current values. Induction disk, induction cup, or static relay types are commercially available for use in a percentage differential protection schemes. The operating time of standard speed induction disk relays is a minimum of 6 cycles, and high-speed induction cup or static relays operate in 1.2–1.5 cycles. Normally, when a percentage differential scheme is used, it is desirable to locate the current transformers so that they include the generator breaker within the protective zone. This gives maximum protection to the circuit breaker and bus duct or cable connecting the generator to the power system.

When the generator neutral is grounded through an impedance limiting the maximum ground-fault current to a value less than the primary rating of the current transformers, a differential ground-fault scheme, ANSI Device 87G, should be considered. When the ground-fault current is limited to values below the current transformer rating the sensitivity of phase differential relays on ground faults is reduced. A differential ground-fault scheme, however, is capable of detecting faults to within 10% from the generator neutral. A current polarized directional ground relay shown in Fig 208 may be used in this application. A neutral current transformer, separate from the ground overcurrent relay current transformer may be necessary for maximum relay sensitivity. The ground differential current transformer should have a primary current rating of 10–50% of the neutral resistor current rating. Figure 209 shows how a single percentage differential relay or a directional product type relay may alternately be used for sensitive ground-fault protection. To match the phase and neutral current transformer ratios, an auxiliary current transformer must be utilized. Again, a careful review of the auxiliary current transformer performance characteristics should be made to prevent misoperation on external faults.

11.4.4 Reverse Power Relay–Device 32. This device function provides backup protection for the prime mover rather than for protection of the generator. It detects the reverse flow of power (watts) that would occur should the prime mover lose its input energy, that is, throttle valve closes, without the accompanying trip of the generator feeder breaker. Under such conditions the generator would start to *motor*, drawing real power from the system. A steam turbine could overheat due to the loss of the cooling effect provided by the steam, a diesel or gas

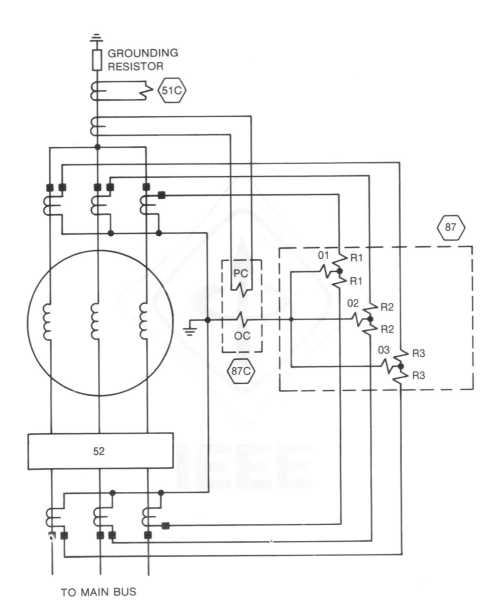

Fig 208
Generator Percentage Differential Relay (Phase Scheme) and
Ground Differential Scheme Using a Directional Relay

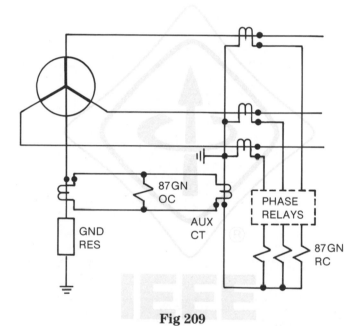

**Fig 209
Percent Differential Relay Used for Ground Differential Protection**

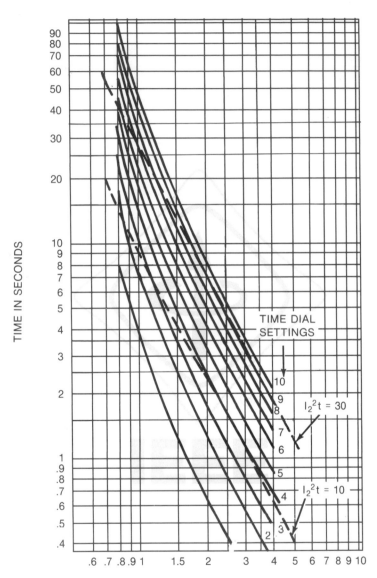

NEGATIVE-PHASE-SEQUENCE CURRENT (I₂) IN PER UNIT VALUES
OF TAP SETTING

Fig 210
Characteristics of Negative Sequence Overcurrent Relay
Showing Generator $I_2^2 t$ Limits

Table 48
Maximum Motoring Power for Prime Movers

Steam turbine	3%
Water wheel turbine	0.2%
Gas turbine	50%
Diesel engine	25%

engine could either catch fire or explode. This relay is a backup for the mechanical devices normally furnishing this type of protection.

The magnitude of motoring power varies considerably depending on the type of prime mover, as shown in Table 48. The reverse power relay should have sufficient sensitivity such that the motoring power provides 5–10 times the minimum pickup power of the relay. An induction disk directional power relay is frequently used to introduce sufficient time delay necessary to override momentary power surges such as might occur during synchronizing. A time delay of 10–15 s is typical. Either a single-phase or a three-phase relay may be used although a single-phase relay calibrated in three-phase watts is frequently selected. Solid-state relays are also available that provide the same functions with equivalent sensitivity.

11.4.5 Phase Balance Current Relay—Device 46. Unbalanced loads, unbalanced system faults, open conductors, or other unsymmetrical operating conditions result in an unbalance of the generator phase voltages. The resulting unbalanced (negative sequence) currents induce double system frequency currents in the rotor that quickly cause rotor overheating. Serious damage to the generator will occur if the unbalanced condition is allowed to persist indefinitely. The ability of a generator to withstand these negative sequence currents is defined by ANSI C50.13-1977 [1][23] as

$$I_2^2 t = K$$

where the negative sequence current is expressed in per-unit of the full-load current and the time is given in seconds. The standard further defines the value of constant K for various types of generators as listed in Table 49.

Generators under 100 MVA are generally able to carry negative sequence currents of up to 8–10% of full-load current continuously without injurious overheating.

A negative sequence overcurrent relay is recommended where protection from this unbalanced condition is desired. It consists of a time overcurrent unit with extremely inverse characteristic matching the generator $I_2^2 t$ curves as shown in Fig 210. The input to the relay is connected through a filter that passes only negative phase sequence currents. The time dial setting determines the level of

---

[23] The numbers in brackets correspond to those of the references listed at the end of this chapter.

Table 49
K Value for Various Generators

Type of Machine	Permissible $I_2^2 t$
Salient-pole generator	40
Synchronous condenser	30
Cylindrical-rotor generators	
Indirectly cooled (air)	30
Directly cooled (H_2)	10

protection offered by the relay and should be set to match the $I_2^2 t$ limit of the generator being protected. The relay is available in both an electromechanical and a static type construction. The static type is generally more sensitive, capable of detecting negative sequence currents down to the continuous capability of the generator. A sensitive alarm unit is offered on most relays that is capable of detecting and alarming for values of I_2, which are near the trip setting.

11.4.6 Loss of Field Protection—Device 40. This device senses when a generator's excitation system has been lost. This protection is important when generators are operating in parallel or in parallel with a utility supply system, although it is not needed on a single isolated unit. Should a generator lose its field excitation, it will continue to operate as an induction generator obtaining its excitation from the other machines on the system. When this happens, the generator rotor quickly overheats due to the slip-frequency currents induced in it. The system itself is also jeopardized because it is forced to supply the lost kVAR output of the machine in trouble, plus provide still more kVAR's in order to excite the unit as an induction generator. There is the danger that the system would not have sufficient kVAR capacity for such a condition (thereby causing instability) or, having the kVAR capacity, the other machines' excitation systems would be operated at dangerously high levels, thus causing overheating. There are at least three types of protective devices that can be used to provide this protection each having different relative cost, complexity of application, and degree of protection offered. The type chosen will be dependent upon the application considering such factors as generator cost, relay cost, and importance of the generator output.

The premium type of protection functions according to the principles of an impedance relay. It detects loss of excitation by the apparent impedance exhibited by the machine under such conditions. Figure 211 illustrates the typical characteristics of such a relay plotted using resistance (R) and reactance (x) coordinates. The generator full-load operating condition is represented by point A, and when excitation is lost, such as from a shorted field winding, the apparent impedance of the generator will trace a locus of points ending inside the

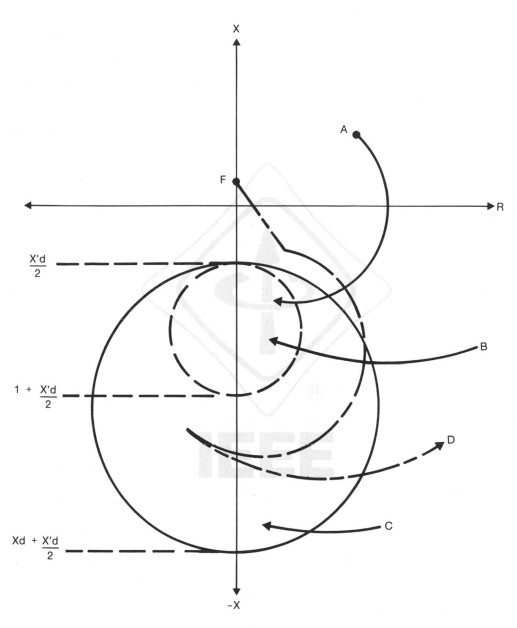

**Fig 211
Loss of Field Impedance Relay Characteristics**

operating characteristic. The curve from B represents a locus of points of a moderately loaded generator, and the curve from C represents the locus of a lightly loaded generator. For more sensitive protection a second impedance element can be added that operates in a more restricted area of impedance values as represented by the dotted characteristics. This impedance element will promptly respond to excitation conditions represented by A and B curves. The first element is then equipped with a time delay relay set for 0.5 s or greater so that upstream fault initiated system swings which may momentarily intrude into the operating area of the relay, illustrated by curve D, will not trip the generator. Loss of field on lightly loaded generators will still be cleared after the preset time delay, but with the inherent adverse effect on the system. This relay can be provided with slightly differing characteristics, but the ultimate quality of protection will be the same.

A second type of loss of field protection can be provided where the expense of the impedance type relay cannot be justified. It is possible to obtain relatively effective loss of excitation protection from a reverse power relay connected so that it operates on reverse kVAR flow. The principle limitation is that it cannot always distinguish between the occasions when the generator is operating at a leading power factor because of system conditions and similar appearing VAR flows due to loss of excitation.

The third protective scheme available involves the application of a dc undercurrent relay that is connected in series with the field. Its low cost makes it attractive for small generators and those supplying noncritical loads where the generator has leads brought out through conventional slip rings. The relay may require the use of a timer in order to ride through momentary interruptions of current that might occur during short circuits in the power system. The relay would not indicate the loss of excitation due to faults in the field winding and may not operate during to the presence of induced ac currents in the field winding during certain operating conditions.

11.4.7 Undervoltage Relay—Device 27. The undervoltage relay can be used to serve any one of several protective functions depending on the voltage tap and time dial setting. The automatic voltage regulator will normally maintain the voltage within specified limits on multiple isolated systems, therefore, a sustained undervoltage could indicate a severe overload condition or the loss of a generator. The relay may be used to initiate the starting of a standby unit. For single machine operation the relay could be used to remove load from the generator should a regulator failure or other malfunction cause the unit to be unable to maintain proper voltage. The relay may also be used to provide a form of single-phase short-circuit protection because close-in or internal faults would normally depress the voltage sufficiently to cause the relay operate. On isolated systems with multiple generators the undervoltage relay may be used to provide a backup to the backup protective devices. In this application a time delay of 15–20 s is necessary in order to give all other relays an opportunity to operate.

11.4.8 Overvoltage Protection—Device 49. Overvoltage protection is normally provided on machines such as hydrogenerators where excessive terminal voltages may be produced following load rejection without necessarily exceeding the V/Hz limit of the machine. In general, this is not a problem with steam or gas turbines since the rapid response of the speed governor and voltage regulator systems preclude this possibility. Many generator excitation systems have integral V/Hz limiters that prevent the overvoltage condition from occurring.

11.4.9 Voltage Balance Relay—Device 60. A voltage balance relay is used to continuously monitor the availability of potential transformer (PT) voltage and to block the operation of protective relays and control devices that will operate incorrectly when a potential transformer fuse opens. This application requires two sets of PT's on the generator circuit, one set supplying potential for the backup overcurrent, directional power, and loss of excitation relays; the other set supplying potential for the voltage regulator, synchronizing relays, and metering devices. In those cases where two sets of potential transformers on the genertor circuit cannot be justified, the bus potential transformers may be used as the second set so long as *dead bus* startup of the generator is not necessary.

When the two sets of PT's have ouput voltage alike, the relay is balanced and both the right and left contacts are open. When a fuse opens in any phase of one set of potential transformers, the unbalance will cause the left contact to close, which may both alarm and block the tripping of protective devices 32, 40, and 51V. When a fuse opens in the second set of PT's, the right contacts close, which may operate an alarm and also switch the voltage regulator to manual operation to prevent it from rising to ceiling voltage.

11.4.10 Generator Field Protection—Device 64F. Generator field circuits are normally operated ungrounded. Thus, a single ground fault will not result in equipment damage, or affect the operation of the generator. If, however, a second ground fault should occur, there will be an unbalance in the magnetic field established by the rotor. This unbalance may be severe enough to develop destructive vibration within the generator. A generator field ground relay is normally used to detect the first field ground and sound an alarm indicating the ground has occurred. Since most generators continue operation with a single field ground, larger generators should be equipped with vibration monitoring equipment so that a second ground will not result in an abnormal amplitude of vibration. The vibration monitor should be arranged to sound an alarm or shutdown the generator and prime mover if a high level of vibration occurs, or both.

For excitation systems using slip rings and brushes, two types of relays are available. The first incorporates a separate grounded voltage source and a current sensitive relay. The ungrounded side of the source is connected in-series with the relay and with one side of the field circuit in a manner, such that any ground occuring in any part of the field will operate the relay. The second type of relay uses a sensitive D'Arsonval dc relay connected between ground and the center of a voltage divider positioned across the field circuit. One side of the voltage divider also includes a voltage sensitive resistor that shifts the null point as the excitation level changes, and thus ensures relay operation for a ground anywhere in the field.

Satisfactory operation of either type of relay requires the generator rotor to be grounded. To accomplish this, a means should be provided to bypass the bearing oil film. This should be done because the resistance of the path may be too large for dependable relay operation and, also, even a small magnitude of current flowing through the bearing may cause its destruction. Bearing seals may be used to provide the necessary bypass in some machines, while others may require an additional brush and slip ring for effective rotor grounding. Generators using a brushless type of excitation system do not have continuous access to the field circuit for ground monitoring. Slip rings with brush lifting mechanisms may be provided for these generators for periodic checking of the field ground status.

11.4.11 Temperature Relays. Generator stator overheating usually results from an overload, a failure of the generator cooling system, or even excessive rise in the ambient air temperature. It is a long-term phenomena not readily detected by other protective devices. In attended stations the generator is rarely tripped on overtemperature, preferring instead to alarm and allow the operator to take the appropriate steps to reduce the generator temperature. In unattended stations the generator may be tripped.

Resistance temperature detectors (RTD's) embedded in the generator stator windings are used to sense the actual winding temperature. Typically, six RTD's, two per phase, are installed and a selector switch provided to connect the thermal relay to the RTD indicating the highest operating temperature. Precaution should be taken to insure that the resistance of the RTD matches the input resistance of the relay.

11.5 Tripping Schemes

11.5.1 Protection Philosophy. Once the task of selecting the desired array of protective relays for the generator has been completed, then decisions must be made that determine how the prime mover-generator set is to be shutdown. It would be very short-sighted if the only consideration given was to disconnecting the generator from the electrical system. The basic operations in initiating shutdown of a prime mover-generator set are:

(1) Trip the generator breaker
(2) Open the excitation source (trip the field breaker)
(3) Remove the prime mover energy source (close the throttle valve)
(4) Initiate an alarm

The precise manner in which these operations are accomplished will be dependent on many factors:

(1) Type of prime mover—diesel/gas engine, gas turbine, steam turbine, or waterwheel
(2) Impact of the sudden loss of output power on the electrical system and the process which it serves
(3) Environmental considerations (if any)
(4) Safety to personnel
(5) Operating experience that focusses on a specific problem requiring special consideration

11.5.2 Classification of Protective Functions. For the purposes of discussing the various methods of shutting down the prime mover-generator set following operation of the protective devices discussed in 11.4, these devices are

Table 50
Classification of Protective Functions

Type	Protective Function	Location of Abnormality
A	Generator differential, 87	Generator
A	Bus differential, 87	Switchgear bus
B	Ground overcurrent, 51G	System
B	Negative sequence overcurrent, 46	System
B	Backup overcurrent, 51V	System
B	Reverse power, 32	Prime mover
B	Loss of excitation, 40	Generator or exciter
B	Undervoltage, 27	System
C	Temperature, 49	Generator auxiliaries
C	Generator field ground, 64F	Generator field

arbitrarily classified as types A, B, or C, as described in Table 50. Type A protective functions are those, such as differential relays, that detect the highly destructive internal equipment fault conditions. Type B protective functions are basically backup relays in which the abnormality is remote from the generator and the urgency for immediate tripping is not as critical as it is for type A functions. Type C protective functions are those, such as temperature relays, in which the urgency for tripping is even less severe and a manual response is preferable.

11.5.2.1 Tripping Schemes for Type A Functions. There are two possible alternatives that are considered feasible.

(1) Trip Generator and Field Breakers, and Close Prime Mover Throttle Valve. The prime mover may tend to overspeed as the generator load is removed and before the steam or hot gases can be dissipated in the prime mover. The throttle valve of a diesel engine or gas turbine will respond very fast and no appreciable overspeed will be experienced. The overspeed of a steam turbine will be somewhat greater, but normally not greater than the overspeed trip setting. This tripping scheme is shown in both Figs 212 and 213.

(2) Trip Prime Mover Throttle Valve Only. This alternative may be considered only for a steam turbine. This scheme may be of value where turbine throttle valve fouling has been experienced, where turbine overspeed following generator breaker opening may be excessive, or where the impact of the sudden loss of load to the process is detrimental. The consequences of sequential tripping is that the greater damage may be realized by the generator than for simultaneous tripping.

11.5.2.2 Tripping Schemes for Type B Functions. There are at least four alternative modes to consider when the backup devices operate.

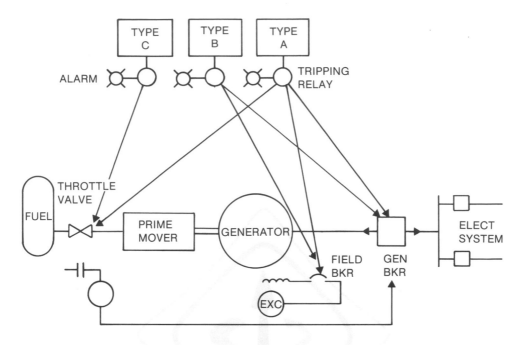

Fig 212
Typical Tripping Scheme — Alternate I

(1) Trip Generator and Field Breakers Only. This scheme, illustrated in Fig 212, promptly clears the generator, but depends of the governor or overspeed trip to sense the load change and close off the fuel supply. This arrangement might be used for small steam turbines, gas turbines, or diesel engines that have high-performance governor systems.

(2) Trip Generator and Field Breakers, and Close Prime Mover Throttle Valve. This scheme disconnects the entire generator-prime mover set and may result in a slight overspeed as the entrapped steam or gases are dissipated.

(3) Trip Throttle Valve Only. This scheme illustrated in Fig 213 is typically used on large steam turbines having back pressure control and extraction stages. After the throttle valve is closed, the valve limit switches then initiate sequential tripping of the generator and field breakers. This procedure eliminates the possibility of any overspeeding of the turbine. When this arrangement is used, the reverse power relay should still be arranged to trip the generator and field breakers since its basic function is to provide backup to the turbine throttle valve. Likewise, the negative sequence overcurrent relay may be regarded by some as a type A protective function since the heating limits of the generator rotor are approached.

(4) Alarm Only, Unload and Shutdown the Prime Mover Set Manually. This scheme might be preferred in those installations having steam turbine drives where there is concern about the ability of governors to respond promptly and the

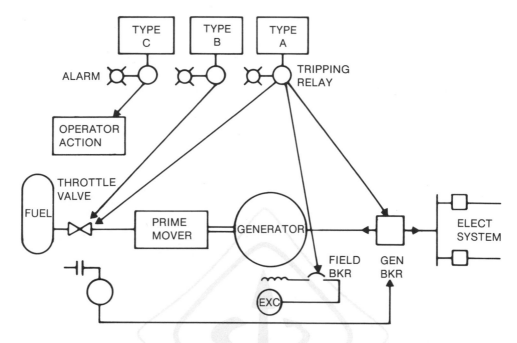

Fig 213
Typical Tripping Scheme — Alternate II

throttle trip valve to trip cleanly, if at all. Such a scheme should not be employed unless the generator is high resistance grounded or ungrounded. The reverse power relay should be connected to trip the generator and field breakers simultaneously.

11.5.2.3 Tripping Schemes for Type C Functions. The alternatives for tripping the less urgent type of protective functions are as follows:

(1) Trip Generator and Field Breaker and Close Prime Mover Throttle Valve. This scheme would be applied only to those installations that operate automatically, without an operator present.

(2) Trip Throttle Valve Only. This scheme, illustrated in Fig 212, will provide an orderly sequential shutdown of the unit.

(3) Alarm Only. Unload the generator manually, and if trouble cannot be corrected, shutdown the unit manually. This scheme, shown in Fig 213, is the generally preferred arrangement for manned installations.

11.6 Recommended Protection Schemes. The recommended protection schemes for generators are given by machine sizes.

(1) Small—1000 kVA maximum up to 600 V, 500 kVA maximum above 600 V

(2) Medium—From small machine sizes up to 12 500 kVA regardless of voltage

(3) Large—From medium machine sizes up to approximately 50 000 kVA

Any recommendation based entirely on machine size is not entirely adequate. The importance of the machine to the system or process that it serves and the reliability required from the generator will be important factors in the selection of the protective devices for the generator.

11.6.1 Small Generators. The basic minimum protection for a single isolated machine, as shown in Fig 214(a), consists of:

1—Device 51V, backup overcurrent relay, voltage restraint, or voltage-controlled type

1—Device 51G, backup ground time-overcurrent relay

Additional protection that should be considered for multiple machines on an isolated system, as shown in Fig 214(b), are:

1—Device 32, reverse power relay for antimotoring protection

1—Device 40, reverse VAR relay for loss of field protection

3—Device 87, instantaneous overcurrent relays providing self-balance type differential protection

For generators having excitation systems that do not have the ability to sustain the short-circuit current, even the basic minimum recommendations will not apply. These machines will typically be single isolated units having very small kVA ratings.

11.6.2 Medium Size Generators. The basic minimum protection for machines rated up to 12 500 kVA, as shown in Fig 215 consists of:

3—Device 51V, backup overcurrent relays, voltage restraint, or voltage-controlled type

1—Device 51G, backup ground time-overcurrent relay

3—Device 87, differential relays, fixed or variable percentage type, either standard-speed or high-speed, or the self-balance type whenever applicable

1—Device 32, reverse power relay for antimotoring protection

1—Device 40, impedance relay, offset mho type for loss of field protection, single element type

In the larger machine ratings, and especially those operating in parallel with a utility company supply, the following additional relay is recommended:

1—Device 46, negative phase sequence overcurrent relay for protection against unbalanced conditions

11.6.3 Large Generators. The recommended protection for the large industrial service generators is shown in Fig 216 and described as follows:

3—Device 51V, backup overcurrent relays, voltage restraint, or voltage-controlled type

1—Device 51G, backup ground time-overcurrent relay

3—Device 87, differential relays, high-speed variable percentage type

1—Device 87G, ground differential relay, directional product type

1—Device 40, impedance relay, offset mho type for loss of field protection, two-element type is recommended for greater sensitivity

1—Device 46, negative phase sequence overcurrent relay for protection against unbalanced conditions

1—Device 49, temperature relay to monitor stator winding temperature

1—Device 64F, generator field ground relay, applicable only on machines having field supply slip rings

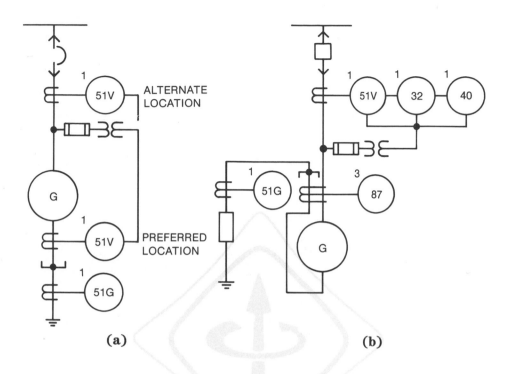

Fig 214
Typical Protective Relaying Scheme for Small Generators
(a) Single Isolated Generator On Low-Voltage System
(b) Multiple Isolated Generator On Medium-Voltage System

1—Device 60, voltage balance relay

Figure 216 shows the bus potential transformers being used to supply potential to the generator relays. This is an acceptable application providing that circuitry is provided to disconnect the relay potential when the generator is out of service. The preferred arrangement is to provide two sets of potential transformers at the generator terminals.

The bus differential relays, 87B, shown in Fig 216, are recommended when the large generators are connected to the system. Although they are not a part of the generator protection scheme, they provide high-speed clearing of bus faults, and thus, the generator backup overcurrent relays are not required to perform this primary protective function. This greatly improves the level of protection and reduces the thermal stresses that would otherwise be imposed on the generator and its components.

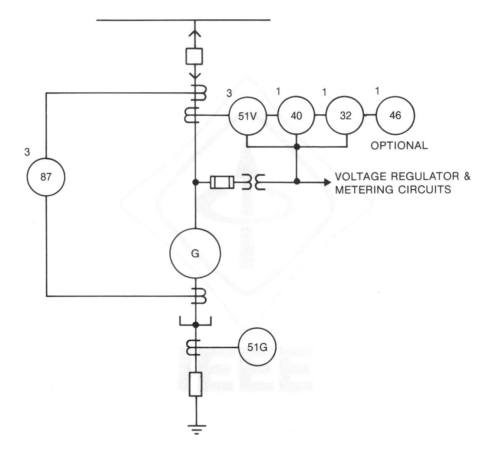

Fig 215
Typical Protective Relaying Scheme for Medium-Sized Generator

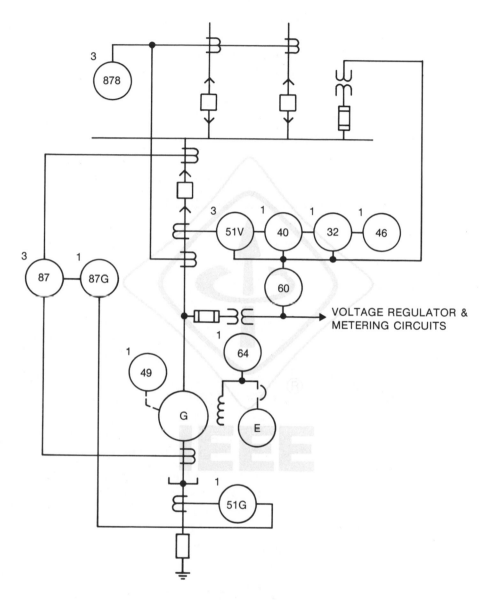

**Fig 216
Typical Protective Relaying Scheme for Large Generator**

11.7 References. The following publications shall be used in conjunction with this chapter.

[1] ANSI C50.13-1977, American National Standard Requirements for Cylindrical Rotor Synchronous Generators.

[2] ANSI C50.14-1977, American National Standard Requirements for Combustion Gas Turbine Driven Cylindrical Rotor Synchronous Generators.

[3] ANSI/IEEE Std 387-1984, IEEE Standard Criteria for Diesel-Generator Units Applied as Standby Power Supplies for Nuclear Power Generating Stations.

[4] ANSI/IEEE C37.2-1979, IEEE Standard Electrical Power System Device Function Numbers.

[5] ANSI/NEMA MG1-1978, Motors and Generators: Part 22, Large Generators.

[6] *Applied Protective Relaying.* Westinghouse Electric Corporation, NJ, Relay-Instrument Division, pub B7235E, 1974.

[7] BAKER, D. S. Generator Backup Overcurrent Protection. *IEEE Transactions on Industry Applications,* vol IA-18, Nov/Dec 1982, pp 632–640.

[8] BARKLE, J. E., and GLASSBURN, W. E. Protection of Generators Against Unbalanced Currents. *AIEE Transactions,* vol 72, pt 111, Apr 1953, pp 282–285.

[9] BERDY, J. Loss of Excitation Protection for Modern Synchronous Generators. *IEEE Transactions on Power Apparatus and Systems,* vol PAS-94, Sept/Oct 1975, pp 1457–1463.

[10] BROWN, P. G., JOHNSON, I. E., STEVENSON, J. R. Generator Neutral Grounding. *IEEE Transactions on Power Apparatus and Systems,* vol PAS-97, May/June 1978, pp 683–684.

[11] *Electrical Transmission and Distribution Reference Book.* Westinghouse Electric Corporation, PA, Central Station Engineers, 1964.

[12] GRAHAM, D. J., BROWN, P. G., and WINCHESTER, R. L. Generator Protection with a New Static Negative-Sequence Relay. *IEEE Transactions on Power Apparatus and Systems,* vol PAS-94, July/Aug 1975, pp 1209–1213.

[13] GROSS, E. T. B., and LE VESCONTE, L. B. Backup Protection for Generators. *AIEE Transactions,* part 111-A, pp 585–589, June 1953.

[14] IEEE Committee Report. Bibliography of Relay Literature, 1974–1975. IEEE Transactions on Power Apparatus and Systems, vol PAS-97, May/June 1978, pp 789–801.

[15] KELLY, A. R. Allowing for Decrement and Fault Voltage in Industrial Relaying, I. *IEEE Transactions on Industry and General Applications,* vol IGA-1, Mar/Apr 1965, pp 130–139.

[16] *Loss-of-Excitation Protection for Synchronous Generators.* General Electric, publication GER-3183, 1980.

[17] MASON, C. R. *The Art and Science of Protective Relaying.* New York: Wiley, 1956.

[18] MCFADDEN, R. H. Grounding of Generators Connected to Industrial Plant Buses. *IEEE Transactions on Industry Applications,* vol IA-17, Nov/Dec 1981, pp 553–556.

[19] POWELL, L. J., JR. Influence of Third-Harmonic Circulating Currents in Selecting Neutral Grounding Devices. *IEEE Transactions on Industry Applications,* vol IA-9, Nov/Dec 1973, pp 672–679.

[20] RAJK, M. N. Ground Fault Protection of Unit Connected Generators. *AIEE Transactions,* vol 77, pt 111, Dec 1958, pp 1082–1093.

[21] ST. PIERRE, C. R. Impact Loading of Isolated Generators, *IEEE Transactions on Industry Applications,* vol IA-17, Nov/Dec 1981, pp 557–566.

[22] ST. PIERRE, C. R. Loss-of-Excitation Protection for Synchronous Generators on Isolated Systems. *IEEE Transactions on Industry Applications,* vol IA-21, Jan/Feb 1985, pp 81–98.

12. Bus and Switchgear Protection

12.1 General Discussion. The substation bus and switchgear is that part of the power system that is used to direct the flow of power and to isolate apparatus and circuits from the power system. It includes the bus, circuit breakers, fuses, disconnection devices, current and potential transformers, and the structure on or in which they are mounted.

To isolate faults in busses all power source circuits connected to the bus are opened electrically by relay action, by direct trip device action on circuit breakers, or by fuses. This disconnection shuts down all loads and associated processes supplied by the bus and may affect other parts of the power system. When bus protective relaying is used, it should operate for bus or switchgear faults only. It is therefore essential that bus protective relaying operate for bus or switchgear faults only. False tripping on external faults is intolerable.

In view of the disastrous effects of a bus fault, the equipment should be designed to be as nearly "fault proof" as practicable, and the high-speed protective relaying or selected fuse should be used to keep the duration of the fault to a minimum, which limits the damage and minimizes the effects on other parts of the power system. When industrial power systems are grounded through resistance to limit fault damage, the current available to detect a ground fault is small; thus the protective relaying should be very sensitive.

To prevent faults the bus and associated equipment should be installed in a location where they are least subjected to deteriorating environmental conditions. A preventive maintenance program is a must to detect deterioration, to make repairs, and to check and test relay performance before a fault occurs [6][24].

12.2 Types of Busses and Arrangements. The substation bus may have many different arrangements depending on the continuity of service requirements for the bus or for essential feeders supplied from the bus. See ANSI/IEEE C37.97- 1979 [2]. The methods of protecting substation busses and switchgear will vary depending on voltage and the arrangement of the busses. The bus arrangements most applicable to industrial power systems are shown in Figs 217–220.

Industrial power system voltages fall into three categories: above 15 000 V, from 15 000 to 601 V, and below 600 V. The industrial power system usually

---

[24] The numbers in brackets correspond to those of the references listed at the end of this chapter.

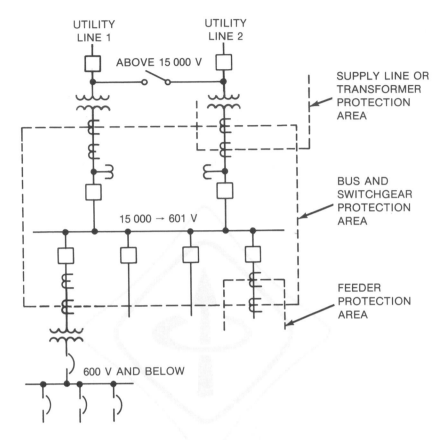

Fig 217
Single Bus Scheme with Bus Differential Relaying

includes only the distribution busses 15 000 V and below. However, it may include the subtransmission substation bus at a higher voltage level. Bus protective relaying at this level may overload sections of equipment supplied by the electric utility. For this reason the high-voltage relaying is usually specified by the utility, and compliance with utility practice is mandatory in most cases. Chapter 13 gives further information on utility service supply line requirements [4].

12.3 Bus Overcurrent Protection. If the system design and operation and the function of the process served do not require fast bus-fault clearance, overcurrent protection is used on each incoming power source circuit. On medium- and high-voltage systems, fuses or overcurrent relays are used. On low-voltage systems circuit breaker trip devices or fuses are used in most applications. Relays require the use of current transformers for fault sensing, and their use in low-voltage switchgear equipment is often not practicable because of physical and equipment limitations. The introduction of solid-state circuitry to perform

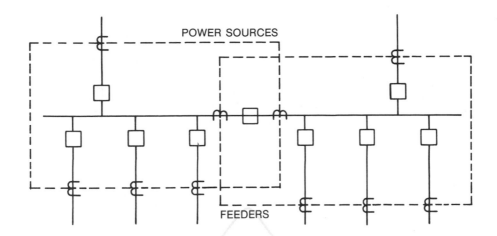

Fig 218
Sectionalized Bus Scheme with Bus Differential Relaying

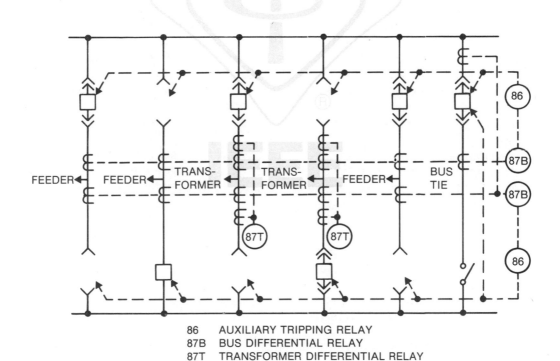

86	AUXILIARY TRIPPING RELAY
87B	BUS DIFFERENTIAL RELAY
87T	TRANSFORMER DIFFERENTIAL RELAY

Fig 219
Double Bus Scheme with Bus Differential Relay

475

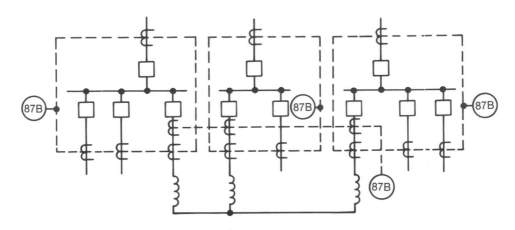

Fig 220
Synchronizing Bus Scheme with Bus Differential Relaying

the sensing and timing functions has provided significant improvements in the quality of protection for low-voltage circuits and apparatus.

Separate circuitry for ground faults detects the faults at much lower levels and clears them much faster than is possible with direct-acting electromechanical phase-overcurrent devices alone. ANSI/NFPA 70-1984 [3] (National Electrical Code [NEC]) requires in Section 230-95 ground-fault protection on solidly grounded wye-connected electric services of more than 150 V to ground but not exceeding 600 V phase-to-phase for any service-disconnecting means rated 1000 A or more. Chapter 6 of this standard describes how to use low-voltage circuit breakers to their best advantage. Chapter 5 covers the application of fuses for protection, and Chapter 3 gives details on relays and procedures for proper settings.

Overcurrent relays and trip devices should have time-delay and high-current settings to prevent opening the source circuit breakers upon the occurrence of a feeder fault. Therefore, overcurrent relays and devices cannot provide sensitive high-speed bus and switchgear protection.

An induction overcurrent relay connected to a current transformer in the power transformer neutral-to-ground circuit will provide good sensitivity for ground faults, but it should be set to be selective for feeder faults. If the feeders have ground-sensor instantaneous protection, only a short time delay is needed on the relay in the transformer grounding circuit. Since most faults are ground faults or eventually become ground faults, good ground-fault protection greatly improves bus overcurrent protection.

12.4 Differential Protection. Differential relaying can provide protection for buses and switchgear. It is high speed, sensitive, and permits complete overlapping with the other power system relaying as indicated in Figs 217–220. The basic principle is that the phasor sum of all measured currents entering and leaving the bus must be zero, unless there is a fault within the protected zone.

Differential relaying is provided to supplement the overcurrent protection. It is frequently used on 15 kV busses, sometimes on 5 kV busses, and rarely on low-voltage busses. The following factors determine whether this relaying should be provided.

Degree of exposure to faults. For example, open-type outdoor busses would have a high degree of exposure, and metal-clad switchgear, properly installed and in a clean environment, would have minimum exposure. Contaminated environments increase the possibilities of faults, and equipment located in these environments needs better protection.

Sectionalized bus arrangements make differential protection more useful and desirable, particularly when secondary-selective low-voltage distribution systems are used. The faulted bus can be isolated quickly and continuity of service maintained to a portion of the load by the other buses.

Effects of bus failure on other parts of the power system and associated processes. If a long down-time period for repairs can be tolerated, differential protection may not be economically justified. On major plant busses the cost of differential relaying is usually insignificant when compared with the reduction in damage to the equipment and the reduced down time of important plant or process facilities.

If there are problems in coordinating the system relay settings, differential relaying is effective in obtaining selectivity. An example would be a system where several busses are required at the same voltage level, with one feeding another. This generally results in unacceptably high current ratings or overcurrent relay settings on the upstream fused switches or circuit breakers to obtain coordination.

On busses fed by a local generator, bus differential relaying is recommended to clear the bus quickly and hold the rest of the system together. The overcurrent relays used to protect generator circuits take considerable time to operate.

The differential protection methods generally used are, in the order of the quality of protection they provide:

(1) Voltage responsive and linear coupler methods
(2) Percentage differential (where applicable)
(3) Current responsive
(4) Partial differential (sometimes not considered a differential scheme)

Since the differential relay should trip all circuit breakers connected to the bus, a multicontact auxiliary relay is needed. This auxiliary relay should be the high-speed lockout type, with contacts in the circuit breaker closing circuits to prevent "panic" manual closing of a circuit breaker on the fault. The lockout relay should be hand reset before any circuit breakers can be closed.

12.5 Voltage Differential Relaying. This method uses through-type iron core current transformers. The problem of current transformer saturation is overcome by using a voltage-responsive (high-impedance) operating coil in the relay. Separate current transformers are required in each circuit connected to the bus

477

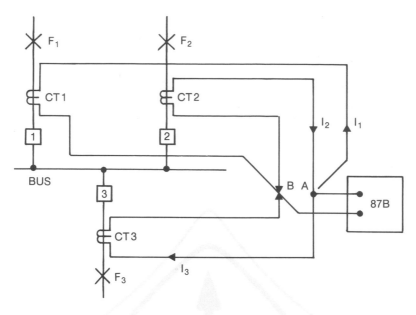

Fig 221
Voltage Differential Relaying

as shown in Fig 221. Voltage differential bus protection is not limited as to the number of source and load feeders and has the following features:

(1) High-speed operation in the order of 1–3 cycles
(2) High sensitivity can be set to operate on low values of phase or ground-fault currents in most installations
(3) Relay operates from all standard bushing current transformers and from switchgear through-type current transformers with distribution windings
(4) Relay is not adversely affected by current transformer saturation, dc component of fault current, or circuit time constant
(5) Discrimination between external and internal faults is obtained by relay settings; no resistant or time delay is required

All current transformers should have the same ratio unless high-impedance relays suitable for this duty are used. Auxiliary current transformers should not be used to match ratios. Current transformers with different maximum ratios can be matched by operating the high ratio transformer as auto transformers using an intermediate tap to obtain a match with the maximum tap of the lower ratio current transformers.

All current transformers should have low secondary leakage reactance; wound-types are generally not suitable. Bushing-type current transformers constructed on toroidal cores with completely distributed windings will generally have negligible leakage reactance. A distributed winding is one which starts and ends at the same point on the core. Through-type current transformers having suitable characteristics are available for use in switchgear assemblies.

The relay must fulfill two requirements. First, it should not trip for any fault external to the zone of protection. Next, it should be capable of operating for all faults internal to the zones of protection. Considering the first requirement, that the relay should not operate falsely for external faults, refer to Fig 221. Assuming a three-circuit-breaker bus with a fault at the location shown, let us consider for simplicity only one of the three phases. For the fault F_3 indicated, the fault current I_3 will flow through circuit breaker 3 with the currents flowing through circuit breakers 1 and 2, each being smaller but summing up to I_3. If we assume for the moment that the current transformers behave ideally, the current transformer secondary current produced at circuit breaker 3 will balance the sum of the currents produced at circuit breakers 1 and 2. This current will circulate in the current transformer secondary circuits and produce little, if any, voltage across points A-B.

If for some reason the current transformer secondary current at circuit breaker 3 does not balance the sum of the currents produced by the current transformers at circuit breakers 1 and 2, excess or difference current will be forced to flow through CT 3, causing the voltage across points A and B to increase considerably to a point where the relay 87B will operate. It thus becomes apparent that the current transformer at circuit breaker 3 has a greater tendency to saturate than those at circuit breakers 1 and 2, for the given fault location, because it gets the total current while the other two each get only a fraction of the total. From the point of view of the relay, the worst condition would be the case where the current transformers at circuit breaker 3 saturated almost completely and hence produced no detectable secondary current, while those at circuit breakers 1 and 2 did not saturate at all and, hence, reproduced the current faithfully. It is important to note that for the condition of complete saturation, the mutual reactance of the bushing-type current transformer approaches zero. If it has no appreciable secondary leakage reactance, then the secondary impedance of the current transformer is just its resistance. Thus for the condition of complete saturation of the current transformer at circuit breaker 3, the voltage developed between points A and B is the product of $(I_1 + I_2)$ and the total resistance in the circuit between points A and B and current transformer 3 at circuit breaker 3, including the current transformer secondary resistance. The differential relay is set so that it does not operate for this voltage. It is obvious that this voltage depends on the magnitude of the fault current, the type of fault, and the total resistance. In the case of internal faults, the secondary currents do not circulate but rather result in a high enough secondary voltage to cause the relay to operate.

A nonlinear resistor or a voltage limiting circuit is connected in parallel with the sensitive high-impedance operating coil to limit the voltage that may be attained during high internal faults. To obtain higher speed operation for high internal faults, the unit is connected in series with the nonlinear resistor.

When offset fault current occurs or residual magnetism exists in the current transformer core, or both, an appreciable dc component in the secondary current is present. This has caused false tripping when simple unrestrained low-impedance relays are used for bus differential. Voltage differential relays are made insensitive to the dc component by connecting the relay sensitive operating coil in series with a capacitor and reactor. The circuit is resonant at the fundamental power frequency and the dc component is blocked by the series capacitor [7].

12.6 Air-Core Current Transformer (Linear-Coupler) Method. This method provides extremely reliable high-speed bus protection. It is highly flexible to future expansion and system changes. The couplers can be open-circuited without any difficulties to simplify switching circuits. The operating time for one type of linear coupler system is 1 cycle or less above 150% of pickup and 1 cycle for another type of linear-coupler system. This scheme eliminates the difficulty due to differences in the characteristics of iron-core current transformers by using air-core mutual inductances without any iron in the magnetic circuit. Therefore, it is free of any dc or ac saturation.

The linear couplers of the different circuit breakers are connected in series and produce secondary voltages that are directly proportional to the primary currents going through the circuit breakers, as shown in Fig 222.

With the simple series circuit shown in Fig 170:

$$I_R = \frac{E_{sec}}{Z_R + \sum Z_C}$$

$$= \frac{I_{pri} M}{Z_R - \sum Z_C}$$

where

E_{sec} = voltage induced in linear-coupler secondary
I_{pri} = primary current, rms symmetrical
I_R = current in relay and linear-coupler secondary
M = mutual impedance, 0.005 Ω, 60 Hz
Z_c = self-impedance of linear-coupler secondary
Z_R = impedance of relay

During normal conditions or for external faults, the sum of the voltage produced by the linear couplers equals zero. During internal bus faults, however, this voltage is no longer zero, and is measured by a sensitive relay that operates to trip circuit breakers and clear the bus.

Linear couplers are air-cored mutual reactors wound on nonmagnetic toroidal cores such that the adjacent circuits will not induce any unwanted voltage. For the conductor within the toroid, 5 V is induced per 1000 A of primary current. Therefore, by design, the mutual impedance M is 0.005 Ω, 60 Hz. In other words, $E_{sec} = I_{pri} M$.

Static voltage differential relays are also available, providing faster operating times than mechanical relays.

12.7 Percentage Differential Relay. Where there are relatively few circuits connected to the bus, relays using the percentage differential principle may be employed. These relays are similar to transformer differential relays that are described in Chapter 10. The problem of application of percentage differential

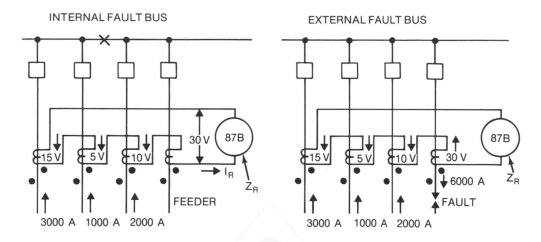

Fig 222
Linear-Coupler Bus Protective System with Typical Values
Illustrating Its Operation on Internal and External Faults

relays for bus protection, however, increases with the number of circuits connected to the bus. It requires that all current transformers supplying the relays have the same ratio and identical characteristics. Variation in the characteristics of the current transformers, particularly the saturation phenomena under short-circuit conditions, presents the greatest problem for this type of protection and often limits it to applications where only a limited number of feeders are present.

12.8 Current Differential Relaying. When voltage or linear-coupler differential protection cannot be economically justified, a less expensive current differential scheme may be considered. This scheme utilizes simple induction-type overcurrent relays connected to respond to any difference between the currents fed into the bus and the current fed from the bus. The current transformer arrangements are the same as shown in Figs 217 to 220. The connections are as shown in Fig 221.

Chapter 3 gives details on these relays. A special form of overcurrent relay is available with an internally mounted auxiliary relay and connections to permit testing the integrity of the current transformer circuits for accidental ground faults and open circuits. The connections are so arranged that while checking on one phase, the relays in the other two phases are still providing protection.

12.9 Partial Differential Protection. This type of protection, sometimes called *summation relaying,* is a modification where one or more of the load circuits are left uncompensated in the differential system (Fig 223). For this reason it may be a misnomer to name it a "differential" scheme. This method may be used as primary protection for busses with loads protected by fuses, as backup to a complete differential protection scheme, and to provide local backup protection for "stuck" load circuit breakers which fail to operate when they should. The phase overcurrent relays are set above the total bus load or the total

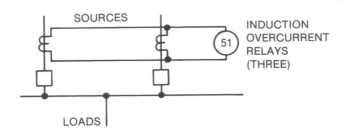

Fig 223
Partial Differential System

rating of all loads supplied from the bus section. The relays should provide enough time delay to be selective with relays on the load circuits. Consequently, the sensitivity and speed of partial differential protection is not as good as for full differential protection.

12.10 Backup Protection. In the event of a failure of the primary protective system to operate as planned, some form of backup relaying should be provided in the industrial power system or in the power supply system.

Bus backup protection is inherently provided by the primary relaying at the remote ends of the supply lines. This is known as remote backup protection. It may not be adequate because of system instability and effects on other power systems, and local backup relaying may be necessary. Kennedy and McConnell [5] analyze the performance of various remote and local backup relaying schemes. Chapter 13 gives further information on utility service supply-line requirements and the backup protection by utility relaying [4].

Circuit breaker failure can cause catastrophic results, such as complete system shutdown. Local circuit breaker failure or stuck circuit breaker relay schemes are available to quickly trip upstream circuit breakers if the circuit breaker on the faulted circuit fails to operate within a specified time. However, those schemes are normally applied only on buses where the extra expense can be economically justified.

12.11 Voltage Surge Protection. Protection against voltage surges due to lightning, arcing, or switching is required on all switchgear connected to exposed circuits entering or leaving the equipment. A circuit is considered exposed to voltage surges if it is connected to any kind of open-line wires, either directly or through any kind of cable, reactor, or regulator. A circuit connected to open-line wires through a power transformer is not considered exposed if adequate protection is provided on the line side of the transformer. Circuits confined entirely to the interior of a building, such as an industrial plant, are not considered exposed to lightning surges, and may not require voltage surge protection.

The protection is provided by surge arresters connected, without fuses or disconnecting devices, at the terminals of each exposed circuit; see ANSI/IEEE

C37.20-1969 [1]. Surge protection connected directly to the bus is not recommended, as the reliability of the bus will be diminished. If it becomes necessary to connect surge protection directly to the bus, it should be connected through fuses or circuit breakers. (Note that circuits will not be protected when the circuit breakers are open.)

The arrestors should be of the valve type, of adequate discharge capacity, and their voltage ratings should be selected to keep the voltage surges below the insulation level of the protected equipment. When the exposed line is directly connected to the switchgear through roof entrance bushings, intermediate or station-type arrestors are recommended. Where an exposed line is connected to the switchgear with a section of continuous metallic sheath cable, the arresters should be installed at the junction of cable and overhead line. If the arresters at the overhead line are intermediate or station type, arresters may not be required at the switchgear. If distribution-type arresters are used at the overhead line, another set may be required at the switchgear terminals, depending on the length of the cable. Cable without a continuous metallic sheath does not reduce the wavefront enough so that valve-type distribution arresters are required at the switchgear. In the latter case a properly installed ground wire in the duct with each three-phase cable provides very nearly the same impedance as continuous metallic sheath cable.

Surge arresters may be required to protect the switchgear at altitudes above 3300 ft, even though the circuits are not connected to exposed circuits. This is due to the voltage correction factors applicable above 3300 ft altitude; see ANSI/IEEE C37.20-1969 [1]. Surge arresters are applied such that the impulse voltage protective level maintained by the arrester is about 20% less than the corrected impulse voltage rating of the switchgear. These arresters should be of the station type.

Low-voltage bus and switchgear are often protected by current-limiting fuses, sized to the full-load rating when bus and switchgear have bus bracings that are less than the available fault current. Current-limiting fuses are often used to limit the fault current to levels that the bus and switchgear can handle.

When the short time-delay setting on a circuit breaker exceeds 3 cycles, the bus and switchgear will need to be tested and specified for the longer time period as standard bracing tests are for 3 cycles only.

To reduce the possibility of destructive arcing ground faults, on 227/480 V systems, the 480 V bus may be insulated. It is far better to prevent a ground fault from occurring than it is to shut down a system or a part of a system after a ground fault has occurred.

12.12 Conclusion. Because of its location and function in the electric power system, the bus and switchgear should be designed, located, and maintained to prevent faults. The preferred practice for bus switchgear protection above 600 V is voltage-responsive or linear-coupler differential relaying with the power system designed with sectionalized busses so that continuity of service can be maintained to a portion of the load. The best protective relaying in a single bus arrangement will operate to cut off power to all circuits supplied by the bus.

Hopefully, the relaying will never be called on to operate. Location of the equipment in a good environment and maintenance on a planned basis will help prevent the need for relays to operate [6]. If a fault does occur, high-speed sensitive relaying will limit the damage so that repairs can be made quickly and service restored in a short time. Fast clearing of faults also can save lives by minimizing explosion and fire aftermath. Furthermore, fast clearing of human contact faults has saved lives or reduced injury. Fuses offer excellent bus and switchgear protection.

12.13 References. The following publications shall be used in conjunction with this chapter.

[1] ANSI/IEEE C37.20-1969, IEEE Standard for Switchgear Assemblies Including Metal-Enclosed Bus.

[2] ANSI/IEEE C37.97-1979, IEEE Guide for Protective Relay Applications to Power System Buses.

[3] ANSI/NFPA 70-1984, National Electrical Code.

[4] BECKMANN, J. J., DALASTA, D., HENDRON, E. W., and HIGGINS, T. D. Service Supply Line Protection. *IEEE Transactions on Industry and General Applications, vol IGA-5, Nov/Dec 1969, pp 657–671.*

[5] KENNEDY, L. F. and McCONNELL, A. J. An Appraisal of Remote and Local Backup Relaying. *AIEE Transactions on Power Apparatus and Systems,* pt III, vol PAS-76, Oct 1957, pp 735–747.

[6] KILLIN, A. M. How Plant Management Evaluates Electrical Performance. *IEEE Transactions on Industry and General Applications,* vol IGA-3, Mar/Apr 1967, pp 75–78.

[7] SEELEY, H. T. and VON ROESCHLAUB, F. Instantaneous Bus Differential Protection Using Bushing Current Transformers. *AIEE Transactions,* vol 67, 1948, pp 1709–1719.

13. Service Supply Line Protection

13.1 General Discussion. This chapter discusses the interface between the supplier of electricity and the consumer. It includes service quality requirements for industrial and commercial power systems, possible system disturbances in utility and consumer systems and their effects, recommended corrective measures in system design and operating techniques, and several protective schemes of typical installations both with and without consumer generation. There are many possible circuit arrangements and protective relaying schemes, and it is not possible to describe and discuss each one. However, a few of the most frequently encountered circuit arrangements are covered.

The basic desire of the power company and the consumer is to provide, and have provided to, respectively, a reliable power supply of adequate capacity to serve the connected load. This needs well-engineered design considering service requirements, system disturbances and protection against such disturbances, and personnel safety.

13.1.1 Design Procedure. A typical sequence in designing an electrical service arrangement for a new plant could be as follows:

(1) Classify and group loads according to their characteristics:

    (a) Power required, real and reactive

    (b) Optimum nominal voltage

    (c) Sensitivity to voltage level

    (d) Sensitivity to voltage dips

    (e) Sensitivity to interruptions

    (f) Sensitivity to frequency variations

    (g) Other unusual service requirements, such as sensitivity to nonsinusoidal wave shapes and harmonics

    (h) Physical location of loads

    (i) Future load considerations

(2) Select, together with the electric utility personnel, a suitable supply service arrangement consistent with the economics of the application. Industrial plant power-system designers together with electric utility personnel should determine all the service requirements, the industrial plant process, and the quality of service expected from the electric utility. This may include system voltage including upper and lower limits and the duration of abnormal voltages, frequency variation and its duration, harmonics and utility system equipment, relaying and reclosing schemes.

(3) Analyze the service line and equipment as to the various electric faults or system disturbances that are likely to occur in order to avoid unnecessary opening of protective devices such as fuses and circuit breakers. It is necessary to obtain an estimate of the number of short-time and long-time outages and statistics of the number of faults on utility service lines.

(4) If necessary restructure the one-line diagram and apply such equipment that will minimize the effect of these faults and disturbances and will improve the system performance and reliability. The consumer should evaluate the cost of such changes against the loss of production, safety, equipment damage, and extra maintenance. This may include:

(a) Addition of power-conditioning equipment, such as capacitors, voltage regulators, or generators

(b) Addition of surge protective equipment

(c) Modification of protective relaying and switching schemes such as the use of time-delay undervoltage relay or latching relays, use of circuit breakers or remotely operated switches and contactors, dc control power instead of ac control, automatic control instead of manual switching, auto-reclosing, loop circuits instead of radial feeds

(d) Use of auxiliary devices or stored energy control systems, such as batteries or capacitive tripping devices

(e) For a plant generating its own power, load shedding may be utilized for those loads that are noncritical during low-frequency and other system disturbances; this may keep the system running by improving system frequency and voltage

(f) Connection to redundant supply sources and internetwork connection for continuity

(g) Addition of current-limiting cable limiters or fuses on 600 V or below systems

A comprehensive engineering liaison between the plant designer and utility personnel should result in a selectively coordinated system, increased reliability, minimum equipment cost by avoiding duplication of equipment, and maximum use of equipment.

13.2 Service Requirements. Consideration of the design, operation, and protection of service lines between a consumer and utility power supplier should be based on deep mutual understanding of each other's needs, limitations, and problems. The electric power supply for an industrial or commercial power system should meet the following basic requirements:

(1) Accommodate normal peak power demand and provide ability to start large-sized motors without excessive voltage drop

(2) Provide minimum deviation from normal frequency and normal voltage

(3) Maintain normal phase rotation in a multiphase system

(4) Provide minimum voltage-wave distortion and harmonics, and maximum freedom from voltage surges

(5) Maintain three-phase supply at all times, avoiding voltage unbalance and single-phasing

(6) Provide high reliability of power supply, that is, a high percentage of time that the system can effectively serve the loads

These requirements are measures of quality of service to a consumer. The quality of electric power supply has become important in operation of many modern electrically supplied systems. The nature of consumer's operation and type of loads sets the requirements of quality of service. The simplicity of these statements of service requirements may tend to obscure the complexity of technical and commercial problems that sometimes arise, but the statements are the true measures of quality of service to a consumer.

Service quality involves two distinct factors, each of which should be considered separately and each of which will have different degrees of importance among consumers. The two factors are power quality and power reliability. Together, these factors make up service quality.

Each load device has specific power quality tolerances within which it will operate normally. Table 51 lists electric service deviation tolerances of various load and control devices. The term *load,* as used in this chapter, means an electric device, such as a motor, capacitor, lighting lamps, heating elements, motor starters, solenoids, computers, communication equipment, annunciators, electronic tubes, inverters, rectifiers, and control circuits. To a consumer, these loads are only a means to an end; they are the muscles and nerve systems needed to operate chemical processes, mines, public buildings, or manufacturing plants.

Reliability requirements for power supply to certain load devices may completely change service considerations. For example, an incandescent lamp will perform satisfactorily on voltage containing myriad abnormalities, but if it is the lamp in an *exit* fixture in a public building, or an operating lamp in a hospital operating room, then it must have power of absolute reliability. By contrast, a computer used for process control or power plant load management must be supplied with power of extremely high quality, but if it is being used to process routine business data then service reliability may become secondary. Power supply requirements for computers will vary depending upon the application of computers. The range of tolerances in voltage, frequency, and rate of change in voltages in the power supply is usually well defined by the manufacturer of the computers.

A study of a consumer's operation and loads can help the utility–consumer team arrive at the required level of service quality. A study of possible system disturbances and their effects (see 13.3) should be made, and where these disturbances exceed tolerances of the load devices for the equipment included in the system, then appropriate steps outlined under 13.4 should be considered. The required reliability should also be kept in mind, and where higher than normal utility reliability is required, suitable measures as outlined under corrective measures and line protection (13.4) should be considered.

13.3 System Disturbances and Their Effects. Many of the control devices and loads that are part of commercial and industrial power systems are sensitive to the magnitude, waveshape, and frequency of the supply voltage. Voltage and

Table 51
Electric Service Deviation Tolerances for Load and Control Equipment

Device	Voltage Level*	Voltage Distortion: Harmonic Content	Frequency	Comment
Alarms, systems operating on loss of voltage	Variable	–	–	
Capacitors for power factor correction	+10% to −110%	**	+0% to −100%	
Communication equipment	±5%†	Variable	–	
Computers, data-processing equipment	±10% for 1 cycle†	5%	+½ Hz to −1½ Hz	
Contactors, motor starters				
ac coil burnout	+10% to −15%	–	–	
ac coil dropout‡	−30% to −40% for 2 cycles	–	–	
dc coil dropout	−30% to −40% for 5–10 cycles	–	–	
Electronic tubes	±5%	Variable	–	
Lighting				
fluorescent	−10% (fluorescent)	–	–	Uncertain starting, shortened life. Lamp will extinguish.
	−25%	–	–	
incandescent	+18%	–	–	10% of normal life. See Fig 173.
mercury vapor	−50% for 2 cycles	–	–	Lamp will extinguish.
Motors, standard induction*	±10%	Variable	±5%	Sum of absolute values of voltage and frequency deviation should be no greater than ±10%.
Resistance loads, furnaces, heaters	Variable	–	–	
Solenoids, shutoff valves for gas or oil fired furnaces, magnetic chucks, brakes, clutches	−30% to −40% for ½ cycle	–	–	

	Voltage			Remarks
Transformers	+5% with rated kVA ≤ 0.80 PF +10% with no load	–	–	Voltage deviations apply at rated frequency. If frequency drops, voltage limits should reduce proportionally.
Inverters (gaseous, thyristor)	+5% with full load +10% with no load –10% transient	2%	±2 Hz	Firing circuits and transformers generally determine tolerances. If supply voltage is +5%, transformer loading should be reduced by 5%.
Rectifiers diode (gaseous)	+5% with full load +10% with no load –10% transient	Sensitive§	–	If supply voltage is +5%, transformer loading should be reduced by 5%.
diode (solid state)	±10%	Sensitive§	–	Some rectifier systems rated by NEMA MG1-1978 [7] for voltage deviation of +5% to –10%.
phase controlled (gaseous, thyristor)	+5% with full load +10% with no load –10% transient	2%	±2 Hz	Firing circuits and transformers generally determine tolerances. If supply voltage is +5%, transformer loading should be reduced by 5%.
Generators	±5%	Sensitive§	–5%	Voltage tolerance generally a function of generator design. Surge protective devices should be applied at generator terminals.
Turbines (steam)	–	–	–1%	

* It is presumed that properly selected lightning arresters and surge protective equipments are installed throughout the system. Deviation tolerance is continuous unless specified. Percent voltage unbalance = (100· maximum voltage deviation from average voltage)/(average voltage).

** Capacitor ratings are based on a nominal voltage distortion caused by overexcited transformers. Silicone-controlled-rectifier phase-controlled loads may cause additional and excessive distortion.

† Voltage tolerances may not be applicable for equipment with integral power supply or voltage regulator.

‡ This type of contactor is not recommended where tripouts due to voltage dips are undesirable.

Phase voltages should be balanced as closely as can be read on the usual commercial voltmeter.

§ Presence of harmonics may raise or lower direct output voltage, balance between phases, or balance between rectifier units operating in parallel.

Systems with rectifiers or inverters may contain harmonics which cause nonsalient pole generator overheating. Generator load reduction may be required as a function of the number of phases in the rectifier system as follows: 24 phases–no reduction; 12 phases–8% reduction; 6 phases–10% reduction.

Table 52
Typical Fault Records for One Utility

	Line Design Voltage		
	23–46 kV	69 kV	138 kV
Temporary line faults per 100 mi/yr	25–40	10–25	5–15
Permanent line faults per 100 mi/yr	0.1–0.2	0.2–0.3	0.1

Table 53
Outage Rates and Average Repair Times
of Electric Equipment

	Outages per Mile (or per Unit) per Year		Average Repair Time (hours)
	Normal	Stormy	
Semipermanent forced outages			
Secondary cable	0.005	0.005	35.0
Network protector	0.0038	0.0038	2.0
Network transformers	0.0049	0.0049	24.0
Primary switch	0.001	0.001	2.0
4 kV open wire	0.045	3.2	2.0
4 kV feeder circuit breaker	0.005	0.005	0.8*
4 kV distribution transformer	0.0017–0.0069	0.0017–0.079	3.5
Maintenance outage			
Network protector	0.2	–	2.0
Network transformer	0.2	–	10.0
Primary switch	0.2	–	2–10
4 kV open wire	2.0	–	4.0
4 kV feeder circuit breaker	0.2	–	0.8*

From [2].
*Time to transfer feeder to spare position.

frequency variation and presence of harmonics in most cases deteriorate the quality of power. Most voltage transients, all deviations in frequency, and most short- and long-term power supply losses originate in the utility system. Further system disturbances are sometimes introduced by the distribution system within the industrial or commercial power system beyond the utility delivery point. Sample fault records for various operating voltages for one utility are given in Table 52, though failure rates differ widely due to design and application variations. Outage rates and average repair time of some sample electric equipment are given in Table 53 and also in [B1][25].

---

[25] The numbers in brackets correspond to those of the references listed at the end of this chapter. The numbers in brackets with the prefix "B" correspond to those of the bibliography listed at the end of this chapter.

The effects of all these possible disturbances are analyzed on the various types of load devices in the following order:

(1) Voltage variation
(2) Frequency variation
(3) Harmonics
(4) Short circuits

These disturbances can take the form of equipment damage, loss of production, production of inferior quality product, damage to plant facilities (fire), and injury to personnel.

13.3.1 Voltage Variations. Voltage variation can be classified under several categories: steady voltage reduction, momentary voltage dip, short-time interruption, voltage transients, voltage flicker, voltage unbalance, and long-time voltage loss. Each type of voltage variation will have a different effect on load devices.

13.3.1.1 Steady voltage reduction results from daily changes in load on transmission lines, transformers, distribution line motors, and is usually in the range of ± 10%. The voltage level, however, may gradually change over a period of minutes or hours, and as a result subject loads to voltages that are either too high or too low to permit continued satisfactory operation. A voltage variation of about ± 10% at the point of load connection is usually allowed for loads other than lighting.

13.3.1.2 Momentary voltage dips are generally caused by transient line faults. Most of these faults are cleared within a few cycles. Most of the load devices in the plant usually remain energized during this condition if the voltage dip is not severe. Under severe voltage dip conditions, ac contactors will drop out and may jeopardize the process. For continuous process plants it is necessary to provide suitable protective measures, and such voltage dips should be considered in initial design stages.

13.3.1.3 Voltage transients of much shorter duration and frequently of much greater magnitude are more common than momentary voltage dips. These transients and swings will result from remote system faults, interruption of faults, switching surges from operating disconnects or circuit breakers in the transmission or distribution system, fuse blowing, impact loads and load dropping, routine operating open-arc or submerged-arc furnaces, SCR controlled loads, and welders. Voltage swings resulting from these causes are transient in nature, but may be severe enough to cause malfunctions of some of the load equipment.

The principal type of transient defect is a high-voltage spike, and the principal source is lightning.

Transient over- and undervoltages can be produced within a plant. Overvoltage spikes result from switching within the plant, particularly switching of the primary side of a transformer, which is coupled to a heavy load. Overvoltage may also occur if some type of line voltage regulating device defectively advances to its full boost position. In some cases, a combination of grounding and an arcing fault condition also produces high voltages. Transient undervoltages are produced by sudden application of large loads, particularly the line starting of large induction and synchronous motors.

13.3.1.4 Transient spikes may be up to 2 kV peak on a 480 V system, with a duration of 10 to 200 μs. Such spikes are considerably attenuated when traveling through cable circuits and transformers.

Voltage flicker is repeated voltage swings of such magnitude that there is a noticeable change in the amount of light produced by lamps. Voltage flicker results from cyclic load variations which are usually single-phase, but may be three-phase. Arc furnaces and electric welding sets are the most common causes of flicker. The effect of this problem is mainly psychological, but it is symptomatic of trouble within the power system. Higher power lamps in common use today are less sensitive to flicker due to a larger mass of the filaments. Various types of lamp performance as affected by voltage levels are given in Figs 224–226.

Voltage flicker may also adversely affect induction motors even though it is considered an illumination phenomenon. Single-phase and small horsepower motors tend to overheat due to voltage flicker. Three-phase flicker causes motor losses to increase, continual variations in torque and speed, and thus an increase in vibration. Motors may even backfeed power for a cycle or two if flicker is severe.

The impact of transient and short-term defects described under 13.3.1, (1)–(3), on industrial loads, such as lighting devices, motors, and heating elements, is not serious. The defect goes unnoticed or produces a minor problem. But for modern electronic devices, such as computers, programmable controllers, and communication systems, the impact could be devastating. Transient voltage spikes entering equipment through supply lines can damage semiconductor devices if proper protection against transients is not taken. Metal oxide varistors, when applied properly, will protect these devices. However, sensitivity of electrical equipment depends on magnitude of voltage dip and its duration. Contactors are sensitive only to magnitude of the voltage dip. They drop out at about 60–70% of system voltage or on 30–40% voltage dip. On the other hand, the duration of the voltage dip is important to other loads. Many electric motor drives are sensitive to the combination of magnitude and duration of voltage dips, both of which play an important role in whether a critical drive will ride through a disturbance.

The effect of voltage variations on standard induction motor characteristics is listed in Table 54. Induction motors, in general, will ride through voltage dips or interruptions where high-speed reclosing is used, without any loss of production. However, under long-term low-voltage conditions, an induction motor may reach an instability condition whereby the motor cannot drive its connected load at rated speed and will therefore slow down. Also during low-voltage conditions, motors tend to draw high (accelerating) current and will cause the voltage to remain depressed at the motor terminals. Depressed motor terminal voltage may prevent proper recovery (reacceleration) and motor overheating will result.

Synchronous motors can be made to ride through voltage dips for several cycles with proper controls, but may be damaged on utility line reclosure [B17]. Continuing low-voltage at the terminals will tend to make a synchronous motor unstable [B11].

Power factor correcting capacitors may be damaged by high voltage, which is still within the apparent tolerable band of the capacitor. This can occur if the high voltage results in overexcitation of transformers or motors. Such over-

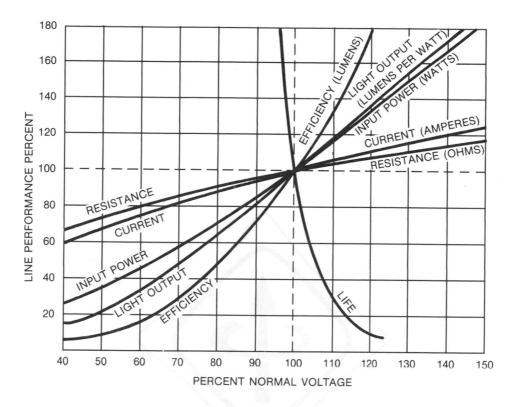

NOTE: These characteristics are averages of many lamps.

Fig 224
Incandescent Lamp Performance as Affected by Voltage

excitations can lead to distorted voltage waveshapes and result in excessive capacitor currents. This consideration is included in capacitor standards. But the compounding of these wave distortions with the distortions caused by solid-state phase-controlled devices has not been included in the standards [B11].

Modern high-speed computers are very sensitive to disturbances in the electrical power supply lines. Overvoltage and undervoltage transients can produce false operation of digital computers and control equipment. Short-term defects, lasting up to one second, may not cause failures in electronic equipment, but can produce serious errors in logic circuits, registers, and other equipment used in computers and digital control systems. These are very sensitive to interruptions of only a fraction of a cycle. An interruption of only a few cycles can cause malfunction of peripherals, such as card reader and magnetic tape units. Most computers are designed to protect the contents of core memory by a

493

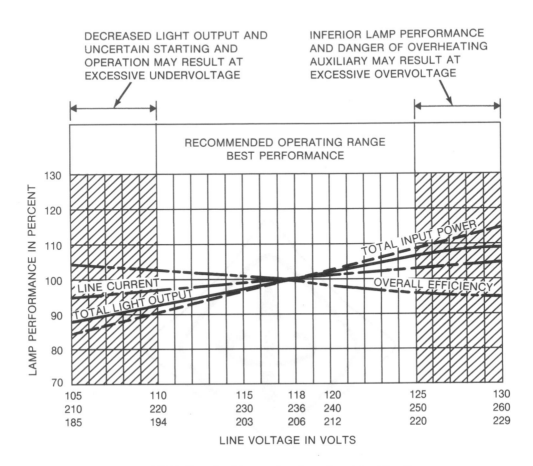

DECREASED LIGHT OUTPUT AND UNCERTAIN STARTING AND OPERATION MAY RESULT AT EXCESSIVE UNDERVOLTAGE

INFERIOR LAMP PERFORMANCE AND DANGER OF OVERHEATING AUXILIARY MAY RESULT AT EXCESSIVE OVERVOLTAGE

RECOMMENDED OPERATING RANGE BEST PERFORMANCE

TOTAL INPUT POWER

LINE CURRENT

TOTAL LIGHT OUTPUT

OVERALL EFFICIENCY

LAMP PERFORMANCE IN PERCENT

LINE VOLTAGE IN VOLTS

NOTE: Energized lamps may be extinguished if voltage drops to approximately 75% of rated line voltage.

Fig 225
Fluorescent Lamp Performance as Affected by Voltage

controlled sequence of dc power supplies in the event of a power interruption. Voltage and frequency transients can cause errors in the computer memory or complete memory loss, or word structure alteration, or nonprogram jumps. Where possible computers should be fed from a supply line other than the one supplying the plant load. If the process controlled by the computer is critical, a UPS system should be used. Recently a complete line of equipment has become available for protecting electronic devices and computers from erratic operation or failure due to power line transients, fluctuations, and interruptions. This equipment should be used for electronic devices which are critical for process and production. Industrial plant load tolerances to power supply variations are listed in Table 51.

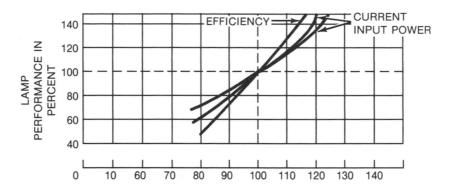

Fig 226
Mercury Lamp Performance as Affected
by Voltage; 120 V Basis

Table 54
Effect of Voltage Variations on the Characteristics
of Standard Induction Motors

Characteristic	Function of Voltage	(95%)	(100%)	(104%)	(110%)
Start and maximum running torque	V^2	90.2%	100%	108%	121%
Full-load speed	Slip α $1/V^2$	98.65%	100%	100.4%	101%
Full-load current	Depending upon design	106%	100%	96%	93%
Starting current	V	95%	100%	104%	110%
Temperature rise, full load	Test	+4 °C	0 °C	−2 °C	−4 °C
Maximum overload capacity	V^2	90.2%	100%	108%	121%

13.3.1.5 Voltage unbalance and loss of a phase (single-phasing) is caused by large single-phase loads, unequal impedances due to untransposed conductors in the supply system, one open fuse, the failure of one pole to close properly in a circuit breaker or contactor, open delta transformer banks, etc. Single-phase condition is an extreme case of voltage unbalance. The voltage unbalance creates negative-phase-sequence components, which cause an increase in motor losses, heating of generator rotors, and heating of motor windings. Severe negative-

Table 55
Effect of Voltage Unbalance on
Motors at Full Load

Voltage unbalance (%)	0	2.0	3.5	5.0
Current, neg. seq. (%)	0	15.0	27.0	38.0
Current, stator (rms %)	100	101.0	104.0	107.5
Increase in losses (%)	0	9	25	50
Stator average (%)	0	2	3	15
Stator maximum (%)	0	33	63	93
Rotor (%)	0	12	39	76
Total motor (%)	0	8	25	50
Temperature rise (°C)				
Class A	60	65	75	90
Class B	80	85	100	120

sequence conditions can lead to motor failures. In NEMA MG1-1978 [7] a voltage unbalance of no more than 1% is allowed in order to avoid excessive temperature rise. A voltage unbalance of 3.5% can result in a 20–25% increase in motor temperature rise, thus shortening the motor insulation life by over one half. Table 55 shows the effect of voltage unbalance on motor losses and temperature rise.

13.3.1.6 An interruption or loss of power supply voltage is defined as complete separation of the consumer system from the utility power system. This is another type of power system disturbance that can vary considerably in time duration, depending upon the cause of trouble and the method used to restore service. An interruption can last from 15–30 cycles if high-speed reclosing is used, 1–60 s or more if delayed automatic reclosing is used, and may continue for many minutes or hours, if remote or manual switching is used, or automatic reclosing was not successful. These disturbances generally result from system faults. When in-plant generation is used, synchrocheck relays should be used to block out-of-phase reclosing of the utility supply or to close the generator breaker.

13.3.2 Frequency Variation. The large interconnected utility systems will provide very constant frequency substantially all of the time. The isolated local generation facility will have a constantly varying frequency within a narrow band centered on the nominal value. Frequency deviation from the nominal system value is a service disturbance of a rather special nature. The change in frequency may be undesirable for computers and motors, but many loads are not frequency sensitive, such as filament-type lamps, resistance heaters, etc. Induction motors and synchronous motors may be over-excited and overheat on low frequency if the voltage is not reduced correspondingly. Volts per hertz is the criterion. When the load control system is designed to maintain the shaft load, the motor may be overloaded during low frequency. The drop in frequency is usually associated with system troubles and a drop in system voltage. Many load-shedding schemes are designed to operate when a drop in system frequency at a rate greater than 1 Hz/s is recognized or if frequency drops to some value below nominal (for example, 58 Hz in a 60 Hz system).

13.3.3 Harmonics. The recent trend toward the use of thyristors and rectifiers for HVDC transmission, variable frequency variable-speed drives, and rectifiers for dc drives has introduced additional harmonics in the electric power system. These harmonics may be propagated over great distances. The presence of harmonic voltages or wave shape distortion may cause problems that require consideration [B12]. Harmonic currents in induction and synchronous machines cause additional losses and heating, and adversely affect their efficiency. Synchronous machines are more sensitive to voltage unbalance and wave distortions than are induction motors. These effects are chiefly attributable to the harmonics of low orders that can have large magnitudes. Harmonics also cause an increase in motor exciting losses because they saturate the iron core. Motor stator I^2R losses increase in proportion to the square of total harmonic current. Induction motors will develop small negative torque due to the presence of odd harmonics thus reducing the torque available from them as compared to the torque at rated speed. Voltage level at the loads may appear to be normal, and yet there may be severe generator or capacitor overheating, or interference with communication or telephone systems due to harmonics. Where large banks of capacitors are used, overvoltage from resonance may be encountered due to harmonics, which may damage the capacitors.

13.3.4 Short-Circuit Current. Abnormalities in current are due to short circuits or faults in the utility supply or plant distribution system. These faults will result in magnetic forces and heat that can cause explosions and fire in switching devices and other equipment if this switching equipment is not selected with adequate interrupting capability and braced to withstand magnetic forces based on the available fault-current magnitude at the equipment location. A short-circuit study is necessary to determine equipment interrupting rating requirements. The flow of short-circuit current is usually of unreasonable magnitude or improper direction. The problems caused by a reversal of power flow may be commercial as well as technical.

Any particular power system disturbance, as described before and noted in Table 56, characteristically originates in a certain part of the power system.

Regardless of the source of these system disturbances, they result in an increase in the total operating cost of a facility due to reduction in motor efficiency and power factor, motor burnouts, power factor capacitor failures, equipment malfunctions, need for complex control schemes and added protection devices, and also increased capital cost required to supply emergency power for long-term disturbances through batteries, inverters, diesel generators, etc. Many consumer processes may suffer only loss of production time; however, some also incur spoilage of product and damage to the production equipment.

13.4 Corrective Measures and Supply Line Protection. After a study of available quality of service and possible system disturbances is made, it may be

Table 56
Electric Power System Disturbances

Disturbance	Duration	Effect on System	Typical Cause*
Voltage level change	Steady	±10% voltage	Normal system voltage variation resulting from load changes.
Voltage swing	10 cycles to 5 min	±30% voltage	Motor starting, shock loads, furnace loads, welders, planers, chippers, roughing drives.
Voltage transients*	Up to 30 cycles	+100% to −50% voltage	Remote system faults, switching surges, lightning strikes, capacitor switching.†
Voltage flicker	Variable	Voltage variations	Repetitive voltage swings or transients.
Voltage loss A	1 s maximum	Down to 0% voltage	Power transmission system or distribution system faults, network system faults.‡
Voltage loss B	1 min maximum	Down to 0% voltage	Power system faults or equipment failure requiring reclosing or resynchronizing operation.‡
Voltage loss C	Extended	−100% voltage	Permanent power system faults, equipment failure; accidental opening of power circuit breaker.
Voltage wave-shape distortion, harmonics, noise	Variable	Fundamental or harmonic voltage up to +200%	Arcing faults, ferroresonance, switching, transients, transformer, iron core reactor or ballast magnetizing requirements, controlled rectifiers, commutators, arc discharges, fluorescent lamps, motors.
Voltage unbalance	Steady	Up to 10% voltage variation among phases of three-phase system	Single-phase or unbalanced loads on three-phase system.
Single phasing	Extended	Down to 0% voltage on one phase of three-phase system	Open conductor, switching with single-pole devices, fuse opening, circuit breaker or contactor failure

Table 56 (*Continued*)

Disturbance	Duration	Effect on System	Typical Cause*
Power direction change, short circuits	Variable	Change of flow of current or power	Supply system faults, loss of transmission lines, synchronizing power surges, switching.
Frequency change	Variable	+1 to −2 Hz	Loss of generation or utility supply line.

*It is presumed that properly selected lightning arresters and surge protective equipment are installed throughout the system.

†Some types of switching transients may be amplified by coincident resonance of power factor capacitors and transformer inductances at the switching frequency.

‡Disturbance may be in either the utility or consumer system. Disturbance may be isolated in 3 to 30 cycles by circuit breakers or 35 cycles by network protectors, after which time service may be restored to disturbance-free portion of system.

found necessary that corrective measures be considered for desired reliability and quality of power. Two routes are available to approach this problem. First, the consumer can take the responsibility for establishing the desired quality by various improvements outlined hereafter. Second, the utility can make changes in its system to provide the consumer with electric power to his quality requirements on a special rate basis. A combination of the above alternatives is also possible [B9].

It is obvious that the use of multiple supply lines (sources), transformers, bus sectionalizing, in-plant generation, use of circuit breakers with proper protective devices, and use of properly rated fusible disconnects or interrupter switches, control schemes for automatic operation, addition of voltage regulating devices, power factor capacitors, and additional protective relays will improve the quality of service.

Each improvement also adds to the cost and space requirements that should be warranted by the process requirements or for personnel safety.

A knowledge of the frequency of outages and the estimated cost of each outage is necessary to determine the need for added equipment.

These costs may include:

(1) Indirect effect on safety
(2) Loss of production
(3) Production of off-tolerance products or scrap
(4) Purging and restart-up expense
(5) Additional personnel needed to be available for restart procedures if not an automatic process
(6) Electric and mechanical equipment damage

Table 57 lists various devices to minimize the effect of power system disturbances. The corrective measures and lines protection are described under the following four headings:

(1) Addition of power equipment
(2) Modification of power system design
(3) Modification of control system
(4) Protective schemes and relaying

13.4.1 Addition of Power Generating Equipment. Addition of power generating equipment in an industrial or commercial plant will generally provide higher reliability and better quality of power supply. Some processes or loads cannot tolerate a shutdown and should be provided with a high reliability energy source to ride through a loss of normal supply voltage.

This can be achieved by installing auxiliary power generating equipment such as diesel generating sets, battery operated generating sets, inplant generation with utility tie, and storage batteries.

The power generating equipment for emergency power (with an automatic transfer scheme) to supply vital and critical process equipment when utility power is interrupted is expensive in initial cost and expensive to operate and maintain.

It may be advantageous for some process plants to generate their own power because it is economical due to available low cost fuel, or the plant has need of large amounts of process steam or a heat requirement for a building is large [B7], or both. In-plant generation provides higher reliability to plant loads and better control of voltage within limits (by excitation of the generator field winding). The plant supply system will be less affected by faults on utility transmission lines and its generating system. Proper protective relaying should be used to disconnect the utility supply when a fault is on utility lines and load shedding devices and it should be provided for optimum use of in-plant generating equipment. In order to improve the quality of power supply, various equipment and devices, such as motor-generator sets, rectifier-inverter sets, rectifiers, automatic voltage regulators, load tap changers on transformers, ferroresonant transformers, power-factor correction capacitors, harmonic filters, reactors, lightning arrestors, etc, can be added.

Equipment such as a motor-generator set, rectifier-inverter set, or rectifier provides electrical isolation of loads from utility supply lines and short-time continuous power to a load during service interruptions. This electrical isolation of the load from the main system thereby eliminates such disturbances as noise, voltage swings, or transients, but the motor-generator set itself will be sensitive to certain disturbances. The inertia of the rotating motor-generator set provides energy to the load for a period of from several cycles to several seconds, depending upon flywheel effect. This type of approach is common in power supplies for computers where it is sometimes essential to have a programmed shutdown of a system so that the computer's memory circuits can retain vital information. These motor generator sets are provided with a flywheel.

The most critical ac loads can be supplied by a rectifier-inverter system incorporating a battery. In this approach the load is supplied by a static or

Table 57
Minimizing the Effect of Power System Disturbances

Device	Minimized Disturbance	Comment*
A. Power Equipment and Power Switching Devices		
Capacitors, shunt connected	Voltage level change	Can reduce normal system voltage variations.
Reactors, shunt connected	Voltage level change	Can reduce system overvoltages during light load periods.
Voltage regulators, induction or load tap changing	Voltage level change, voltage swing, voltage unbalance	Will restore voltage level on load. Operating range usually ± 10%. Response too slow to correct for voltage transients. Single-phase sensors and regulators required to correct for voltage unbalance.
Generators, synchronous motors, or condensers used with automatic voltage regulators	Voltage level change, voltage swing	Will restore system voltage within limits imposed by maximum and minimum excitation on machine field.
Circuit breakers used with protective relays, network protectors, reclosers, fuses	Voltage transients, voltage single phasing, power flow change, frequency change	Device isolates cause of disturbance. Restoration of service required. See Table 58
B. Control Circuit Modification and Small Power Equipment		
Battery, battery charger system	All	Provides uninterrupted power for control, instrumentation, communication. Alternate power source for emergency lighting, critical loads. Limited to about 50–300 A·h capacity during loss of voltage.
Cable shielding in control or communication circuits	Voltage waveshape, harmonics, noise	Reduces or eliminates induced voltages which interfere with correct transmission of intelligence or control signals.

* The solution to any disturbance may result in "side effects" which are equally disturbing. For example, capacitors switched without preinsertion surge-limiting resistors may create damaging transient overvoltages; improperly installed or grounded shield communications cable may cause other types of noise or burnout due to power-fault induced currents, etc.

Table 57 (*Continued*)
Minimizing the Effect of Power System Disturbances

Device	Minimized Disturbance	Comment*
Circuit breaker, close/trip control circuits	Voltage loss	
Capacitor stored energy		Provides energy for releasing circuit breaker or bolted pressure switch tripping mechanism for several seconds.
Compressed air storage		Provides energy for 2–3 circuit breaker or bolted pressure switch closing operations.
Spring energy		Provides energy for 1 circuit breaker or bolted pressure switch closing operation.
Control relays	Voltage loss	
Time delay dropout		Primarily used in motor control circuits to maintain starter circuit in "run" position for 2 to 4 s after loss of voltage.
Mechanically latched relay		Retains control circuit condition during loss of voltage.
Engine generator set	Voltage loss B and C (Table 56)	Provides emergency power for lighting or critical loads. Normally requires minimum of 5–60 s to start if automatic.
Filters, tuned circuits	Voltage distortion, harmonics, noise	
Motor-generator set†	All except voltage loss B and C (Table 56), extended frequency change	Loss of voltage ride-through capability with flywheel: (1) 0.3 s with deviation less than −3% voltage, −½ cycle, or (2) 1.8 s with deviation less than −3% voltage, −1½ cycles.

* The solution to any disturbance may result in "side effects" which are equally disturbing. For example, capacitors switched without preinsertion surge-limiting resistors may create damaging transient overvoltages; improperly installed or grounded shield communications cable may cause other types of noise or burnout due to power-fault induced currents, etc.

† Improved performance results from addition of a flywheel.

rotating inverter that receives its power from a system consisting of a battery in parallel with a generator or a rectifier, which is in turn supplied by the main power system. Such an array of equipment could provide uninterrupted services to a load of up to 400 kVA for 30 min or more.

Other equipment such as an automatic voltage regulator, a transformer

Table 57 (*Continued*)

Device	Minimized Disturbance	Comment*
Uninterruptible power system (UPS), static system, or motor-generator set	All	Total isolation from power system disturbance for loads up to about 400 kVA. Provides power for 15–30 min after loss of voltage. A UPS system contains a rectifier, inverter, and battery (or motor-generator sets with battery) to supply 60 Hz power.
Voltage stabilizer‡	Voltage level change, voltage swing, voltage transient, voltage unbalance, voltage flicker	Static device. Holds constant output voltage with voltage input variations up to ±30%. Full correction in ½ cycle. Limited to loads below 50 kVA.
C. System Line Diagram	All	Use of reactors, bus ties, etc, can minimize flicker on critical loads. Emergency separation of critical and noncritical loads.

*The solution to any disturbance may result in "side effects" which are equally disturbing. For example, capacitors switched without preinsertion surge-limiting resistors may create damaging transient overvoltages; improperly installed or grounded shield communications cable may cause other types of noise or burnout due to power-fault induced currents, etc.

‡May contain harmonic distortion in regulated voltage.

load-tap changer, or ferroresonant transformers will reduce voltage fluctuations and will restore voltage level on load and practically maintain constant voltages. Their operating range is usually ± 10%, but response is too slow to correct voltage transients. Power factor correction capacitors help in reducing demands of reactive power from the utility in addition to reducing normal system voltage variation and voltage dip. Harmonic filters are required where high-power rectifiers and inverters are used. Reactors are generally used for reducing short-circuit currents, but are also useful in reducing system overvoltages during light load periods, smoothing distorted wave shapes and reducing harmonics. Surge arresters at the point of interconnection will protect equipment from insulation breakdown due to lightning or switching surges.

Storage batteries are normally used to provide reliable control power for the switchgear and other control devices, but are sometimes used as an emergency source of power for a motor that drives an extremely critical equipment, such as vacuum pumps or lubrication pumps for large rotating machinery.

13.4.2 Modification of Power System Design. Modification in power system design will increase reliability as well as power quality. Some of the examples of modification are: changing simple disconnect switches and contactors with power circuit breakers or interrupter switches, using a synchronous motor instead of a squirrel cage induction motor for large drives, power supply from relatively large capacity (stiff) system, dual source supply and bus sectionalizing, and underground power distribution. Power circuit breakers, network protectors, reclosers, and bolted pressure disconnect switches, though more expensive than simple disconnect switches, might provide better protection, because they can be opened or closed against all magnitudes of current up to their short-circuit ratings and can be operated locally or remotely. Circuit breakers are usually provided with protective relays for selective operation to minimize interruption of power and provide fast clearing of faults. Most disconnect switches generally do not provide all such functions and are normally operated locally. Disconnect switches provided with fuses for overload and short-circuit protection may be a source of single-phasing, although now switches are available with single-phase protection and remote operation. These switches will open if one fuse is blown. Circuit breakers always open all the three-phases on a fault. Manufacturers' data and recommendations should be consulted when these devices have operated at or near their interrupting ratings.

Voltage dips can be reduced by using a large capacity power supply system, or starting large motors by reduced voltage starting methods, or segregating the power supply to impact loads from other critical loads that might be affected, or using 0.8 or 0.9 leading power factor synchronous motors of low starting current (3.5 to 4 times motor full-load current) or a wound rotor motor instead of a squirrel cage induction motor. The synchronous motor can improve the power factor of the system by regulating the excitation of the field thus reducing the voltage variation.

In most cases dual-circuit supply from the utility and sectionalizing of the substation bus provide much higher reliability and better quality of power than that obtained by segregating critical plant loads from the larger loads responsible for the voltage dips.

In lightning-prone areas, it may be desirable to use underground distribution circuits to shield the circuit from lightning and physical abuse. Also, provision of a distribution transformer at the interconnection of utility and plant bus will attenuate the surges from utility transmission lines to a certain degree.

To protect for power supply loss from such causes as fire, tornado, or earthquake, it is necessary that all important substations be physically separated as far as possible, so that not more than one substation is affected due to such causes.

13.4.3 Modification of Control System. Modification of the industrial control and its power supply can provide fast removal of faults and minimize the effects of voltage dips and interruption of power, thus providing higher reliability. These modifications could be the use of a time delay drop out scheme, reaccelerating schemes for motors, dc instead of ac control, regulated or stabilized control buses, latching relays, automatic reclosing schemes, etc. Automatic reclosing should not be attempted where the protected equipment is cable or transformer(s). Faults in this type of equipment are more likely to be permanent rather than transient in nature. Automatic reclosing, in such cases, will cause further equipment damage and personnel hazard.

Motors generally are little affected by short interruptions and voltage dips, but ac contactors and solenoids may drop out on reduced voltage. A common approach here would be to use a time-delay dropout scheme for motor control. Sometimes latching devices are used for the same purpose.

Another approach is to supply the control system from a battery or other reliable dc source. Disturbances in the ac portion of the system are thus not reflected into the control circuit, and power devices will resume their normal operating condition after restoration of service. Also stored energy tripping mechanisms, such as capacitor trip devices, compressed air storage, separate batteries for circuit breaker, or bolted pressure switch control may be used to provide energy for tripping such equipment after ac voltage has disappeared.

13.4.4 Protective Schemes and Relaying. Protective devices and relays are applied to the utility interconnection circuits in the same manner and employing the same principles as in all other locations in industrial plant power systems and utility systems. The basic purpose is to protect all circuits and equipment from abnormal electrical conditions as well as to minimize the effects of faults. Protective relays, fuses, and other devices provide this function by initiating the proper switching at the designated time under abnormal electrical conditions. A prerequisite to the use of protective relays is the presence of suitable sensing devices (instrument transformers) and suitable switching devices (circuit breakers or remotely operated switches, etc). Therefore, selection of a protective relaying scheme is inherently dependent on the circuit arrangement as well as the equipment and processes to be protected and the service continuity needed.

Relays and other protective devices are used to detect faults, and abnormalities in voltage and frequency such as overvoltages, undervoltages, single-phasing, and under-frequency. Protective relay settings and fuse ratings are determined and coordinated with utility protective devices to isolate the faulted segment of the system as quickly as possible to permit the unfaulted part of the system to continue to operate. Methods of restoration of service after loss of voltage are tabulated in Table 58.

The details of protecting specific devices are covered in other chapters of this recommended practice. The discussion of protective devices in this chapter centers on the specific applications in service supply lines as indicated on the sample circuit arrangements. Surge arresters are also important to incoming circuit protection. The application of surge arresters for equipment protection is covered in respective chapters. Typical applications are shown on typical systems illustrated in this chapter.

The application of protective devices is accomplished by a division of the electrical circuit into zones of protection which can be accommodated in an economical manner using various protective schemes. These zones of protection are selected on the basis of the individual characteristics of each installation. Some of the considerations are:

(1) Electrical characteristics of the utility supply circuits, especially fault-current distribution

(2) Load continuity requirements and capacities

Table 58
Restoration of Service after Loss of Voltage

Method	Minimum Time to Restore Service	Comments
Reenergize circuit*		If motor loads exist which support plant voltage after loss of system voltage, then reclosing should be delayed either for a definite time or until residual plant voltage has decayed to less than 25% normal, or as recommended by manufacturer to prevent damage to motors. The reclosing should be in accordance with ANSI C50.41-1982 [1] (the impressed voltage should not exceed 1.33 V/Hz of rated value).
Automatic reclosing after temporary fault	25 to 70 cycles	
Remote-controlled reclosing of circuit breakers or switches	Up to 1 min	
Manual or remote-controlled reclosing after manual isolation of cause of disturbance; replace fuses	Up to 1 h or longer	
Transfer incoming line to alternate power source†		Intentional time delay may eliminate unnecessary transfer under some conditions.
Automatic transfer	15 to 30 cycles	
Manual transfer	Up to 30 min	
Start generators in consumer system	Variable	Standby generation may be sufficient to supply emergency or critical loads.

* Reclosing should include resynchronizing if consumer generators are operating in parallel with utility system.

† May include transfer of emergency lighting and loads to a battery source or engine-driven generators.

CAUTION: Do not apply automatic or remote reclosing on circuits consisting of cables or transformers where reclosing will reinitiate the permanent faults associated with such equipment.

(3) Probability of system disturbances due to exposure, circuit length, type of equipment, etc

(4) Utility standard requirements established to ensure maximum quality service to their users

(5) Available protective equipment with due consideration to economics

(6) Motor stability and other pertinent load characteristics

(7) Reclosing requirements

(8) Fault-locating requirements

(9) Physical layout

Protective schemes used in an industrial plant in combination with service supply lines are described according to their locations in an industrial utility tie line under seven groups (see Fig 227).

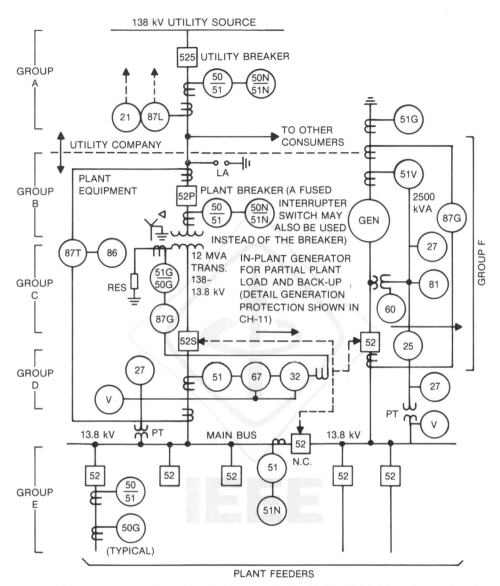

PLANT FEEDERS

NOTE: If fused interrupter is used, relaying associated with the 12 MVA transformer is not used.

**Fig 227
Grouping of Protective Schemes**

Group A — Supply circuit protection
Group B — Service entrance protection
Group C — Supply transformer protection
Group D — Transformer secondary protection
Group E — Plant feeder circuit protection
Group F — In-plant generator protection
Group G — Bus relaying (not shown)

These protective device locations are not always well defined due to differences between installations, and in some cases one or more groups may be nonexistent. For example, when the utility supplies power at 208Y/120 V, 480Y/277 V, or even 4.16 kV or higher, no transformer is required and group C is omitted. Group D is omitted when there is no secondary protection circuit breaker, and group F is present only with in-plant generation. Protective devices which often are supplied for protection of each group, will be discussed in the following paragraphs. A portion of the protective equipment described is shown in Figs 228–232.

While protective relaying is emphasized in this chapter, it is not intended to exclude series tripping or fusible protective devices from consideration. In circuits of 600 V and less, these two types of protection are more common than protective relays. However, even at these voltages ground-fault protective relaying is now used extensively to protect against arcing ground faults. In many plants with low and medium voltage (480 V–34.5 kV) distribution, fused interrupter switches are used because of their simplicity, lack of dependence on battery control power, and low maintenance.

A broader connotation of protective schemes is better expressed as system or service protection. This carries the additional responsibility of automatic service restoration when warranted, separation of the more critical loads from less critical loads that can be deliberately dropped when the main service from the utility is lost for any reason, and similar sophisticated control schemes designed to make the maximum utilization of the utility service even during periods of partial inadequacy. Identification of the general area of a fault through relay targets or open fuse indication so as to minimize operating confusion and down time is also part of this concept.

13.4.4.1 Group A—Supply Circuit Protection. The protective equipment at this point is on utility premises and is usually a utility property. Its primary purpose is to protect the main utility circuit from the adverse effect of faults between the utility circuit breaker and the service entrance equipment. The main goal is to clear faults on the feeder quickly, rather than jeopardize the service of all users supplied from the source bus. Another function is to back up the service entrance relaying and prevent an in-plant disturbance from affecting the utility source bus. Relays commonly employed at this point are as follows:

(1) Inverse time overcurrent phase and ground fault relays (Devices 50/51 and 50N/51N) are used whenever possible because of simplicity and economy.

The more inverse relays are sometimes used when coordination with fuses is needed or when it is required to ride over high inrush currents upon restoration of power after a service outage. Standard inverse or minimum time inverse

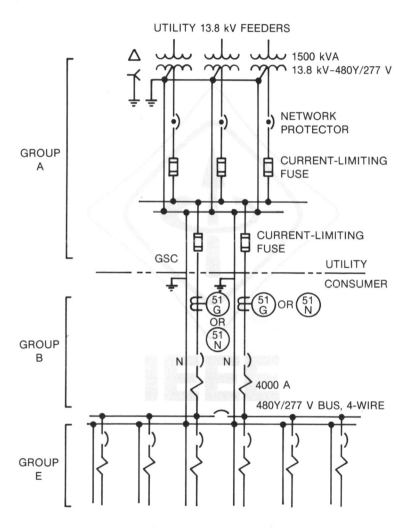

Fig 228
Low-Voltage Network Supply (480Y/277 V or 208Y/120V)

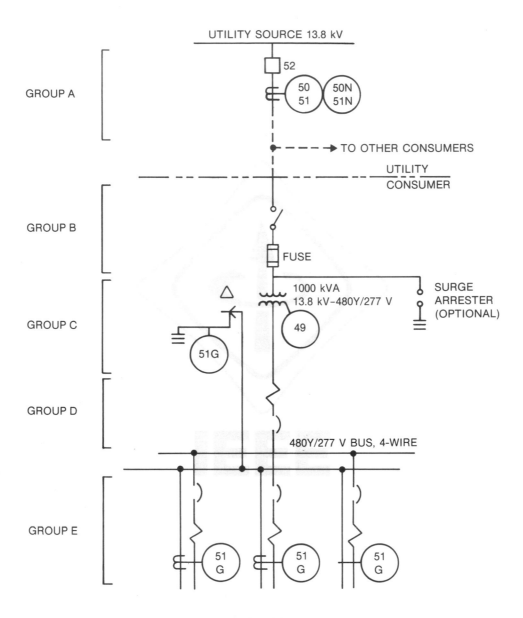

**Fig 229
Fused Primary and Low-Voltage Plant Bus**

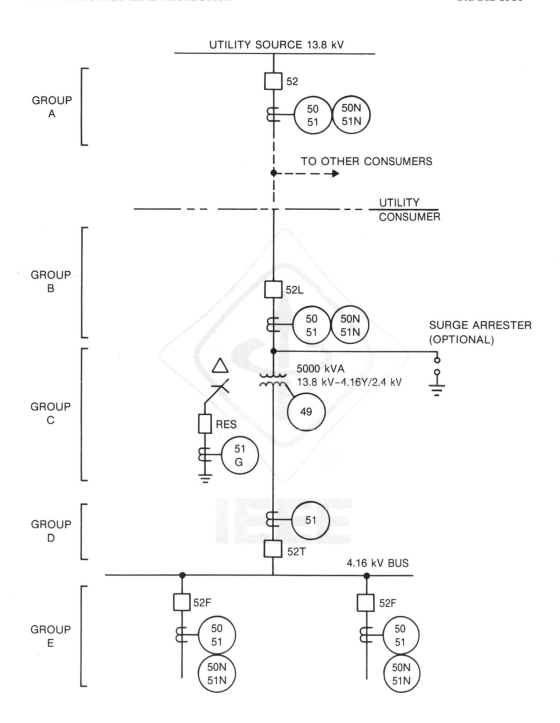

**Fig 230
Single Service Supply with Transformer
Primary Circuit Breaker**

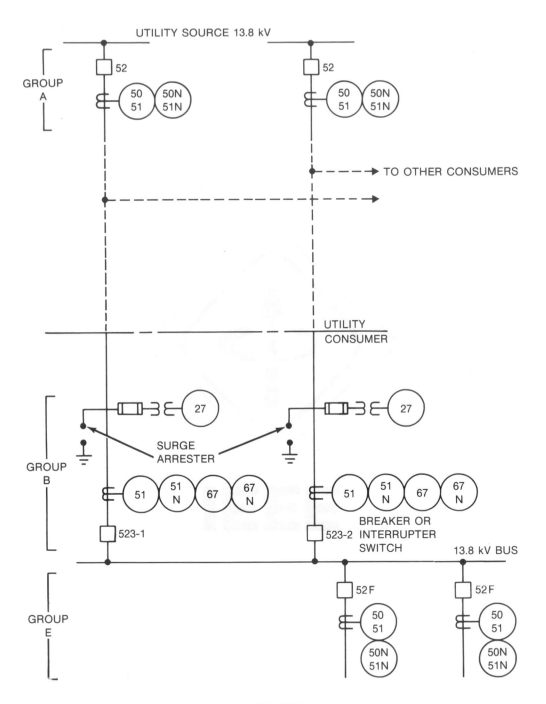

Fig 231
Dual Service without Transformation

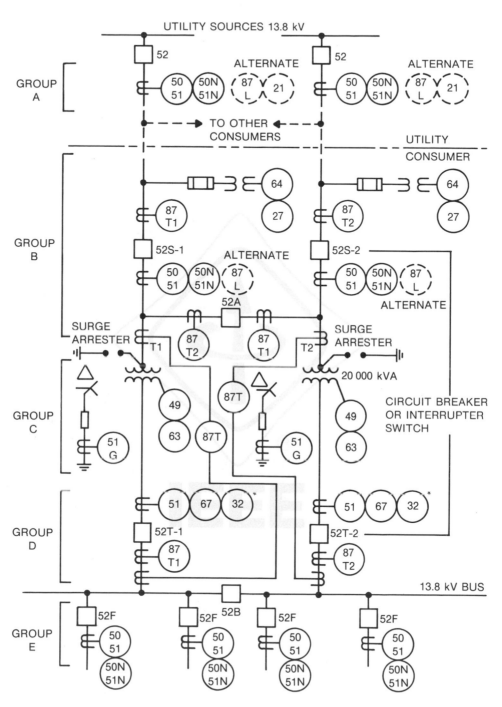

*Use Device 67 or 32.

**Fig 232
Dual Service with Transformation**

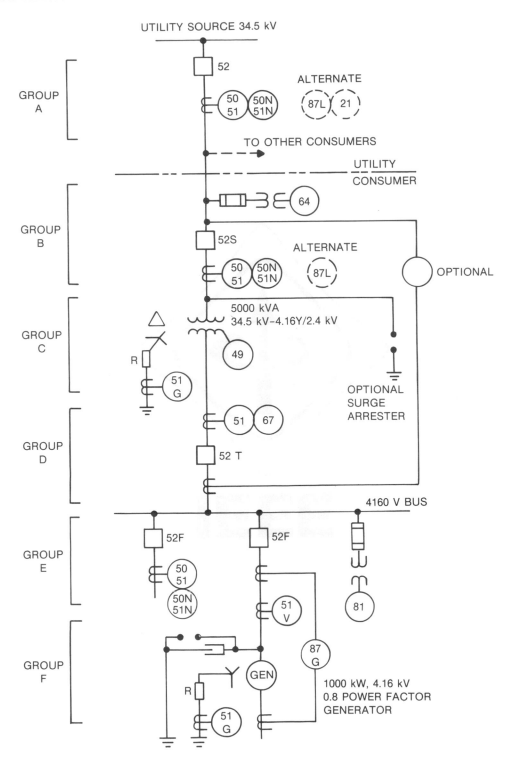

Fig 233
Single Service with In-Plant Generation

characteristics are also used for a variety of reasons, one condition being, for example, when the short-circuit current magnitude is dependent largely on system generating capacity at the time of fault. When the utility uses a fixed-time type relay on the industrial service line, a less inverse relay characteristic on the industrial system may be preferred.

Instantaneous attachments (Devices 50 and 50N) should be employed only when they will coordinate with the load-side protective devices.

When instantaneous units are used, they are set high enough so that they do not detect faults beyond about 80% of the distance to the next load-side overcurrent protective device. In some applications they must be omitted or immobilized in order to achieve coordination.

(2) Distance Relays (Device 21)—These relays are generally used by a utility company on their tie line to an industrial plant. There are several types of distance relays, including impedance, reactance, offset distance, and mho. These relays are needed when fast tripping is desired over most of the protected line length. These relays are considerably more expensive than overcurrent relays. A distance relay compares the current and voltage of the power system to determine whether a fault exists within or outside its operating zone. These relays can be obtained with one to three zones of operation. The first zone provides instantaneous protection for up to about 90% of the protected line. The second and third zones, if used, are time delayed and extend backup protection into the area protected by the service entrance relays.

The zone of operation is a function of the protected line impedance, which is a fixed constant and is relatively independent of current and voltage magnitudes. Distance relays are sometimes necessary when it is impossible to get selective, reasonably fast protection with overcurrent relays because of the long time needed to get selective tripping over a wide variation of fault-current magnitudes. They may also be needed for conditions where fault currents are low and are difficult to distinguish from load currents. For long lines, impedance-type distance relays are most commonly used. Distance reactance and certain other newer relays can be applied on medium to short lines. Where very short lines are involved, pilot relay systems can be considered.

Distance-controlled overcurrent relays (Devices 51 and 21) can be used in combination to provide fast tripping for faults on the primary of supply transformers, plus backup time delay for low-side faults with some limitations. This combination is useful where overcurrent relays alone cannot be set to respond to transformer low-side faults.

(3) Pilot Relay System (Device 87L)—This relay system employs a communication channel in conjunction with protective relays in order to compare the circuit conditions at both ends of the protected line simultaneously. This system can thus provide fast protection over 100% of the line. This relay system determines whether fault is within the protected line or external to it.

Additional relaying should be considered for backup protection.

The majority of pilot relay systems used on lines to consumer plants are those of the pilot-wire-type (Device 87L). These relays measure the current in both ends of the line differentially and detect an internal fault by differences between "in" and "out" current. A metallic telephone type communication channel is required for this relaying. Pilot wire relays are limited in distance to a maximum of roughly 10–15 miles, depending on the pilot wire used. Complications arise if the line serves more than one customer, although multiterminal installations are possible. Recently, fiber optics became available as an alternative to telephone-type channels. Other pilot channels, such as powerline carrier, microwave, and audio tones utilizing wire communication, are not covered in this publication. Additional information can be obtained from manufacturers and [B13] and [B2].

When selecting the basic relay types, it is well to consider the competence of the available personnel. This includes the ability to calculate the proper settings to attain the desired results as well as the qualifications of test personnel who are going to start up and maintain the relays.

13.4.4.2 Group-B—Service-Entrance Protection. Where service-entrance relaying is used, it normally operates the main interrupting device. However, when fuses are used, the fuse provides the functions of both sensing and interrupting. This protective equipment is sometimes provided by the consumer, in which case the required characteristics and settings should be selected based on selective operation of the consumer and utility protective devices, and when applicable, by the pertinent governmental codes. Refer to Chapter 7 for ground-fault relaying, and requirements of ANSI/NFPA 70-1984 [6] (National Electrical Code [NEC]) on systems below 600 V.

The protective devices function to disconnect the supply from the consumer's system for certain faults within the supply transformer primary and secondary connections and serve as a backup to the protective devices associated with the transformer secondary.

There are several schemes which can be used to open the utility supply circuit breaker within group A when there is no transformer primary circuit breaker and the fault currents are not sufficient to operate the relays of group A. This is accomplished by employing transferred tripping schemes through the use of pilot wire, grounding switches, or power-line carrier audio tone or microwave signals.

Typical protective devices associated with group B are as follows:

(1) Overcurrent phase and ground-fault relays (Devices 50/51 and 50N/51N) are applied similarly to the overcurrent relays described in 13.4.4.1,(1).

(2) Directional phase overcurrent relay (Device 67)—these relays differ from nondirectional relays in that they operate for faults in one direction. When applied for tie line protection they look into the utility system and detect faults on incoming lines, thus preventing overloading of in-plant generators and parallel lines. Their principal function is to prevent tripping when fault current flows in nontripping direction, thus preventing unnecessary tripping—a most important consideration for higher reliability. Also their directional selectivity allows for sensitive settings that would not be possible with overcurrent relays.

Directional relays are recommended at the industrial plant when two or more radial feed circuits are operating in parallel, or when there is in-plant generation in parallel with the utility supply lines. These relays are located at the service entrance location when there is no transformation at the industrial plant. The directional relays are normally located at the secondary circuit breaker, but the relays for group B and D are combined in this case. Some utilities accept consumer surplus power from the in-plant generators. In such cases, voltage restrained feature enables the relay to distinguish between load and fault current. A close coordination with the utility company engineers is necessary in the selection of directional relay characteristics and settings so that the protection system will be compatible with both the utility system protection and plant operating requirements [B16].

Directional overcurrent ground-fault relays (Device 67N) are used in addition to directional phase overcurrent relays (Device 67) for the complete protection of incoming lines in grounded systems.

These relays are required at the industrial plant when paralleling incoming lines without wye–delta transformers or when in-plant generation is paralleled with the utility supply line. However, when supply transformers are involved, these relays (Device 67N) are considered optional in view of the reduced ground fault possibility and possible use of ground relays (nondirectional) (Device 51N or 51G), which will operate to clear the ground fault. Also Device 67N requires polarization and added cost of polarization is not justified.

(3) A ground-fault detector relay (Device 64) may be needed to clear a supply line ground fault by disconnecting the supply line at the consumer end. This situation can occur when there are parallel supply lines or in-plant generation, and the utility end is opened by overcurrent relays. The system ground is thus removed, and the supply line is then operating as a normally ungrounded system. An over- and undervoltage relay connected to a line-to-ground voltage transformer or an overvoltage relay connected to the broken delta secondary of three line-to-ground voltage transformers can be used. Alternately, a sensitive power directional relay (Device 32) may accomplish the same purpose.

(4) The consumer terminal of pilot relay systems (Device 87L) as described in 13.4.4.1,(3), can be used here and functions to trip the service entrance circuit breakers as well as the utility supply circuit breakers.

(5) 600 V and below service reliability may be improved by using current-limiting cable limiters on each end of each service cable where there are three or more cables per phase.

13.4.4.3 Group C—Supply Transformer Protection. Transformer protection is listed separately even though some of the protection is provided by the circuit breakers and relays or fused devices covered in groups B and D, and a chapter in this book is devoted entirely to transformer protection. Nevertheless, the basic protective devices are described briefly:

(1) Overcurrent phase and ground-fault relays (Devices 50/51 and 50N/51N) are as described in 13.4.4.1,(1). Instantaneous ground-fault relays (Device 50G) should be connected to core-balanced (ground-sensor)-type current transformers for best results, although they can be used with conventional current transformers in a residual connection with Relay 50N with good results on solidly grounded systems.

(2) Transformer differential relays (Device 87T) provide fast clearing for phase-to-phase plus phase-to-ground faults and can be obtained as regular or

high-speed devices. Differential protection is almost universally applied to large (above 5000 kVA) or important transformers when a primary circuit breaker or circuit switcher is used. Harmonic restraint relays are used to allow greater sensitivity and yet not operate on magnetizing inrush currents. These differential relays are arranged to cause both the primary and the secondary circuit switching devices to trip and lock out through a lockout relay (Device 86).

(3) Pressure relays (Device 63) have gained acceptance as reliable devices for large power transformers. They are applicable to all transformers that have a sealed gas chamber above the oil level. They are more sensitive than differential relays for internal faults and are particularly useful for faults in tap-changing equipment. They can be used in lieu of or to supplement the transformer differential relays.

(4) Transformer temperature protective devices (49) are usually provided with all power transformers. These devices are essentially for overload protection and in many applications merely indicate alarm or activate cooling apparatus. They protect against excessive transformer temperatures. Refer to ANSI C57.92-1981 [3].

13.4.4.4 Group D—Transformer Secondary Protection. Where there is no transformation, these relays are not involved. Relaying/protection at this location also varies with the number of transformers used, whether or not there is in-plant generation or for other similar considerations. Basic protection here includes:

(1) Overcurrent phase relays (Device 51) are required to protect against bus faults and to back up the in-plant feeder overcurrent relays. When there is a primary circuit breaker with associated overcurrent relays [see 13.4.4.2,(1)], it is possible to eliminate these overcurrent relays at the secondary circuit breaker without a great sacrifice in protection. Instantaneous attachments (Device 50) cannot be used here successfully without loss of selectivity between main and feeder relaying [B14]. Therefore, if high-speed protection against bus faults is desired, bus differential relays (Device 87B) should be used. Residually connected overcurrent ground relays (Device 51N) are not needed because the transformer neutral ground relay (Device 51G), described in the following paragraph, is more effective.

(2) A transformer neutral ground relay (Device 51G) is connected in the grounded neutral lead of the transformer secondary. This relay is very sensitive and will back up feeder ground-fault relays. This relay usually trips the secondary breaker. A time-delay relay may be connected in series that will be activated at the same time to trip the primary breaker (62P) if the fault continues after set time delay (usually 1–2.5 s). This additional time-delay relay will protect the windings and terminals from ground faults. It is the only protection against bus ground faults when bus differential relays are not specified. A high set Device 50G can provide additional protection for turn-to-ground faults in the transformer secondary winding. In a resistance grounded system, the tranformer differential relay sensitivity may be such as to be unable to recognize the low-side line-to-ground fault. In such cases differential ground relay scheme (Device 87G)

should be used; see chapters on ground-fault protection and transformer protection.

(3) Directional overcurrent phase and ground relays (Devices 67/67N) can be useful in those dual-service arrangements where the dual services are exclusively used by a single consumer. Otherwise the general comments of 13.4.4.2,(2) apply to this location equally well. In addition, installations with in-plant generation utilizes directional relaying at this location to clear faults in the supply circuit or supply transformer from in-plant generator and bus sources.

(4) A directional power relay (Device 32) can be used to disconnect the incoming line and the supply transformer upon utility tie loss in case of parallel operation or when there is in-plant generation. For example, a sensitive relay can detect core power loss when the transformer is reversed magnetized from the secondary side. For an alternate method of accomplishing this, see 13.4.4.2,(3) (Device 64).

13.4.4.5 Group E— Feeder Protection. This relaying is selected on the basis of the type of load, type of circuit, and general degree of protection required. The detailed considerations for load circuits are fully covered elsewhere in this publication in respective chapters.

13.4.4.6 Group F— In-Plant Generation Protection. The presence of in-plant generation adds flexibility and reliability to the consumer's electrical supply sources, but it also adds certain complexities in protection and control. Additional protection often needed includes standard generator protective relaying (Chapter 11) as well as certain of the following. Some of the following are standard generator protection.

(1) Overcurrent relays with voltage control (Device 51V) are used. Low voltage usually results from faults rather than overloads. Thus when the overcurrent is accompanied by reduced voltage, the relay will operate. However, if the circuit voltage is maintained, the relay operation is blocked (or in some designs, restrained) since the condition is probably an overload that requires that the generator remain connected. This characteristic allows setting to be lower than generator full-load rating.

(2) Generator neutral ground relay (Device 51G)—if the generator is grounded, a time overcurrent relay in the neutral circuit can be used in much the same manner as the transformer neutral ground relay [see 13.4.4.4,(2)]. If the in-plant generator is to parallel the transformer, the type of grounding could be either the same as the transformer or it could be high resistance grounding. In the latter case, care should be exercised in selecting line-to-line voltage rating on the potential transformers used for protection and synchronizing. Furthermore, a voltage relay would be used for ground detection on a high resistance grounded neutral. Where necessary, surge protection should be provided in the generator.

(3) Generator differential relays (Device 87G) protect against faults in the generator or generator leads.

(4) Underfrequency relays (Device 81) can be used to operate at preselected frequencies to drop load or sectionalize buses in order to keep remaining generation and load in operation during disturbances. Time delays of 6–30 cycles

are often used to prevent operation on transient or temporary conditions. Static frequency relays which operate faster on frequency change are most useful for this purpose. Frequency relays are sometimes needed to protect some generators that may be subject to overspeeding, though turbine manufacturers prefer to use governors for protection against overspeeding.

In case of a fault in the utility system, frequency relays may be used to separate local generation plus critical loads from a utility system. They may also be used in the simplest case to drop unessential load during system overloading. The theory in this latter case is that if each consumer drops some load, the utility may not disconnect the total feeder.

Frequency relays may also be needed to disconnect the utility service in certain cases where the supply line is equipped with automatic reclosing. In this case the consumer's main circuit breaker is tripped to protect synchronous or large induction motors and generators.

(5) Directional relays (Device 67) are usually needed, but are normally located elsewhere [see 13.3.4.4,(3)].

13.4.4.7 Group G—Bus Relaying. Bus and switchgear relaying has been covered in Chapter 12 of this recommended practice.

13.5 Supply-System Protective Schemes—Examples. Six typical schemes have been selected for detailed analysis, and one-line diagrams showing protective relays are prepared to illustrate these schemes. The protective relaying is discussed with cross reference to protection groupings A–F, which appear to the left of the diagrams. These schemes should not be considered as preferred or recommended schemes, they are merely examples of various utility tie arrangements with suggested relaying. Additional relaying or different configuration of utility tie may be used depending on consumer requirements and serving utility.

The electric supply system for an industrial plant should be designed to fulfill the particular needs of the individual plant [B12]. For example, duplicate service equipment may be needed for an industrial process requiring a constant supply of electric power.

13.5.1 Network Supply Systems Below 600 V. In many sections of the country, the utility tariffs permit supply of power at secondary voltages only, and all higher voltage equipment is owned and installed by the utility on the consumer's premises. An example of this type installation is shown in Fig 228. The arrangement is a network system with the sources supplying network transformers operated in parallel on a common service bus. With this type installation, network protectors are required and are supplied by the utility to segretate a faulted line or transformer from the service bus. The network protectors perform three basic functions in the system. They automatically isolate the fault in the primary feeder or network transformers without dropping load, also isolate ground fault in the primary feeders in a single-ended grounded delta–wye network transformer, and on predetermined voltage conditions they automatically close the network protector breakers.

In the example (Fig 228), since the short-circuit current is extremely high, current-limiting fuses are necessary to limit this short-circuit current on the customer's equipment. The plant engineer should review very carefully with the utility the type and design of approved service entrance equipment.

The protection required for the consumer's service-entrance equipment is largely regulated by national codes. Circuit breaker, service protector, bolted pressure switch and fuse selections are governed by the continuous-current rating and number of service main circuits allowed. In addition to phase fault protection, the NEC [6], Section 230-95, requires ground-fault protection for solidly grounded wye service entrances. Furthermore, in a grounded power system, the NEC [6], Section 250-23, requires connection of a service conductor specified in Section 250-25 to a grounding electrode at the service switchboard. With few exceptions, the connection is required to be made on the source side of service disconnecting means. Ground-fault protection is often recommended on grounded circuits since the phase protective devices usually provide little or no protection against low-to-medium-magnitude arcing ground faults.

Air circuit breakers or other interrupting switches (Device 52S) in group B may be provided with ground-fault protective device in conjunction with phase overcurrent device or separate ground sensing relays (Device 51G or 51N) may be provided as shown in Fig 228.

13.5.2 Fused Primary and Low-Voltage Plant Bus. Many utilities permit the customers to own and operate higher voltage service entrance equipment and the associated transformers. Figure 229 shows a simple single service of this type installation adequate for small industrial customers. The fuse size on the transformer is limited to the maximum size that will provide selectivity with the power company's overcurrent relays (group A, Devices 50/51 and 50N/51N), secondary breaker (Device 52S) protective devices, and the NEC [6] requirements for transformer protection. This selectivity is important to the utility to prevent interruption of service to other customers on the same line as well as to the plant loads.

No special relaying equipment is provided at location B. A ground relay (Device 51G) is recommended at location group C to protect against low-magnitude arcing ground faults. Since the transformer secondary is low voltage (480Y/227 V), the protective devices at group D are integrally mounted on the low-voltage power circuit breakers or bolted pressure switches. Where the transformer is not close coupled to the secondary circuit breaker, protection for this connecting circuit is dependent on fault clearing by the primary interrupting device.

Transformer primary fuses should be selected with due consideration to the reduced magnitude of secondary phase-to-ground arcing faults. It is now possible to select fuses which give this protection.

13.5.3 Single-Service Supply with Transformer Primary Circuit Breaker. This arrangement (Fig 230) is more expensive, but provides more protection than described in 13.5.2. The fused disconnect switch is replaced by a circuit breaker (Device 52L) and protective relays. This is required when the transformer full-load requirements exceed the maximum permissible size fuse. The high-side circuit breaker may also be justified to clear low-side ground faults, especially when resistance is grounded as shown in this arrangement. Some interrupter switches are also rated to handle this duty.

The service entrance overcurrent relays (group B, Devices 51/50 and 51N/50N) are often set and tested in cooperation with the utility to coordinate with relays of group A. The service entrance relays serve a dual purpose since they also are the backup protection for the transformers. Circuit breaker or switch 52L may not be considered the service circuit breaker or switch by the NEC [6], and a main secondary circuit breaker or switch may be required if more than six feeder circuit breakers or switches are installed. Instantaneous relay (Device 50) in group A can be set only if there is sufficient impedance in the circuit between utility circuit breaker (Device 52S) and consumer circuit breaker (Device 52L), and should be set to see faults along 80% of the line; otherwise it should be disconnected [B14]. Instantaneous relay (Device 50) in group B should be set to coordinate with the device 50 of group E. Overcurrent relay (Device 51) in group D is optional since it provides overcurrent protection to 4.16 kV bus and as a backup for feeder faults.

A time-delay relay (Device 62) may be added to ground fault relay (Device 51G) of group C to trip primary circuit breaker (Device 52L) if the ground fault continues after tripping secondary circuit breaker or switch (Device 52T).

13.5.4 Dual Service without Transformation. This arrangement (Fig 231) illustrates a method of utilizing a dual service from the utility. With this arrangement, one service circuit breaker or interrupter switch (Device 52S-1) may be normally open and the other service circuit breaker or interrupter switch (Device 52S-2) normally closed. The reasons for not operating in parallel in this instance are to minimize the short-circuit duty on the plant 13.8 kV bus and to eliminate the need for directional relays (Device 67). Automatic transfer (Device 27) should be delayed to coordinate for faults on other portions of the system. The voltage will be depressed due to a fault until a circuit breaker or a fuse operates to clear it. Because of the setting requirement placed by the utility on the overcurrent relays for circuit breakers (Devices 52S-1 and 52S-2), selectivity between the service circuit breakers (Device 52F) may be difficult to obtain. Overcurrent blocking is utilized to lock out both source lines in the unlikely event of a bus fault or a lack of coordination with fuses protecting feeder circuit. Selectivity would depend on the instantaneous overcurrent relay (Device 50) tripping only the feeder circuit breaker (Device 52F) for feeder faults.

13.5.5 Dual Service with Transformation. This arrangement (Fig 232) illustrates a preferred dual-service scheme. With this arrangement the industrial plant is protected not only from the loss of a supply line but, in addition, against the failure of a primary bus or one transformer. The secondary sides of both transformers are tied together through a normally closed tie breaker or switch (Device 52B). This scheme could also be provided with automatic transfer on the primary side of the transformer by closing tie breaker or switch (Device 52A) after fault is cleared when one of the lines is faulted. A fault in the transformer and on the secondary side of the transformer is seen by one or more of the following protective devices: the sudden pressure relay (Device 63), the transformer differential relay (Device 87T), the overcurrent relays (Device 51), and ground relays (Devices 51G and 87G). Directional overcurrent relay (Device

67) provides backup protection by looking at current flowing to the transformers from the low voltage side. Partial differential relaying is provided for only the source circuit buses, using an overcurrent relay with time delay (Device 51). The relays protecting the feeders or circuits are not in the differential. This scheme provides protection against bus faults and isolates the faulted section by opening secondary circuit breakers or switches (Devices 52T-1 or 52T-2). Transformer differential relay (Device 87T) should be considered for transformers of 5000 kVA or more. Since each installation has a wide variety of individual conditions, the relaying complement of Fig 232 is used only to illustrate certain protective schemes and is not intended to show all the devices that may be required. The transformer differential relay connection on all these low-side grounding schemes needs particular attention for proper functioning on internal and external ground faults.

When a tie circuit breaker or switch is normally closed, special relaying may be necessary to allow for the possibility of circulating current between bus sections.

If the system shown in Fig 232 is operated with either primary or secondary bus tie device normally closed, faults on the utility system are normally cleared by the user's protective devices. Many utilities find this unacceptable as they do not wish to accept the liability for fault interruption unless calibration and setting of relays and maintenance of protective devices is performed by utility personnel. It may be preferable to operate the secondary bus tie device normally opened. This greatly simplifies relaying by eliminating bus differential and directional relaying on the low voltage side. In such cases, interrupter switches can be used for the medium-voltage bus source and tie switches.

The utility's point of connection to the plant distribution scheme will determine which protective relaying scheme will be required. Each installation would have to be examined individually, but the main criteria of the protective relaying scheme should be to separate the two sources upon sensing a disturbance.

13.5.6 Single-Service Supply with In-Plant Generation. This scheme (Fig 233) is the same as shown in Fig 230, except in-plant generation is added. Some manufacturing plants find it economical to generate a portion of their electrical requirements in conjunction with their manufacturing process. The generator, when operated in parallel with the utility system, presents many additional protection and control problems. The system will probably have to be designed to shed load in the event the utility source is lost. This can be difficult since accurate control is needed over the generator governor to ensure that excessive loading does not occur during this load-shedding period. Another problem concerns disconnection of the generator from the utility system when the utility source becomes disconnected under other than fault conditions. Under these circumstances, the industrial plant's generator may attempt to pick up loads of other consumers as well as noncritical plant load normally intended to be supplied by the utility system only.

To attempt to solve these problems, a ground detection relay (Device 64) would recognize a ground fault on the 34.5 kV side of the transformer and would cause device 52S to trip. However, a directional power relay (Device 32) may be provided which would measure the current flow toward the utility source. The relay pickup current should be set with sufficient margin above the magnetizing current of the transformer. In some cases calculations may show that the fault-sensing relay (Device 67) may provide this function. But actual setting of Device 67 for both functions should be determined before this conclusion is reached. The under-frequency relay (Device 81) could be set to recognize a slowdown of the generator due to overload. The relay could trip the transformer secondary circuit breaker switch or preselected feeder circuit breakers switch to shed load to within the rating of the generator.

When the main step-down transformer is out of service, an alternate neutral ground should be established. This consideration frequently results in a separate source of neutral ground, through either a zig-zag ground transformer or the local generators. The inplant generators can safely carry the system ground because the main transformer delta winding prevents any interaction between the utility system ground faults and the local generator neutral system.

13.6 References. The following publications shall be used in conjunction with this chapter.

[1] ANSI C50.41-1982, American National Standard Polyphase Induction Motors for Power Generating Stations.

[2] ANSI/IEEE C37.95-1973, IEEE Guide for Protective Relaying of Utility-Consumer Interconnections.

[3] ANSI/IEEE C57.92-1981, IEEE Guide for Loading Mineral-Oil-Immersed Power Transformers Up to and Including 100 MVA with 55 °C or 65 °C Winding Rise.

[4] ANSI/IEEE Std 141-1986, IEEE Recommended Practice for Electric Power Distribution for Industrial Plants.

[5] ANSI/IEEE Std 446-1980, IEEE Recommended Practice for Emergency and Standby Power Systems for Industrial and Commercial Applications.

[6] ANSI/NFPA 70-1984, National Electrical Code.

[7] NEMA MG1-1978, Motors and Generators.

13.7. Bibliography
[B1] ALACHI, J. Reliability Considerations in Cement Plant Power Distribution. *IEEE Transactions on Industry Applications*, vol IA-15, Mar/Apr 1979.

[B2] *Applied Protective Relaying*. Westinghouse Electric Corporation, Relay-Instrument Division, NJ.

[B3] BRACKEN, J.F. Effects of Momentary Interruptions on Distribution Circuits. *Proceedings of the American Power Conference*, vol 28, 1966, pp 1028–1033.

[B4] CONRAD, R.R. and DALASTA, D. A New Ground Fault Protective System for Electrical Distribution Circuits. *IEEE Transactions on Industry and General Applications*, vol IGA-3, May/June 1967, pp 217–227.

[B5] DICKINSON, W.H. Report on Reliability of Electric Equipment in Industrial Plants. *AIEE Transactions (Applications and Industry)*, pt II, vol 81, July 1962, pp 132–151.

[B6] DONELLY, G.J. Computer Protection Offsets Supply Line Transients. *Electrical World*, May 30, 1966.

[B7] FAWCET, D.V. The Tie Between a Utility and an Industrial, When Industrial Has Generation. *IEEE Transactions on Power Apparatus and Systems*, July 1958, pp 136–143.

[B8] HIGGINS, T.D. and PEACH, N. A Proposed Publication on System Coordination and Protection for Industrial and Commercial Systems. *IEEE Transactions on Industry and General Applications*, vol IGA-1, Nov/Dec 1965, pp 410–416.

[B9] KUSKO, A. Quality of Electric Power. *IEEE Transactions on Industry and General Applications*, vol IGA-3, Nov/Dec 1967, pp 521–524.

[B10] LANGENWALTER, D.F. Protection for an Industrial Plant Connected to a Utility System with Reclosing Breakers. *Distribution*, vol 15, Nov 1963.

[B11] LINDERS, J.R. Effects of Power Supply Variations on AC Motor Characteristics. *IEEE Transactions on Industry Applications*, vol IA-8, July/Aug 1972, pp 383–400.

[B12] LINDERS, J.R. Electric Wave Distortions: Their Hidden Costs and Containment. *IEEE Transactions on Industry Applications*, vol IA-15, Sept/Oct 1979.

[B13] MASON, C.R. *The Art and Science of Protective Relaying*. New York: Wiley, 1956

[B14] MATHUR, B.K. A Closer Look at the Application and Setting of Instantaneous Devices. Conference Record, *1979 IEEE Industrial and Commercial Power System Technical Conference*, IEEE 79 CH 1460-51A, pp 107–118.

[B15] POTOCHNEY, G.J. and STEBBINS, W.L. The Application of Statistical Reliability Data in the Selection of a Utility Power Supply Scheme for an Industrial Plant. *IEEE Transactions on Industry Applications*, vol IA-15, Sept/Oct 1979.

[B16] SMITH, DAVID H. Problems Involving Industrial Plant-Utility Power System Interties. *IEEE Transaction on Industry Applications*, vol IA-11, Nov/Dec 1975.

[B17] WALSH, G.F. The Effects of Reclosing on Industrial Plants. *Proceedings of the American Power Conference*, vol 23, 1961, pp 768–778.

14. Overcurrent Coordination

14.1 General Discussion. Overcurrent coordination is a systematic application of current-actuated protective devices in the electrical power system, which, in response to a fault or overload, will remove only a minimum amount of equipment from service. The objective is not only to minimize the equipment damage and process outage costs, but also to protect personnel from the effects of these failures. The coordination study of an electric power system consists of an organized time-current study of all devices in series from the utilization device to the source. This study is a comparison of the time it takes the individual devices to operate when certain levels of normal or abnormal current pass through the protective devices.

The objective of a coordination study is to determine the characteristics, ratings, and settings of overcurrent protective devices that will ensure that the minimum unfaulted load is interrupted when the protective devices isolate a fault or overload anywhere in the system. At the same time, the devices and settings selected should provide satisfactory protection against overloads on the equipment and interrupt short circuits as rapidly as possible.

The coordination study provides data useful for the selection of instrument transformer ratios, protective relay characteristics and settings, fuse ratings, low-voltage circuit breaker ratings, characteristics, and settings. It also provides other information pertinent to the provision of optimum protection and selectivity in coordination of these devices.

In new installations, electrical equipment ratings often change prior to plant startup, but after protective devices have been ordered. These changes should be anticipated when selecting a protective device, so that the type and range are sufficiently flexible so as to protect the individual load or branch circuit. A preliminary coordination study should be made during the early stages of a new system design in order to verify that the protective device ratings can be selective, and that the supplying electric utility protection practices have been considered. The actual settings for the protective devices are determined after the design has been completed and all load/fault currents have been calculated.

Backup protective devices and settings are selected to operate at some predetermined time interval after the primary device. Thus a backup device should be able to withstand the fault conditions for a greater time period than the primary protection. For most applications, the operation of the backup device will isolate other circuits, in addition to the faulted or overloaded circuit that had inoperative primary protection.

In applying protective devices, it is occasionally necessary to compromise between protection and coordination. While experience may suggest one alternative over another, the preferred approach would be to have protection rather than coordination if a choice must be made.

A coordination study should be made when the available short-circuit current of the source to a plant is increased. A coordination study or revision of a previous study may be needed for an existing plant when new loads are added to the system or when existing equipment is replaced with higher rated equipment. This study determines settings or ratings necessary to assure coordination after system changes have been made.

A coordination study definitely should be made for an existing plant when a fault on the periphery of the system shuts down a major portion of the system. Such a study may indicate a need to change or replace devices.

14.2 Primary Considerations

14.2.1 Short-Circuit Currents. In order to obtain complete coordination of the protective equipment applied, it may be necessary to obtain some or all of the following information on short-circuit currents for each local bus.

(1) Maximum and minimum momentary (first cycle) short-circuit current

(2) Maximum and minimum interrupting duty (5 cycle to 2 s) short-circuit current

(3) Maximum and minimum ground-fault current

These values are obtained as described in Chapter 2.

The momentary currents are used to determine the maximum and minimum currents to which instantaneous and direct-acting trip devices respond.

The maximum interrupting current is the value of current at which the circuit protection coordination interval is established. The minimum interrupting current is needed to determine whether the circuit protection sensitivity of the circuits is adequate.

14.2.2 Coordination Time Intervals. When plotting coordination curves, certain time intervals must be maintained between the curves of various protective devices in order to ensure correct sequential operation of the devices. These intervals are required because relays have overtravel and curve tolerances, certain fuses have damage characteristics, and circuit breakers have certain speeds of operation. Sometimes these intervals are called margins.

Coordination can be easily achieved with low voltage current-limiting fuses that have fast response times. Manufacturer's time current curves and selectivity ratio guides are used for both overload and short-circuit conditions, precluding the need for calculating time intervals.

When coordinating inverse time overcurrent relays, the time interval is usually 0.3–0.4 s. This interval is measured between relays in series either at the instantaneous setting of the load side feeder circuit breaker relay or the maximum short-circuit current, which can flow through both devices simultaneously, whichever is the lower value of current. The interval consists of the following components:

circuit breaker opening time (5 cycles)	0.08 s
relay overtravel	0.10 s
safety factor for CT saturation, setting errors, etc	0.22 s

This safety factor may be decreased by field testing relays to eliminate setting errors. This involves calibrating the relays to the coordination curves, adjusting time dials to achieve specific operating times.

A 0.355 margin is widely used in field-tested systems employing very inverse and extremely inverse time overcurrent relays.

When solid-state relays are used, overtravel is eliminated and the time may be reduced by the amount included for overtravel. For systems using induction disk relays, a decrease of the time interval may be made by employing an overcurrent relay with a special high-dropout instantaneous element set at approximately the same pickup as the time element with its contact wired in series with the main relay contact. This eliminates overtravel in the relay so equipped. The time interval often used on carefully calibrated systems with high-dropout instantaneous relays is 0.25 s.

When coordinating relays with downstream fuses, the circuit opening time does not exist for the fuse and the interval may be reduced accordingly. The total clearing time of the fuse should be used for coordination purposes. The time margin between the fuse total clearing curve and the upstream relay curve could be as low as 0.1 s where clearing times below 1 s are involved.

When low-voltage circuit breakers equipped with direct-acting trip units are coordinated with relayed circuit breakers, the coordination time interval is usually regarded as 0.3 s. This interval may be decreased to a shorter time as explained previously for relay-to-relay coordination.

When coordinating circuit breakers equipped with direct-acting trip units, the characteristic curves should not overlap. In general only a slight separation is planned between the different characteristic curves. This lack of a specified time margin is explained by the incorporation of all the variables plus the circuit breaker operating times for these devices within the band of the device characteristic curve.

14.2.3 Delta–Wye Transformers. When protecting a delta–wye transformer, an additional 16% current margin over margins mentioned in 14.2.2 should be used between the primary and secondary protective device characteristic curves. This helps maintain selectivity for secondary phase-to-phase faults since the per-unit primary current in one phase for this type of fault is 16% greater than the per-unit secondary current which flows for a secondary three-phase fault. This is illustrated in Fig 234 and 14.5.2.

14.2.4 Load Flow Currents. In addition to short-circuit and voltage drop studies, a load flow study can be made to determine the normal and emergency load curents at each load center and through each branch circuit. The load current data is used to establish cable, equipment, and protective device continuous ratings. Such data are valuable when setting protective devices to protect both the equipment and the installed cable.

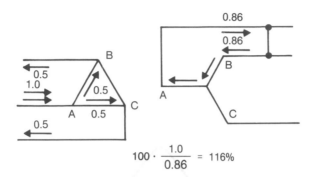

$$100 \cdot \frac{1.0}{0.86} = 116\%$$

Fig 234
Currents in Delta-Wye Transformer for Secondary Phase-to-Phase Fault

Another factor in protecting the circuit cable is the maximum short-circuit current available at the extremity of the cable circuit. The conductor insulation should not be damaged by the high conductor temperature resulting from current flowing to a fault beyond the cable termination. As a guide in preventing insulation damage, curves of conductor size and short-circuit current based on temperatures that damage insulation are available from cable manufacturers. In coordinating system protection, the cable should be able to withstand the maximum through short-circuit current for a time equivalent to the tripping time of the primary relay protection or total clearing time of the fuse. Many times this will determine the minimum conductor size applicable to a particular power system. See Chapter 8 on conductors.

14.2.5 Protection Guidelines. Before proceeding with overcurrent coordination, the individual load or branch circuit protection should be applied in accordance with accepted guidelines recommended or mandated by NEC, ANSI, CEC, IEEE, or a similar code or body. ANSI/NFPA 70-1984 [7][26] (National Electrical Code [NEC]) is primarily an installation code; enforcement of it is the prerogative of the enforcing authority.

To meet the requirements of the NEC [7] and maintain system selectivity, some applications require installation of larger cable than originally anticipated or installation of supplementary overcurrent devices. The alternative to these remedies is a compromise of system selectivity.

---

[26] The numbers in brackets correspond to those of the references listed at the end of this chapter.

14.2.6 Pickup. The term "pickup" has acquired several meanings. For many devices, pickup is defined as that minimum current that starts an action. It is accurately used when describing a relay characteristic. It is also used in describing the performance of a low-voltage power circuit breaker trip device. The term does not apply accurately to the thermal trip of a molded-case circuit breaker, which deflects as a function of stored heat.

The pickup current of an overcurrent protective relay is the minimum value of current that will cause the relay to close its contacts. For an induction disk time overcurrent relay, pickup is the minumum current that will cause the disk to start to move and ultimately close its contacts. For solenoid-actuated devices, tap or current settings of these relays usually correspond to pickup current.

For low-voltage power circuit breakers, pickup is defined as that calibrated value of minimum current, subject to certain tolerances, which will cause a trip device to ultimately trip the circuit breaker. A trip device with a long-time delay, short-time delay, and an instantaneous characteristic will have three pickup values, all given in terms of multiples or percentages of trip-device rating or the long-time delay pickup.

For molded-case circuit breakers with thermal trip elements, continuous ampere ratings, not pickups, are used, since a properly calibrated molded-case circuit breaker carries 100% of its rating at 25 °C in open air. The instantaneous magnetic setting could be called pickup in the same way as that for low-voltage power circuit breakers.

For fuse applications, continuous ampere ratings are used instead of pickup ratings, because low-voltage power fuses can withstand 110% of their continuous rating indefinitely under controlled laboratory conditions. Both single and dual element type fuses are available up through 600 V, with the dual-element fuse utilizing one element for overload and the second element for short-circuit protection. Medium- and high-voltage power fuses will typically not operate for currents below 200% of their nominal ampere ratings.

14.2.7 Primary Pickup Device Coordination. To be effective, the design ratio of the back up device minimum pickup (or continuous) rating to the primary device pickup (or continuous) rating should be as small as possible.

As an example, a 600 A trip setting on a low-voltage circuit breaker can hardly be expected to backup a motor control center (MCC) 20 A branch circuit protected by a molded case breaker, its ratio being 30 (600 ÷ 20 = 30).

Another example is determining the setting of such a 600 A trip device that protects a MCC, where one or more large loads may predominate. Assume that the MCC has only four motors, each 100 hp with a full load motor current of 120 A each and a locked rotor current of 720 A. If three motors are running and the fourth motor starts, a total current of 1080 A passes through the low-voltage trip coil. (Total $I = 3 \cdot 120 + 1 \cdot 720 = 1080$ A.) Depending upon the motor acceleration time, this total current could affect the settings of either or both trip elements (long time, short time) of the low-voltage circuit breaker. This 1080 A current should be permitted to persist, and no false trips should prevent the motor from reaching rated speed. Allowance should also be made for longer acceleration time that may occur during allowable low-voltage conditions.

(Because a starting motor PF is close to 20% and a running motor PF is over 90%, the 1080 A total is not quite true.) A 600 A trip setting on the supply breaker serving the MCC would coordinate well with each of the above loads which would normally be protected with a long-term pickup of approximately 140–150 A (ratio $600/150 = 4$).

14.2.8 Current Transformer Saturation. The function of a current transformer is to produce a secondary current that is proportional in magnitude and in phase with the primary current, this secondary current being applied to protective relays of compatible range and load (burden) characteristics.

When checking coordination, the effect of transformer saturation is to slow the induction disk relay operation. When the current transformer becomes saturated the actual secondary relay current is less than it should be, its waveshape is distorted, and the relay operates more slowly.

Saturation can occur in current transformers used to measure low voltage ground fault current, especially in underdesigned core balance current transformers, and often in the current transformer for the backup ground-fault relay. Saturation has also occurred in the solid-state low-voltage trip devices that use current sensors (not to be confused with current transformers, except for the fact that they reduce phase currents to a value compatible with their devices electronic circuitry). These current sensors form a residual circuit for the measurement of ground-fault current; normal equipment starting current or downstream phase faults may produce an unbalanced current that can cause a false ground-fault current trip.

Actually, a burden value often applies only to a particular value of secondary current. Most devices have a magnetic circuit in which their burden decreases due to saturation as the current increases. Thus the impedances of the applied devices should be known for several values of overcurrent so that the CT secondary circuit burden can be approximated for several values of primary current.

In most industrial systems, current-transformer saturation is significant only on circuits with relatively low-ratio current transformers. In most cases these circuits feed utilization equipment; relays with instantaneous settings below the current transformer saturation point can be applied. As one progresses back toward the source, the current-transformer ratios get larger at the same voltage level, the transformers have more turns, developing higher voltages, and therefore they are less likely to saturate when normal burdens are applied. Usually current-transformer saturation problems occur only in coordination between the two lowest-setting time-element relays.

Saturation of current transformers due to the dc component of an asymmetrical fault current can cause a delay in the operation of some instantaneous relays. It can also cause false tripping of residually connected instantaneous ground fault relays.

14.2.9 How to Construct Curves. A basic understanding of time-current characteristics is essential to any study. On an ordinary coordination curve, time 0 is considered as the time at which the fault occurs, and all times shown on the curve are the elapsed time from that point. The curves that are drawn are response times since, for a radial system, all the devices between the fault and

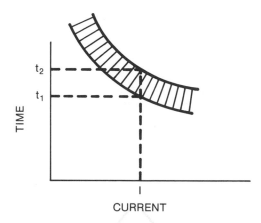

**Fig 235
Time-Current Curve Band
Including Resettable Allowance
(Impulse Characteristic)**

the source experience the same fault currents (except for motor contributions) until one of them interrupts the circuit.

A coordination curve is arranged so that the region below and to the left of the curve represents an area of no operation. The curves represent a locus of a family of paired coordinates (current and time) which indicate how long a period of time is required for device operation at a selected value of current. Protective relay curves are usually represented by a single line only. Circuit breaker tripping curves that include the circuit breaker operating time as well as the trip device time are represented as bands. The bands represent the limits of maximum and minimum times at selected currents during which circuit interruption is expected. The region above and to the right of the curve or band represents an area of operation.

Figure 235 shows a time-current curve represented as a band. Time t_2 is the maximum time from the initiation of the current flow I within which operation of the device and circuit breaker is assured. Time t_1 is the maximum time from initiation of the current flow I within which the current must be normalized to assure the device under consideration will not operate due to the impulse characteristic of the trip device.

Reading current along the abscissa, the time or range of times in which any device is expected to operate, corresponds to the ordinate or ordinates of the curve plotted. Usually, circuit breaker curves begin at a point of low current close to the trip device rating or setting and an operating time of 1000 s; relay curves begin at a point close to 1½ times pickup and the corresponding time for this point. Curves usually end at the maximum short-circuit current to which the device under consideration will be subjected. A single curve can be drawn for any device under any specified condition, although most devices (except relays) plot an envelope within which operation takes place. This envelope takes into consideration most of the variables which affect operation, such as ambient temperature, manufacturing tolerances, and resettable time delay.

Contact-closing characteristics of time-overcurrent relays have historically been defined as either inverse or definite time. The definite time relay will operate within the same elapsed time regardless of the magnitude of the current, as long as the current is above the minimum pickup and below the saturating limit. The inverse time relay operating time is inversely proportional to the current magnitude, that is, the inverse relay operates in a short time for a high current magnitude, and operates slower for low current magnitudes above the minimum pickup. The relay may have characteristics which are inverse, very inverse, extremely inverse, or have some similar designation. These terms mean that a small value of overcurrent will require more time for the relay to operate than would a larger current magnitude, with the extremely inverse relay operating faster than the very inverse relay, which in turn operates faster than the inverse relay for the higher current magnitudes. Manufacturers of solid-state and electromechanical time-overcurrent relays have individual standards for the relay inversity, so the manufacturer's data should be scrutinized before attempting coordination. See Fig 249 for time-current curves for a very inverse relay.

The relay having an inverse or very inverse characteristic is most commonly used, and often the same inverse characteristic can be used throughout the system. A relay with more inverse characteristics is used closer to the load with the backup relay characteristics being less inverse. For instance, a very inverse relay is used to coordinate with fuses (fuses have extremely inverse characteristics). Relays with definite time characteristics are used in medium-voltage motor protection circuits, ground fault protection, and any similar application where there may exist a wide range of fault currents. Selection of curve relay characteristics is often based upon preferences or standardization, with the application being more of an art than a science.

14.3 Initial Planning. There are six possible steps to follow in planning a coordination study.

(1) Develop a one-line diagram including the data indicated in 14.4. Adhering to the principles of protection outlined in other chapters of this recommended practice will minimize the number of modifications of the one-line diagram necessary to achieve a coordinated system.

(2) Determine the load flow.

(3) Collect other data indicated in 14.4.

(4) Determine the level of short-circuit current at each location in the system. Section 14.2.1 gives the fault currents necessary for a study.

(5) Select characteristics of protective devices and CT ratios, and collect the time-current characteristic curves on standard log-log paper.

(6) Collect utility equipment rating and overcurrent device settings.

Section 14.5 indicates the procedure to follow and gives examples of a coordination study.

14.4 Data Required for a Coordination Study. The first requisite for a coordination study is a one-line diagram of the system or portion of the system involved in the study. This one-line diagram should show the following data:

(1) Power and voltage ratings as well as the impedance and connections of all transformers (include tolerances if equipment has not been delivered).

(2) Normal and emergency switching conditions.

(3) Short-circuit data, such as transformer impedances, subtransient reactance of all major motors and generators, as well as transient reactances of synchronous motors and generators, plus synchronous reactances of generators.

(4) Conductor sizes, types, configurations, and temperature ratings.

(5) Current transformer ratios.

(6) Relay, direct-acting trip, and fuse ratings, characteristics, ranges of adjustment.

The second requirement is a complete short-circuit current study as described in 14.2.1 for both momentary and interrupting duties where medium-voltage breakers are involved. This study should include maximum and minimum expected duties as well as available short-circuit current data from all sources for all involved voltage levels, including transformer through-fault currents.

The third requirement is the thermal limit of the device being protected. This can be stated as an I^2t rating (ampere square-seconds) or simply given as one time period for a certain magnitude of current, for example, a safe stall time at a motor locked rotor current. The fourth requirement is the expected maximum loading on any circuit considered. Any limiting devices such as utility settings on relays should be noted.

14.5 Procedure. Time can be effectively used by developing and following a procedure similar to the following one:

(1) Select a convenient current scale (see 14.5.1)

(2) On the log-log graph paper, indicate

   (a) Available short circuit currents in A

   (b) Ampacities or load flow, or both

   (c) I^2t damage points or curves

(3) Start plotting at the lowest voltage level and largest load.

The principle of using overlays for making coordination curves removes much of the tedium from coordination studies. Once a specific current scale has been selected, the proper multipliers for the various voltage levels considered in the study are calculated. Characteristic curves for the various protective devices are then placed on a smooth bright surface such as a white sheet of paper, or a glass-topped box with a lamp in it. The sheet of log-log paper on which the study is being made is placed on top of the device characteristic curve, the current scale of the study lined up with that of the device characteristic. The curves for all the various settings and ratings of devices being studied may then be traced or examined.

14.5.1 Selection of Proper Current Scale. Considering a large system or one with more than one voltage transformation, the characteristic curve of the smallest device is plotted as far to the left of the paper as possible so that the curves are not crowded at the right of the paper. The maximum short-circuit level on the system is the limit of the curves to the right, unless it seems desirable to observe the possible behavior of a device above the level of short-circuit current on the system under study. A minimum number of trip characteristics should be plotted on one sheet of paper. More than four or five curves on one sheet tend to be confusing, particularly if the curves overlap. Refer to Figs 238 and 239.

All relay characteristics must be plotted on a common scale even though they are at different voltage levels. As an example, consider a system on which a 750 kVA transformer with a 4160 V delta primary and a 480 V wye secondary is the largest piece of equipment. Assume that this transformer is equipped with a primary circuit breaker and a main secondary circuit breaker supplying some feeder circuit breakers. On this system, full-load current of the transformer at 480 V is $(750 \cdot 10^3)/(480 \cdot \sqrt{3}) = 902$ A. When 902 A is flowing in the secondary of the transformer, the current in the primary of the transformer will be this same value of current (902 A) multiplied by the voltage ratio of the transformer $(480/4160 = 0.115)$. In the case under consideration, the primary current will be $902 \cdot 0.115 = 104$ A. If we establish the full-load current to be 1 per unit, then 902 A at 480 V · 1 per unit = 104 A at 4160 V. As far as the time-current coordination curve is concerned, both 104 A at 4160 V and 902 A at 480 V represent the same value of circuit current: full load of the 750 kVA transformer and 1 per unit current. Plotting current on the time-current plot, 902 A at 480 V is the same as plotting 104 A at 4160 V. This type of manipulation permits the study of devices on several different system voltage levels on one coordination curve if the proper current scales are selected for the plot.

14.5.2 Example of Step-by-Step Phase Coordination Study

(1) One-Line Diagram. Draw the one-line diagram of the portion of the system to be studied with ratings of all known devices shown (Fig 236).

(2) Short-Circuit Current Study. Calculate the short-circuit current values available at different points in the system:

34.5 kV system	500 MVA (from utility or system study)
4160 V bus	55.5 MVA (from system only—motor contribution excluded for simplification of calculation)
480 V bus	12 800 A, symmetrical, through 750 kVA transformer
480 V substation feeder	12 800 A, symmetrical, through 750 kVA transformer plus 3600 A, symmetrical, motor contribution from 480 V system
480 V, 100 A panelboard	11 000 A, symmetrical (by calculation)

(3) Protection Points. Determine the protection points desired for certain large system components (see Table 59).

The ANSI point was formerly shown on time-current graphs and was a point in Table 56 on the curve "short-time loads (following full load), oil immersed transformers" in the appendix to ANSI/IEEE C37.91-1985 [1] and corresponded to the maximum through-fault current based solely on transformer impedance. However, its use has been abandoned, and it is listed herein for purposes of

Table 59
Protection Points Summary

3750 kVA Transformer	750 kVA Transformer
(a) ANSI Point*	
16.6 · I_{fl} · 0.58 = I_{ANSI} for 4 s	17.6 · I_{fl} · 0.58 = I_{ANSI} for 3.75 s
16.6 · 520 · 0.58 = 5000 A at 4160 V	17.6 · 902 · 0.58 = 9300 A at 480 V
16.6 · 63 · 0.58 = 606 A at 34.5 kV	17.6 · 104 · 0.58 = 1060 A at 4160 V
(b) Inrush Point	
12 · I_{fl} = I_{inrush} for 0.1s	8 · I_{fl} = I_{inrush} for 0.1 s
12 · 520 = 6250 A at 4160 V	8 · 902 = 7216 A at 480 V
12 · 63 = 755 A at 34.5 kV	8 · 104 = 832 A at 4160 V
(c) 6 Times Full Load (NEC [7] Rule)	
6 · 520 = 3120 A at 4160 V	6 · 902 = 5412 A at 480 V
6 · 63 = 378 A at 34.5 kV	6 · 104 = 624 A at 4160 V

continuity. The through-fault protection curve shown in Figs 237–241 inclusive are based upon ANSI/IEEE C57.109-1985 [4].

When a delta–wye transformer is involved in a system, a line-to-ground fault producing a 100% fault current in the secondary winding will produce only 58% fault current ($1/\sqrt{3}$) in each of two phases of the incoming line to the primary of the transformer. This means that the indicated current of the ANSI point must be decreased to 58% of the value for three-phase faults.

(4) Scale Selection

(a) Examine the range of currents to be depicted at different voltages (Table 60). The range of currents shown in the table extends over four cycles of log paper to completely depict all the devices under consideration.

(b) Select a scale that will minimize multiplications and manipulations on devices where a range of settings is available. Since the load-end device is fixed, settings will be selected for two devices at 480 V and two at 4160 V in addition to determining cable sizes, only during preliminary coordination study. Since relays have current transformer multipliers, an even-digit scale at 4160 V appears to offer the easiest working scale. Using a multiplier of 10 for 4160 V currents, multipliers of 87 for 480 V currents and of 1.21 for 34 500 V currents follows.

(5) Basic Tripping Characteristics. Plot the following on log–log paper.

(a) Through-fault protective curve, inrush point, and 6 times full-load current of transformer (OA or AA).

(b) Short-circuit currents.

(c) Largest fuse or molded-case circuit breaker continuous rating. Generally, the largest low-voltage device is plotted first, whether or not it is a 100 A molded-case breaker or a 100 A fuse. The protective device characteristic curve is taped to a light box, and the graph paper is placed over the protective device curve for accurate tracing (see Fig 237).

Table 60
Range of Currents at Various System Voltages

| System | 34.5 kV Scale | | | 4160 V Scale | | 480 V Scale |
	Full-Load Current for 3750 kVA Transformer	500 MVA Short-Circuit Capacity	55.5 MVA Short-Circuit Capacity	100 A Load	16 400 A Short-Circuit Current
34.5 kV	63 A	8400 A	930 A	1.4 A	229 A
4160 V	520 A	69 200 A	7700 A	11.5 A	1890 A
480 V	4500 A	600 200 A	66 500 A	100.0 A	16 400 A

(6) Low-Voltage Circuit Breakers

(a) 480 V feeder circuit breaker. A 750 kcmil cable has an ampacity of about 500 A; hence a trip device set at 500 A adequately protects this cable. A short-time-delay trip device rather than an instantaneous element is chosen to be selective with the downstream molded-case circuit breaker. Select a 600 A long-time-delay trip element set at 80% (480 A) and a short-time trip element set at 4 times (2400 A) with a minimum time characteristic (Fig 238). Also see Fig 241 for electromechanical devices.

(b) 750 kVA transformer secondary main circuit breaker. The next circuit breaker to be selected is the 750 kVA transformer secondary circuit breaker. For a 902 A full load, a 1200 A trip is selected with the maximum time band on both long- and short-time-delay elements. Set the short-time setting at 3 times (3600 A).

While in this example liberal curve separation is effected for clarity, in actual practice many engineers prefer to use faster bands more closely stacked in order to provide tighter coordination.

(7) Medium-Voltage Feeder Relays. By examination of the through-fault protection curve and the inrush point, the limits of the curve for primary protection of the 750 kVA transformer can be determined. Keeping in mind that a low pickup for this device is desirable for good cable protection, it is good practice to keep the characteristic curve as far to the left as possible in order to operate faster on faults. Allowing 16% current margin between the short-time setting of the main secondary circuit breaker (3600 A at 480 V), select a pickup for the medium-voltage feeder overcurrent relay. This should be less than 624 A at 4160 V and more than $3600 \cdot (480/4160) \cdot 1.16 = 480$ A. The 624 A limit equals 6 times the 750 kVA transformer full load current at 4160 V, and is the maximum permitted by the NEC [7], Section 450-3. Since this will also protect the cable supplying the substation, the lower value is preferred.

In this example a 300/5 A current transformer has been selected. The 300/5 A rating current transformer has a 60:1 ratio, and the tap setting times 60 will produce the relay minimum pickup in terms of the primary current ($60 \cdot 8$ A = 480 A primary current). The 8 A tap allows for addition of future load, although tap settings of 5 A or lower are often desirable and can be obtained by

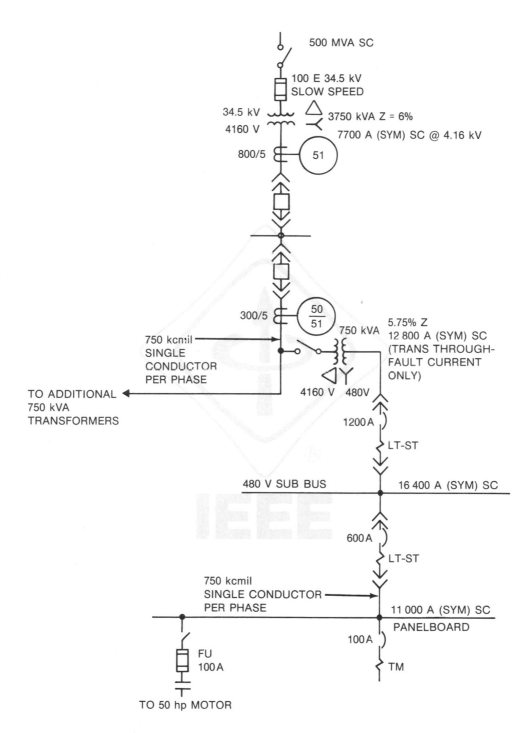

500 MVA SC

100 E 34.5 kV
SLOW SPEED

34.5 kV
4160 V

3750 kVA Z = 6%

7700 A (SYM) SC @ 4.16 kV

800/5

51

300/5

50
51

750 kVA

5.75% Z
12 800 A (SYM) SC
(TRANS THROUGH-
FAULT CURRENT
ONLY)

750 kcmil
SINGLE
CONDUCTOR
PER PHASE

4160 V   480V

TO ADDITIONAL
750 kVA
TRANSFORMERS

1200 A

LT-ST

480 V SUB BUS

16 400 A (SYM) SC

600 A

LT-ST

750 kcmil
SINGLE CONDUCTOR
PER PHASE

11 000 A (SYM) SC

PANELBOARD

100 A

FU
100 A

TM

TO 50 hp MOTOR

**Fig 236
One-Line Diagram for Coordination Study of Fig 240**

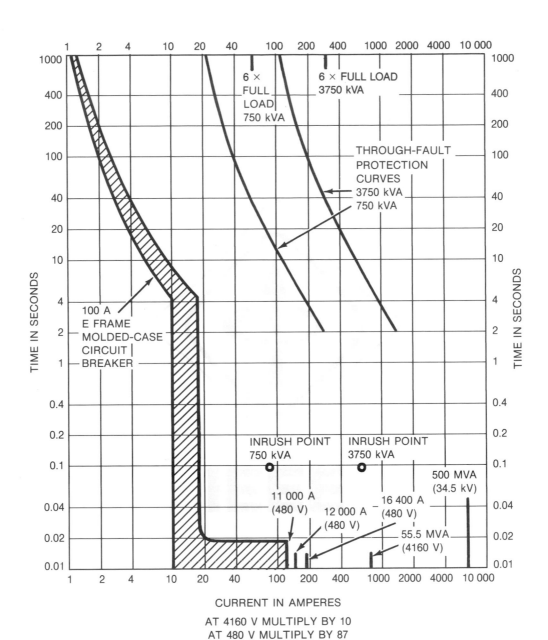

Fig 237
Plot Showing Fixed Points, Transformers Protection Curves,
Maximum Short-Circuit Currents, and 100 A Circuit Breaker Characteristics

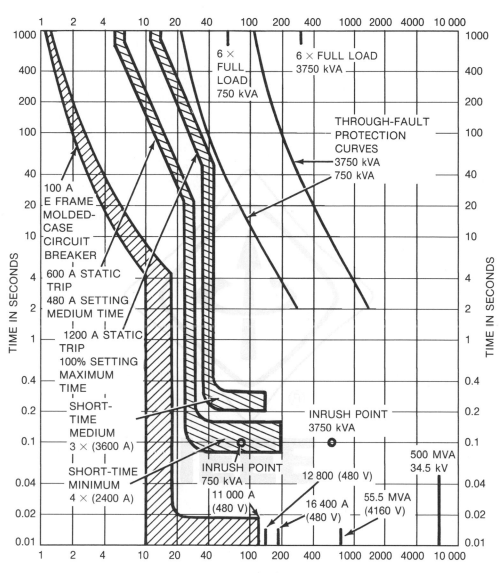

CURRENT IN AMPERES
AT 4160 V MULTIPLY BY 10
AT 480 V MULTIPLY BY 87
AT 34.5 kV MULTIPLY BY 1.21

Fig 238
Selection of Main and Feeder Low-Voltage Circuit Breakers

closer stacking of curves or tolerating some overlap with the transformer secondary circuit breaker curves.

When choosing this relay's characteristics, it is better to be tentative because there may be difficulty with the main 4160 V circuit breaker relays. Hedging slightly, pick a very inverse relay instead of the inverse characteristics sometimes recommended. Figure 249 illustrates the very inverse relay time-current characteristic curves for time dial settings (tripping delays) from ½ to 11, and for multiples of tap setting from 1 to greater than 40. With a 480 A primary current pickup, the tap multiples of the 8 A tap are also multiples of 480 A, that is, a relay tap multiple of 2 is equal to 960 A primary current and a relay tap multiple of 4 is equal to 1920 A primary current in this example.

To place the relay time-current curve on the coordination chart, line up the relay 1 vertical line at 480 A, and the other multiples of the tap setting will automatically align, that is, the 2 A relay tap multiple will align at 960 A, and the 4 A relay tap multiple will align at 1920 A. The time scale (horizontal lines) should also be aligned before tracing the relay time-current curve. (Because transparencies may not align throughout the ordinate, use a convenient horizontal time line near the most critical coordination point, such as the 0.1 s line.) Selecting the appropriate time dial setting is the next step, and a No 1 time dial provides adequate margin (greater than 16%) above the 3600 A (3 · 1200 A) short-time-delay element. Note that 12 800 A (480 V) is the maximum transformer through-fault current, and for this reason the 3600 A short-time-delay element does not extend beyond 12 800 A on the coordination chart. A higher time dial setting would not be appropriate because it would privide less margin to the through-fault protection curve, and would reduce the margin to the 3750 kVA transformer secondary circuit breaker relay curve.

If the 8 A, No 1 time dial had not produced a coordinated tripping curve, it would be necessary to try some alternatives, such as:

(a) Select a different relay tap (this would shift the curve to the right or left).

(b) Adjust the relay minimum pickup setting between taps (if so available) or adjust time dial settings between calibration marking or do both. Calibrating the relay can verify the more refined settings often required.

(c) Select a different relay characteristic (inverse or extremely inverse—see Fig 249).

(d) Use a different current transformer ratio or auxiliary current transformer. (Many transformer bushing current transformers are multiratio.)

(e) Change devices or settings of adjacent devices.

Note that a 4160 V feeder current transformer ratio and overcurrent relay settings have been selected to protect the cable to the substations. The pickup of the relay is no greater than the cable ampacity. (The cable ampacity is more than 480 A.) One 750 kcmil cable per phase meets the requirements of coordination and protection.

The instantaneous element is set above the available asymmetrical transformer through-fault current to preclude tripping on 480 V faults. The value used is $12\ 800 \cdot (480/4160) \cdot (5/300) \cdot 1.6 = 39.4$ A (Fig 240). ANSI/IEEE C37.91-1985 [1] suggests using a 1.25 multiplier times the maximum symmetrical through-fault current instead of the 1.6 multiplier used. An exact multiplier would

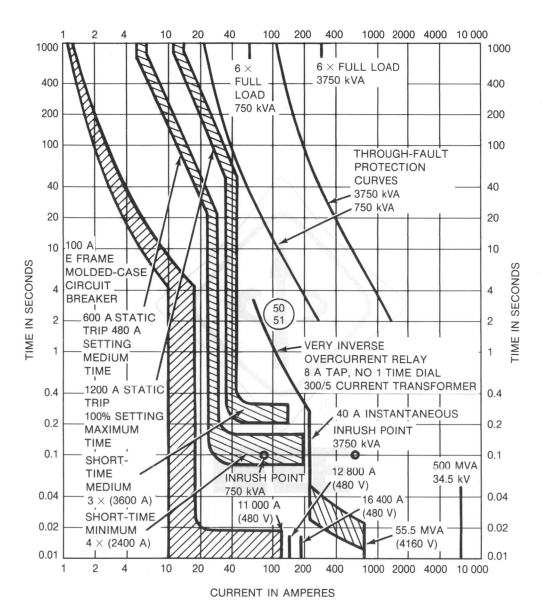

**Fig 239
Selection of Overcurrent Relay Curve and Instantaneous Setting**

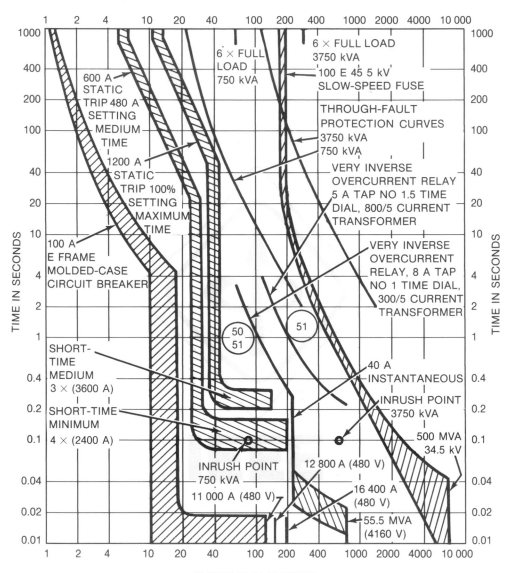

CURRENT IN AMPERES

AT 4160 V MULTIPLY BY 10
AT 480 V MULTIPLY BY 87
AT 34.5 kV MULTIPLY BY 1.21

Fig 240
Selection of Main Overcurrent Relay Curve

depend upon the X/R ratio of the calculated short-circuit current at the transformer secondary.

(8) Medium-Voltage Main Relays. Select a pickup for the 3750 kVA transformer secondary circuit breaker relays no lower than 125% of full load $(520 \cdot 1.25 = 650$ A) and no higher than 300% full load $(520 \cdot 3.00 = 1560$ A) (per NEC [7], Section 450-3).

A good selection is 800 A with an 800/5 current transformer. Do not use an instantaneous attachment on this relay since it cannot be made selective with the feeder instantaneous element.

Select a time dial setting such that 0.3–0.4 s delay is obtained at the theoretical 100% fault-current point. This margin will assure coordination between the main secondary circuit breaker and the 4160 V feeder breakers.

(9) High-Voltage Fuse. According to published tables, a standard speed 100 E fuse will protect the 3750 kVA transformer. However, a slow-speed characteristic is selected, since it will protect the transformer according to the established criteria and will provide more space to fit in all the devices necessary between this fuse and the largest load device. A larger fuse would also protect the transformer, but a tentative selection of the smaller fuse provides the transformer with better protection. Some engineers provide a margin of 0.2–0.4 s between the primary fuse minimum melting time curve and 3750 kVA transformer main secondary circuit breaker relay characteristic at the maximum 4160 V value of short-circuit current. This value is 55.5 MVA or 7708 A. The 100 E 34.5 kV fuse shown in Fig 240 illustrates this point.

(10) The Art of Compromise. Normally, selective coordination starts with the lowest voltage, working up to the highest voltage level. All the lower voltage or primary protective device characteristics should be below and to the left of the backup protective device curve. If the lower voltage device or primary noncurrent protective device curves cannot be made to fit under the backup (same voltage or higher voltage) device curve, an attempt should be made either to raise the backup device or to compromise the coordination. When selectivity must be compromised, the sacrifice should be made at the location in the system with the least economic consequences. This location will vary from system to system. Likely candidates are:

(a) Sacrificing coordination between a transformer primary protection and its secondary main circuit breakers. Loss of selectivity here is usually not detrimental to system security.

(b) Sacrificing selectivity between a load protective device and the next upstream protective device (typically, a feeder circuit breaker and a motor control center main protective device). The economic consequence of loss of selectivity here is usually more acceptable than at locations other than described in this section.

When closer spacing between curves is required, advantage can be gained by utilizing an extremely inverse relay as primary protection for a downstream device, being backed up by a very inverse relay.

545

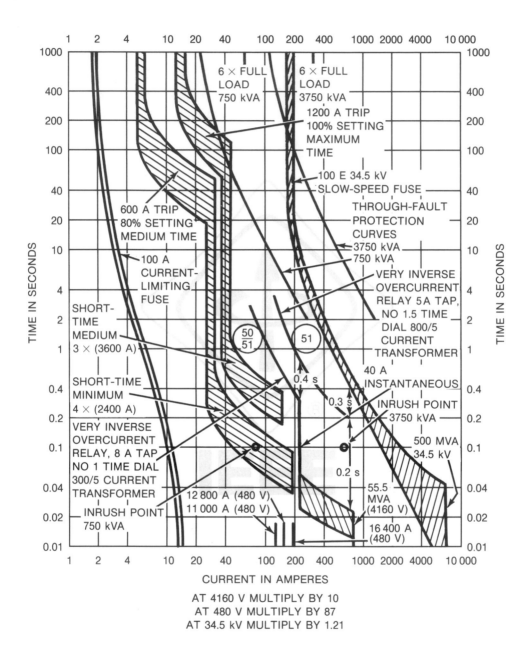

Fig 241
Fig 240 Replotted, Using Electromechanical Trip Device Characteristics

14.6 Ground-Fault Coordination on Low-Voltage Systems. While many argue over the merits of the different types of ground-fault protection, two factors are commonly accepted: arcing ground faults are the more destructive type because the arc limits the fault current sensed by phase overcurrent devices, and selectivity can be achieved only by including more than one level of ground-fault relays. The NEC [7] requires only one ground-fault relay at the service equipment set no higher than 1200 A with a maximum time delay of 1 s for ground-fault currents equal to or greater than 3000 A for solidly grounded wye electrical systems of more than 150 V to ground, but not exceeding 600 V phase-to-phase for each service disconnecting means rated 1000 A or more. No such specifics are defined for service voltages of 1 kV and over.

Several tests have been made to determine the magnitude of the 480 V arcing fault damage. An 1800 kW cycle fault can be considered the minimum magnitude of perceptible fault damage, sufficient to melt $\frac{1}{20}$ in³ of copper. A 10 000 kW cycle fault is the maximum that can be contained within an 11 gauge enclosure under worst-case conditions. Using these as guides, primary and backup ground-fault protection can be coordinated.

Whereas self-sustaining arcing faults are sometimes stated to exist only above a magnitude equal to 38% of the available bolted line-to-ground fault current, exceptions have been known. However, a criterion for minimum circuit protection should also consider selectivity as well as damage. In addition, small branch circuits have low X/R circuit ratio and tend not to be the arcing type. In Fig 242 the ground-fault protection is omitted for all branch circuits for 15 hp motor loads or 30 A feeder circuits, because phase protection coordinates with ground-fault protection.

In Fig 242, the MCC ground-fault protection used an extremely inverse relay Device 51G2, and is backed up at the source by a very inverse relay Device 51G1. For ground faults within the MCC, individual branch circuit phase overcurrent or ground-fault protection should operate. In the event of failure of these primary protective devices, the MCC Device 51G2 would operate to trip the feeder breaker to the MCC. The source Device 51G1 would be the ultimate backup and trip the bus supply breaker.

Other approaches to ground-fault overcurrent coordination are shown on Figs 243 and 244. These designs utilize the ground-fault protection option that is integral within the LVPCB feeder. For coordination, it is necessary to purchase separate ground-fault devices on MCC feeder breakers 100 A and larger on fusible motor starters with dual element fuses of 60 A and larger.

When applying bolted pressure contact switches, there is a possibility of blowing the fuse in only one phase during a ground fault. This problem is called *single-phasing* and it can be injurious to motors because many older motor overload protection devices do not react in time to protect the motor for this condition. In addition, when a single fuse isolates a ground fault, the fault can still be fed from the other phases through the motor windings, even though the current magnitude has been greatly reduced. Thus, the switch should be purchased with the antisingle phasing option and an electrical shunt trip. At the same time, a ground fault trip unit can be purchased to trip the switch.

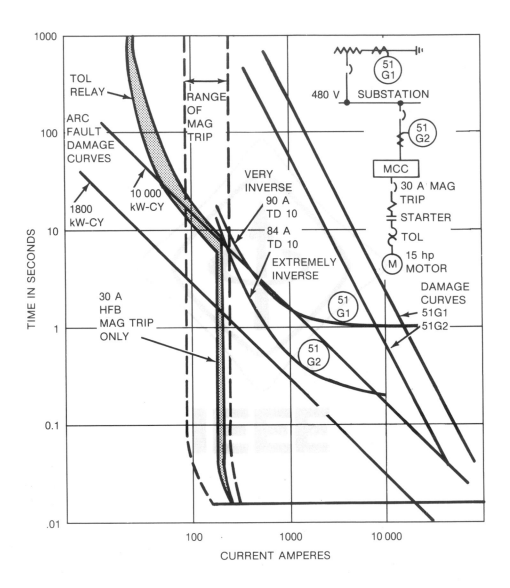

Fig 242
Ground-Fault Protection — 15 hp Motor Load Time-Overcurrent Relays

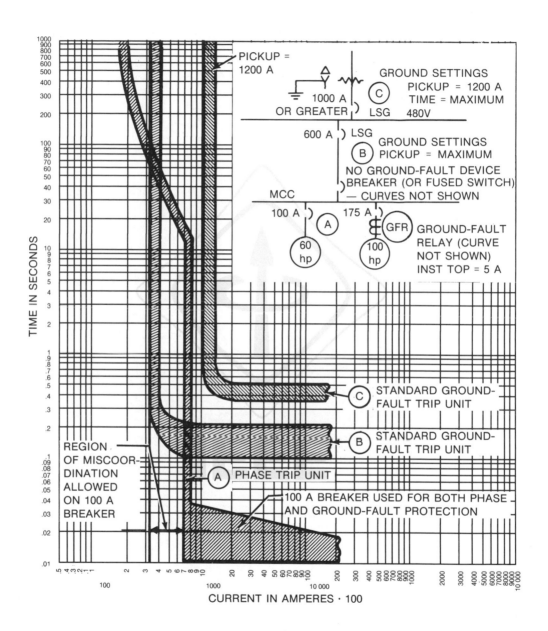

Fig 243
Ground-Fault Time-Current Characteristic Curves

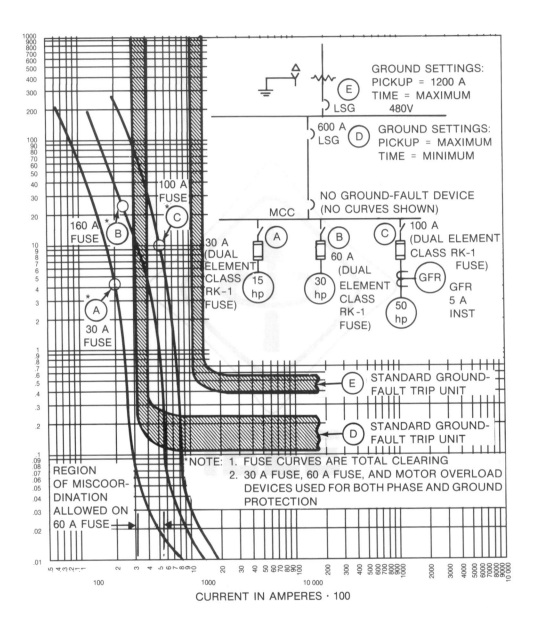

Fig 244
Ground-Fault Time-Current Characteristic Curves

Coordination of the ground fault unit follows the established procedures so that the faulted circuit is isolated from the system by the nearest protective device, and the other parts of the systems are not affected (shown on Fig 245).

14.7 Phase Fault Coordination on Substation 600 V or Less. There is a multitude of protective device options that can be utilized in low-voltage systems. Coordination of three of the most popular options is discussed in this section.

In the first option, shown in Fig 246, fused bolted pressure contact switches are used in the substation main and substation feeders, and a fused switch is used for an MCC motor load. Current-limiting fuses can be selectively coordinated by maintaining at least a minimum ampere rating ratio between the main fuse and feeder fuses, and between the feeder fuse and the branch circuit fuses. These ampere rating ratios are provided by the fuse manufacturers, and a chart is given in Chapter 5, Table 27. The size of the fuse is also chosen to protect the individual circuit components: conductors, motors, buses, and transformers.

The second option, shown in Fig 247, utilizes circuit breakers with direct-acting trip devices in the substation main and MCC feeder, and a fused-switch at the MCC main. To have the best coordination, select circuit breakers with the short-time rather than instantaneous option on the substation main and feeder. It should be recognized that this will not provide the fastest protection because the fault is cleared in the minimum time with instantaneous tripping when compared with the time-delayed tripping obtained from a short-time element. A current-limiting fuse is often used on the MCC main to limit the fault current to a standard, economical short-circuit rating of the MCC bus, circuit breakers, and motor contactors, which is 22 000 A. (This fuse could have been incorporated into the LVPCB feeder, and thus preclude the possibility of single-phasing.) By looking at Fig 247, it can be seen that there is a lack of coordination between the MCC main fuse and the MCC feeder breaker for fault currents above 8000 A. This is a compromise made between protection and coordination.

The third option, shown in Fig 249, utilizes circuit breaker protection at each level. Due to high fault current levels, the molded-case circuit breakers have current limiters for fault currents above 6000 A; there is a miscoordination between the LVPCB feeder to the MCC and the current limiter on each molded case circuit breaker. This is also a compromise between protection and coordination. Another option would be to use current-limiting fuses, since they would operate in approximately ¼ cycle at high fault levels, and would coordinate for overcurrents, as well.

14.8 References. The following publications shall be used in conjunction with this chapter.

[1] ANSI/IEEE C37.91-1985, IEEE Guide for Protective Relay Applications to Power Transformers.

[2] ANSI/IEEE C37.95-1973, IEEE Guide for Protective Relaying of Utility-Consumer Interconnections.

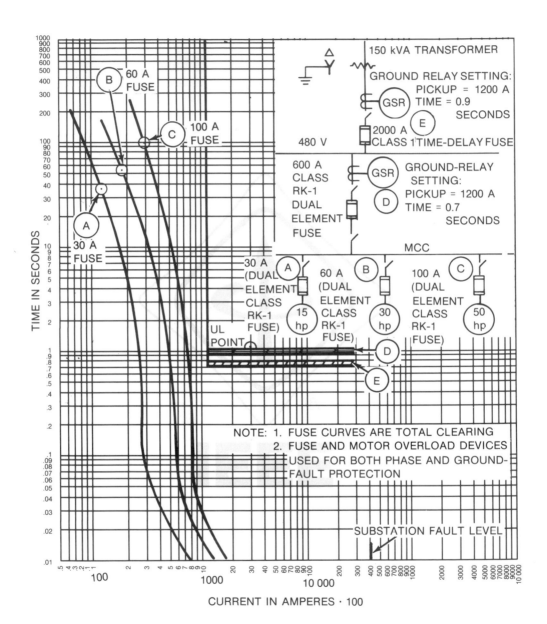

Fig 245
Ground-Fault Time-Current Characteristic Curves

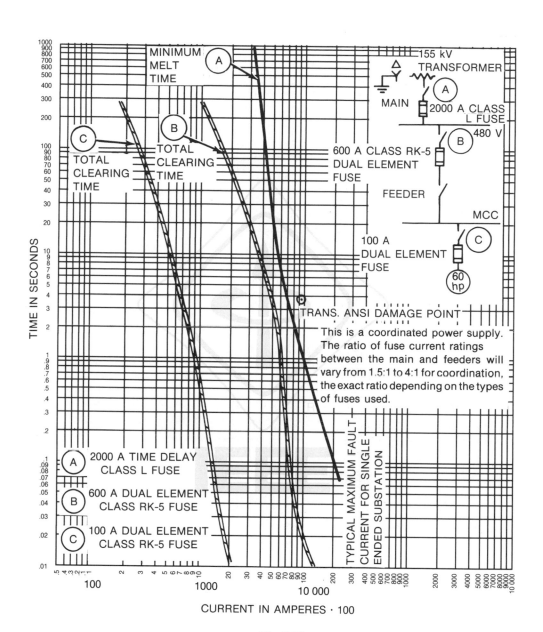

Fig 246
Ground-Fault Time-Current Characteristic Curves

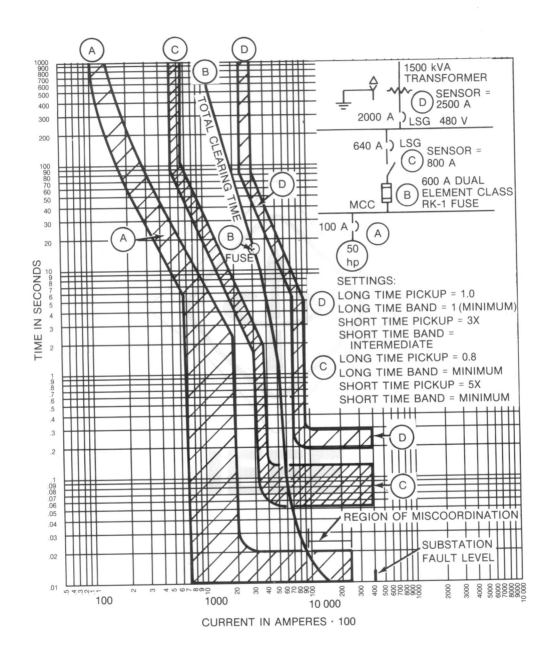

Fig 247
Phase Time-Current Characteristic Curves

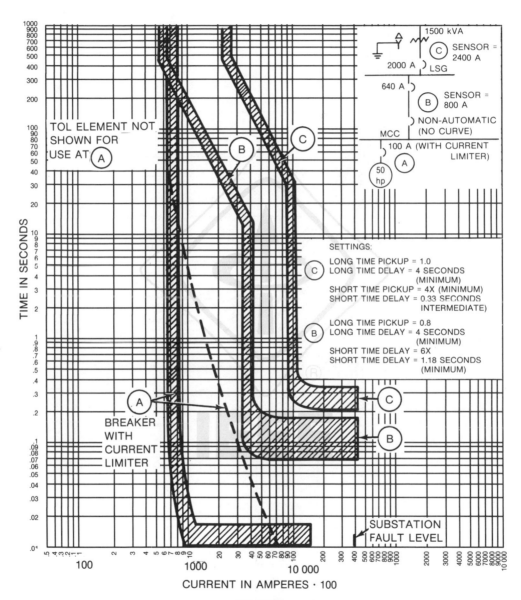

Fig 248
Phase Time-Current Characteristic Curves

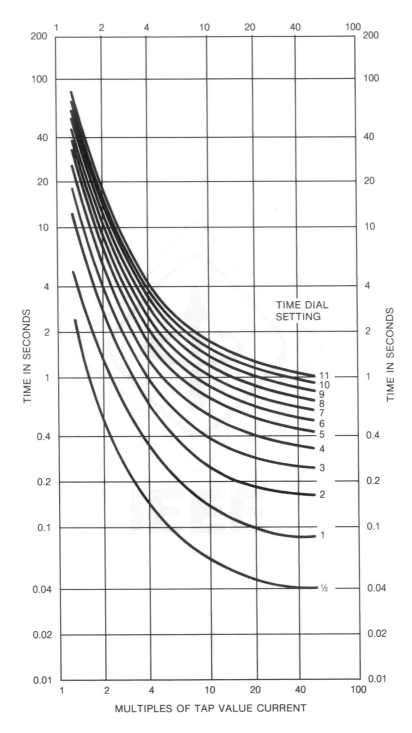

**Fig 249
Typical Time-Current Curves for Very Inverse-Time Relay**

[3] ANSI/IEEE C37.96-1976, IEEE Guide for AC Motor Protections.

[4] ANSI/IEEE C57.109-1985, IEEE Guide for Transformer Through-Fault-Current Duration.

[5] ANSI/IEEE Std 141-1986, IEEE Recommended Practice for Electric Power Distribution for Industrial Plants.

[6] ANSI/IEEE Std 241-1983, IEEE Recommended Practice for Electric Power Systems in Commercial Buildings.

[7] ANSI/NFPA 70-1984, National Electrical Code.

[8] IEEE Std 346-1973, IEEE Standard Definitions in Power Operations Terminology Including Terms for Reporting and Analyzing Outages of Electrical Transmission and Distribution Facilities and Interruptions to Customer Service.

[9] BEEMAN, D. L., Ed. *Industrial Power Systems Handbook.* New York: McGraw-Hill, 1955.

[10] REICHENSTEIN, H.W., and CASTENSCHIOLD, R. Coordinating Overcurrent Protective Devices with Automatic Transfer Switches in Commercial and Industrial Power Systems. *IEEE Transactions on Industry Applications,* vol IA-11, Nov/Dec 1975, pp 620–635.

15. Maintenance, Testing, and Calibration

15.1 General Discussion. It is impossible to predict when an abnormal electrical condition will occur, and therefore, the entire electric protective system should be ready to operate properly at all times. To reduce the possibility that the electric protective devices will not perform, they should be maintained [2], [7], [9][27].

Preventive maintenance should not be confused with repairs after a breakdown. The definition of maintenance implies that we inspect this equipment or system to discover its weaknesses and then repair or replace these areas before a breakdown occurs. A maintenance program for protective devices and the electric system in general could be broken down into six simple steps.

15.1.1 Clean. Remove or reduce the concentrations of all contaminants that are incompatible with the proper operation of electrical components. These contaminants include dust, moisture, rust, and various vapors or gases, or both, that reduce the integrity of an electric system.

15.1.2 Tighten. Loose electrical connections are heat sources that reduce the efficiency of the electric system and will ultimately result in the destruction of a portion or portions of the system. An energized ac system is subjected to constant vibration, and, therefore, connections may become loose.

15.1.3 Lubricate. Direct-acting low-voltage power circuit breakers, relay-actuated power circuit breakers, and switches contain many pivoting or sliding mechanical parts. These parts require lubrication so that they do not become inoperable because of nonuse. However, care should be exercised when lubricating so that improper types of grease and lubricants do not come in contact with current-carrying parts of the system, and lubricants which collect dust and dirt are to be avoided. Consult manufacturers for the proper lubricants.

15.1.4 Inspect. From time to time it becomes necessary to check the ratings and settings of all electric protective devices to assure that they are in accordance with good system design. Conditions change, loads shift, and human beings tamper in an effort to solve a problem.

15.1.5 Test. This requires that the system or protective device be subjected to abnormal electrical conditions and the operation of the system or devices compared to manufacturers' specifications for these conditions.

15.1.6 Record. All test results should be recorded upon completion. A sound

---

[27] The numbers in brackets correspond to those of the references listed at the end of this chapter.

recordkeeping system helps to establish reliable purchasing practices and future testing schedules.

The primary scope of this chapter will be to deal with the maintenance, testing, and calibration of the following types of electric protective devices:

(1) Motor overload relays
(2) Molded-case circuit breakers
(3) Low-voltage power circuit breakers
(4) Protective relays
(5) Medium-voltage circuit breakers
(6) Fuses and switches

In addition to the electric protective devices mentioned, the protective system may include such accessory items as:

(1) Battery chargers
(2) Storage batteries
(3) Current and potential transformers
(4) Control circuitry
(5) Auxiliary or control relays

15.1.7 Safe Working Practices. Safe work habits and use of good common sense cannot be overemphasized. Metal ladders or step stools should never be used around switchgear. Safety helmets are recommended; a fiber or plastic helmet should be worn. Loose key chains, tool pouches, or pieces of wire should never be allowed to hang from the body. Rings and metal watchbands should be removed. Whenever possible, power circuits should be de-energized and properly tagged to conform with approved procedures. It must not be assumed that circuits are de-energized.

Testing and maintenance of protective equipment should be undertaken only by qualified personnel. Alternately, services of manufacturers service organizations or qualified testing companies can be obtained [10].

The equipment or device tester has a responsibility not only for his own safety, but for the safety of other personnel. If inadvertent operation occurs, the equipment or devices should be immediately de-energized and testing or inspection not resumed until safe conditions are reestablished and clearances are reissued.

15.1.8 Scheduling. Operating personnel should be fully informed on the maintenance schedule and the work to be performed. They should be notified in advance and, if possible, operations should be shifted so that the circuits may be de-energized.

Proper communication with all affected parties, protective barriers, warning signs, and tags and locks are some of the suggested means that could allow safe performance of specified tests or inspections on time.

15.1.9 Initial Field Testing. While practices may vary, certain procedures are recommended for testing and inspection prior to energizing equipment and placing protective devices in service.

Control power and control circuit checkout is necessary to detect abnormal conditions or wiring errors. Circuit breakers, switches, and protective relays should be inspected and tested, and specified settings verified. Proper grounding of circuits should be ascertained.

Polarity and performance tests of instrument transformers and sensors are needed to verify proper operation of protective devices, especially in differential and directional schemes.

NEMA recommends field testing of the ground fault protection systems [5], [6]. ANSI/NFPA 70-1984 [1] (National Electrical Code [NEC]) requires performance testing of the ground-fault protection system when first installed on site for installations that are under jurisdiction of regulating authorities. In addition to circuit breaker switch and relay testing, these tests may include evaluation of the interconnected system, sensor location, grounding points, and insulation tests of neutral conductors or primary injection tests (ANSI/UL 1053-1982 [3]).

Special operating or environmental conditions may require reevaluation of protective device settings or effects on performance [8].

Settings of protective devices are often a compromise to satisfy possibly conflicting requirements of maximum continuity of service (absence of nuisance trips) and minimum damage (high sensitivity, shortest time).

Availability of sensitive and accurate protective devices does not necessarily mean that minimum settings can be used in any application or power system.

Constraints that are imposed by the system should be considered. They may include, but are not limited to, normal phase current; voltage or phase angle differences; inrush capacitive or inductive currents; transient power reversals or voltage dips; conductor positioning; stray magnetic fields; CT saturation; harmonics and associated noise, vibration, and fatigue; surges and interference; etc.

Thus, if an abnormal condition or operation is encountered during initial phases of the system startup, record keeping and supplementary testing may assume additional importance.

15.2 Motor Overload Relays. Motor overload protection and how it is applied are covered in Chapter 9. These principles should be understood by the persons responsible for maintaining motor overload protection.

There are five basic types of motor overload relays:

(1) Solder pot
(2) Bimetal strip
(3) Winding-temperature sensor
(4) Hydraulic-magnetic relay
(5) Solid-state

Small motors may be protected by a direct-acting bimetal or solid-state temperature-detecting device adjacent to the motor winding. When such a device opens at a predetermined temperature, the motor is shut down. These units are not adjustable.

15.2.1 Selecting Heater Size. The most widely used motor overload relays are the bimetal or solder-pot types, both of which incorporate heater elements.

Proper heater size is essential for the effective operation of these relays and should be regularly checked.

All manufacturers of heater-type thermal motor overload relays have heater selection tables. The user of these tables should read all the relay manufacturer's instructions before selecting a heater. This is necessary so that the basis of the tables will be known. For instance, one manufacturer prepares his tables so that the heater sizes may be used directly if the service factor of the motor is 1.15 with motor and controller in same ambient temperature. If these conditions are not met, appropriate steps should be taken to either uprate or derate the heater.

15.2.2 Testing Motor Overload Relays. Trip tests of overload relays can be conducted with minimum downtime. Normally, one test point is needed to establish whether the device is operating correctly, that is, whether its time of operation falls within its manufacturer's band of tripping time. It is suggested that the value of test current be equal to three or four times the normal rating of the relay.

To test the motor overload relay, it is necessary to disconnect the relay from the electric system, pass the test current through the relay heater circuit, and determine the time required for the device to trip. This time should be compared with manufacturer's recommended time of operation for the value of current used in some tests. Generally speaking, motor overload relays are nonadjustable, and if the test reveals an improper operation, it may be necessary to replace the relay heater with one of the proper size. Sometimes it is necessary to replace the entire relay. Adjustable designs of thermal relays are available.

15.3 Maintenance and Testing of Molded-Case Circuit Breakers. The circuit breaker should be kept clean and dry, and the line and load connections should be checked occasionally for tightness. If located in a dirty or dusty atmosphere, it should frequently be removed and blown out with clean, dry, low-pressure air. Do not blow the dirt into the recesses of the unit.

Check for excessive heating. Heating may be due to poor electrical connections, improper trip unit alignment, or improper mechanical connections within the circuit breaker housing or at the terminals. Also, certain environments may contaminate the silver contacts, but manufacturers can usually provide modified designs to compensate for those conditions.

Should heating be abnormal, as evidenced by discoloration of terminals, deterioration of molded material, or if there is nuisance tripping, measurement of load current is advised. If the current in each of the individual poles is less than the rating of the circuit breaker, all bolted connections and contacts should be examined and tightened in accordance with the manufacturer's recommendations.

Various types of temperature measuring devices can be employed. Some are quite mobile, for example, hand-held devices having thermally sensitive detectors with digital readouts.

Other reasons for the circuit breaker to operate improperly include:

(1) Excessive friction or binding of the trip bar so that the bimetal cannot exert sufficient pressure to move the trip bar

(2) Defective bimetal or dashpot assembly disabling its operating characteristics

(3) Open turn or thermal of a trip coil

(4) Foreign matter preventing movement of one or more of the following: trip bar, armature, latch release, trigger, bimetal, bellows or dashpot, or spring assemblies

If the circuit breaker has a removable trip unit or is of the design that incorporates fuses, the trip function can be readily checked. On units with removable trip units, the loosening of the trip unit mounting hardware is usually sufficient to cause tripping. Under no circumstances should a molded-case circuit breaker remain closed and latched when the trip unit is physically removed. When replacing the trip unit, the manufacturer's torque requirements should be maintained.

Fused units have an added feature that should be checked: a spring-loaded plunger, which releases when a fuse blows and strikes a trip bar extension on the trip unit. This action causes the circuit breaker to trip on any damaging fault current above the rating of the fuse. A safety feature requires that the circuit breaker be in the *open* position when checking or replacing the fuses. If it does not trip when the fuse blows, either the circuit breaker or the fuse is defective. A defective fuse will not eject its plunger when the link melts. The defective unit should be replaced. In no instance should it be possible to remove the fuse cover without the circuit breaker automatically tripping.

To determine if the circuit breaker mechanism allows its closure under a blown-fuse condition, one of the fuses is removed and a blown fuse substituted for the good fuse; then the cover is replaced. It should not be possible to latch and close the circuit breaker with a blown fuse in any of its fused legs. If the circuit breaker will latch and close with a blown fuse in any of the phases, it is defective and should be repaired or replaced.

If a molded-case circuit breaker does not trip under abnormal load conditions, or within the manufacturer's specified times, and mechanical troubles are not found, the circuit breaker is either labeled incorrectly or it is out of calibration. Recalibration of molded-case circuit breakers can be accomplished only under controlled test conditions by the manufacturer. It is unwise to attempt to make changes in the calibration of the circuit breaker trip unit except for any adjustable feature such as pickup of the instantaneous unit. This adjustment is readily made with a calibrated dial, which is accessible without opening the case.

Periodically, these circuit breakers must be electrically tripped to assure proper operation. Experience has indicated that if they are allowed to remain in service for an extended period of time without an electrical operation, the internal mechanism and joints may become stiff so that the circuit breaker operates improperly when subjected to abnormal current. Therefore, each pole of the circuit breaker should be electrically exercised.

In testing a molded-case circuit breaker, several points should be remembered.

(1) Maintain rated voltage at the terminals if the circuit breaker has an undervoltage attachment

(2) Portable power supplies for field testing have low voltage at their output terminals and it is advisable to use connections having the shortest possible length and largest possible cross section between the test unit and the circuit breaker

(3) The connection to the circuit breaker should be tight

(4) The circuit breaker is tested one pole at a time

(5) Trip devices should be allowed to fully cool before performing a current timing test

The recommended tests for a molded-case circuit breaker are timing and instantaneous pickup.

It is suggested that these circuit breakers be subjected to a test current equivalent to 300% of the circuit breaker trip unit rating and that the tripping time be measured. This time should be compared to the manufacturer's specified values or curves. Repeated testing of the time-delay element without a cooling period will raise the temperature of the thermal element, thereby decreasing the tripping time noticeably on successive tests.

Molded-case circuit breakers may be relatively precise; however, the published time-delay characteristic indicates a wide band of operation. The electrical test will reveal circuit breakers that will not trip, those that take abnormally long to trip, and those that have no time delay. If the test reveals that the circuit breaker is tripping within ± 15% of the outside limits of its published curves and this tolerance does not affect the electric system coordination or stability, the circuit breaker should be considered satisfactory.

In some molded-case circuit breakers, the instantaneous element is not adjustable and it is set and sealed at the factory. Other molded case circuit breakers have an adjustment that permits changing the pickup of the instantaneous unit. This type of circuit breaker may be shipped from the factory with the instantaneous unit pickup set at its maximum if a setting has not been specified by the purchaser. Therefore, it is necessary to check these adjustable instantaneous settings before putting the circuit breaker in service to be assured that the instantaneous pickup is not above the available short-circuit current at its location in the electric system. In many electric systems, one purpose of initial startup tests is to determine optimum settings for instantaneous units.

An electrical test for pickup of the instantaneous unit should be run to verify that the circuit breaker is tripping magnetically. Testing at one of the lower calibration marks is satisfactory. The adjustment may be set to the lowest calibration point and tests made at the point to verify that the unit will pick up. If the instantaneous unit picks up at the minimum calibration point, it may be reasonably assumed that, when reset to the desired calibration point, pickup will be within manufacturers' tolerances. Published tolerances based on factory calibration for standard adjustable instantaneous trip units usually are as follows:

"LO" position   ±25%
"HI" position   ±10%

NEMA AB2-1984 [4] may be used as an alternate guide for testing molded-case circuit breakers.

15.4 Low-Voltage Power Circuit Breakers

15.4.1 Reasons for Maintenance. Causes of malfunction of the circuit breaker are dependent upon time and severity of duty. Common causes of malfunction of low-voltage power circuit breakers generally fall into four categories.

(1) Loss of oil or air seal due to physical damage to dashpot; aging of seals or physical wear, causing incorrect operation

(1a) Normal oscillation (vibration) caused by alternating current may cause wear

(2) Clogging of orifices with foreign matter or oil sludge that forms due to atmospheric and environmental conditions and aging of oil (contamination)

(3) Freezing of components in plunger assembly due to corrosive atmospheres and extensive periods of inoperation

(4) Failure of solid-state components or assemblies

Of all the possible faults, improper delay in opening automatically under overload conditions is the most dangerous to safety. Normal operation procedures and careful maintenance inspections will reveal most of the other conditions that are likely to remove protection from the circuit. However, unless overload tests are run, there is no way to predict whether the circuit breaker will recognize an abnormal circuit condition and operate properly.

There are a number of conditions that may render a low-voltage power circuit breaker unfit for service until it is corrected. Some of the more common ones are:

(1) Frozen contacts of mechanism (circuit breaker may not open automatically or manually)

(2) Improper calibration, that is, trips too fast or too slow

(3) Trip element improperly set

(4) High contact resistance

(5) Trip element armature fails to strike trip bar

(6) Open contacts or damaged series element

(7) High resistance or arcing fault often due to loose or improper fit between primary disconnect assembly and bus stabs

(8) Broken or cracked arc chutes

(9) Loose parts

(10) Excessive force required to operate circuit breaker

(11) Dirt

These conditions can be initiated by a variety of causes including moisture, corrosion, abuse, wear, and vibration.

15.4.2 Maintenance and Test Procedures. A complete test program should include checks to determine the condition of the circuit breaker with respect to each of the foregoing items. The following recommendations are made concerning action to be taken on circuit breakers undergoing overhaul. Most of the conditions can be determined easily.

(1) Broken, cracked, eroded, or contaminated arc chutes are revealed by visual inspection

(2) Frozen or open contacts, as well as excessive force required to operate circuit breakers, may be determined by manually opening and closing the circuit breaker

(3) Determination of leakage current requires an insulation resistance tester and indicates contaminated or degraded insulation

(4) Improper calibration and high contact resistance can be determined by electrical tests; ductor tests will indicate damaged contacts or insufficient contact pressure

Safety rules and cooperation with operating personnel must be strictly observed.

When working on a low-voltage power circuit breaker, it should be de-energized and removed from the cubicle. It should be kept in mind that the stabs on the line side of the cubicle are still energized unless the entire substation or section of the bus is out of service.

15.4.3 Mechanical Inspection. The following steps should be taken in servicing and testing these circuit breakers.

(1) Schedule work.

(2) Ascertain that the circuit is de-energized and the circuit breaker can be removed for servicing.

(3) Open the circuit breaker, rack it out, and remove it from the cubicle. Note that these circuit breakers are heavy and should be handled with some type of lifting device. If no lifting device is available, at least two men or more will be needed to lower the circuit breaker to the floor.

(4) Inspect primary disconnect assembly in back of the circuit breaker. Check that no springs are missing or broken. Check for excessive wear. Clean out dust or dirt.

(5) Remove arc chutes; inspect for cracked, broken, or burned parts. Replace defective parts and clean.

(5a) Remove dust and dirt from the cubicle and arc chutes. Use a high-suction industrial vacuum cleaner to clean cubicle. The circuit breaker and arc chutes may be removed to a cleared area and cleaned with dry low-pressure compressed air.

(6) Inspect and clean main and arcing contacts in accordance with manufacturer's instruction. Note that commercial products are available to resilver these contacts, but replacement may be preferred.

(7) With the arc chutes removed, mechanically close the circuit breaker to inspect contact action and alignment. Deactivate undervoltage releases if applicable. When the arc chutes are removed, moving parts of the circuit breaker are exposed. Be extremely careful to keep all parts of the human body clear. These exposed circuit breaker parts are actuated by stored energy devices, and serious injury will result if any part of the body is caught by the contacts or moving mechanism.

(8) Open and close the circuit breaker several times to determine that operation is smooth and there is no binding.

(9) Lubricate and inspect the "racking-in" device located both in the cubicle and on the circuit breaker. Make sure that there is no evidence of binding.

(10) Lubricate in accordance with the manufacturer's instructions all mechanical joints and mechanisms used to close and open the circuit breaker.

(11) Check for signs of overheating and tighten all screwed or bolted connections that are not pivotal.

(12) Determine that proper movement of the trip bar will trip the circuit breaker.

(13) Determine that the trip arms on the trip devices of each pole properly engage the trip bar.

(14) Check that the rack-in interlock functions in tripping the breaker.

(15) Determine that the circuit breaker position indicator is showing proper position of the circuit breaker.

(16) If any adjustments are necessary, consult the manufacturer's instruction bulletin.

15.4.4 Electrical Test. Since 90% of all the low-voltage power circuit breakers manufactured contain the long-time delay and instantaneous trip unit combination, a discussion of a typical test program covers these units.

Most of these circuit breakers are equipped with one series overcurrent trip device per phase. The electrical test should be run on each individual trip device. The operation of any one of these devices will trip all poles of the circuit breaker.

In testing a low-voltage power circuit breaker equipped with series overcurrent trip devices, several points should be remembered.

(1) Nameplate rated voltage should be available at the input terminals of the test set throughout the test.

(2) The values of current are high and the voltage is low. Therefore, it is advisable to use connections having the shortest possible length and largest possible cross-sectional area between test unit and circuit breaker.

(3) The connection to the primary disconnect assemblies of the circuit breaker must be tight.

(4) The circuit breaker is tested one pole at a time.

(5) Trip devices should be allowed to fully cool before performing a current timing test.

(6) Variations in test results when compared to manufacturer's curves may be caused by nonsinusoidal test currents resulting from low-voltage high-current test sources.

The recommended tests for a low-voltage power circuit breaker are timing (long- and short-time delay units if the circuit breaker has both type trip units) and instantaneous pickup.

The recommended values of test current for long-time delay is three times the trip unit setting, and for short-time delay it is one and one-half times the circuit breaker short-time setting. If the circuit breaker does not operate within the tolerances shown by the manufacturer's time-current curves, then suitable adjustments should be made as recommended by the manufacturer.

15.4.5 Static Trip Devices for Low-Voltage Power Circuit Breakers. Recently, all major manufacturers have brought out low-voltage power circuit breakers equipped with a static trip device. These trip devices eliminate the need for a dashpot to obtain a time-delay trip of the circuit breaker (see Chapter 6). Maintenance of the circuit breaker equipped with a static trip device is the same as any other low-voltage power circuit breaker.

15.4.6 Electrical Tests for Static Trip Devices. Electrical tests for a low-voltage power circuit breaker are performed one pole at a time; therefore, for a complete test, all three poles must be checked. These tests may utilize a high-current low-voltage source connected to the primary disconnect assemblies on the circuit breaker, namely primary injection; or a low-current source connected to the static trip device terminals, which are normally connected to the secondary terminals of the circuit breaker input signal sensor (current transformer), namely, secondary injection. Regardless of the method used, pickup of long-time, short-time, instantaneous, and ground-fault units should be obtained. In addition, a timing point should be checked on the long-time and short-time characteristics. Suggested values of test current are three times pickup for the long-time characteristic and one and one-half times short-time unit pickup for its time characteristic. The test values obtained should be compared to the manufacturer's specifications and suitable adjustments made as recommended by the manufacturer.

The ground trip feature of a static trip device should be disconnected or bypassed when performing tests to determine the pickup and timing of the long-time, short-time, or instantaneous features, because the ground-fault function will otherwise incorrectly interpret the single-pole test current as a ground fault and trip the circuit breaker.

Some static trip devices employ a separate power sensor (current transformer) to energize the circuit breaker trip circuit. If precision results are desired when performing electrical tests on this type static device, it is recommended that the power sensor and signal sensor be energized from separate sources. This action should eliminate any discrepancies in the test results introduced by saturation of the power sensor.

15.5 Protective Relays. The basic types and applications of protective relays are described in Chapter 4.

15.5.1 Knowledge of Control Circuit. It is important to know the peculiarities of the control circuit associated with the relay being maintained, such as the trip circuit being interlocked with a cubicle door, or that removal of potential from the relay operating coil before opening the trip circuit results in tripping the power circuit breaker.

Another point that should be worked out with the operating personnel is the action to be taken in case a circuit is inadvertently tripped. It is the responsibility of the operating personnel to determine whether or not the circuit should be re-energized.

If the relay technician makes a mistake and trips a circuit, he should acknowledge it immediately. This will save many hours of work, trouble-shooting the control circuit looking for nonexistent problems. It should be remembered that one of the reasons for preventive maintenance and testing is to build confidence in the protective system.

The protective relay is the brain of the electric protective circuit. It is the relay that senses an abnormal condition and then sends the message to other devices on the system. Therefore, it is imperative that any relay work be done in a very thorough manner. A job should not be rushed or left half done.

15.5.2 Identification of Circuits. Double-check the identification of the relays on the panel. This can be complicated when work on relay terminals must be done behind the board and there are a large number of similar boards. One way to solve this problem is to have an assistant turn the ammeter switch from the front of the panel while you observe which switch arm turns in back of the board.

15.5.3 Importance of Target or Operation Indicator. One device often overlooked on a protective relay is the operation indicator or target. When a circuit breaker trips under abnormal conditions, this device indicates the relay element that initiated the tripping action. This information is important, and therefore, a target should never be reset under operating conditions without the supervisor's knowledge and permission.

In addition, the target unit's operation closes a set of contacts that are in parallel with the relay's trip circuit contacts and spiral spring, thus preventing burnout of the relay's trip circuit due to high tripping current and opening of the circuit breaker trip circuit through opening of the relay induction disk trip circuit contacts.

15.5.4 Current Transformer. The current transformer is a vital part of the electric protective system. It is this device that reduces the current in the power circuit to a value that can be handled by the protective relay and insulates the relay from the power circuit. There is a danger associated with a current transformer. Consider a 600/5 current transformer. As long as 600 A is present in the primary, the secondary will try to put out 5 A. Appylying Ohm's law to the secondary circuit, $E=IR$ (voltage equals current multiplied by resistance). Inasmuch as there is current, the higher the resistance of the secondary, the higher the voltage across the secondary. This means that if the secondary circuit opens, a very high voltage is obtained. Therefore, *under no circumstances should the secondary of a current transformer be opened while the primary circuit of the transformer is energized.*

15.5.5 Inspection. The manufacturer's instructions for the specific relay should be on hand; they will supply much useful information concerning connections, adjustments, repairs, timing data, etc. The technician should make sure to obtain proper instruction leaflets.

15.5.6 Visual Inspection
(1) Remove cover from relay case.
  (a) The trip circuit is a live circuit and on some relays it is possible to cause an instantaneous trip while removing the relay cover.
  (b) Inspect the cover gasket.
  (c) Check glass for tightness in the frame, cracks, etc.
  (d) Clean glass inside and out.
(2) Remove relay from case.

(a) To eliminate uncertainties, short the current transformer secondary by jumping the relay-operating coil terminals on the back of the relay case. This jumper should be clipped on with square jaw clips such as crocodile clips.

(b) Open trip circuit by opening the red-handled switch or removing the test block, depending on the type of relay involved.

(c) Open the rest of the switches, on all black-handled relays.

(d) Open the latches that hold the relay in the case and carefully remove relay from case. Remember that the switch blades attached to the case as well as the bars in the bottom of the cases are still energized. If extreme care is not exercised, the circuit breaker may be tripped. With capacitor trip, voltages as high as several hundred volts are present in the trip circuit.

(3) Foreign material such as dust or metal particles should be removed from the relay case and the relay. This foreign material can cause trouble, particularly in the air gaps between the disk and magnet.

(4) Dust can be removed by blowing air from a small hand syringe.

(5) Remove any rust or metal particles from disk or magnet poles with a magnet cleaner or brush.

(6) Hold relay up to the light to make sure that disk does not rub and has good clearance between magnet poles.

(7) Inspect relay for the presence of moisture. If free moisture is present or rust spots are noted, it may indicate that the relay is in the improper atmosphere and presents a design problem.

(8) Connections, especially taps, should be checked for tightness. Tighten all screws, nuts, and bolts that are not pivotal joints.

(9) Sluggish bearings may be detected by noting smoothness of relay reset. Rotate the disk manually to close the contacts and observe that operation is smooth. Allow the action of the spiral spring to return the disk to its normal de-energized position. If the bearings appear bad or operation of the relay is questionable, the relay should be returned to the manufacturer for repair. On an instantaneous plunger-type relay, occasionally a burr or groove may develop on the plunger which would cause the relay to hang up.

(10) Check mechanical operation of targets by lifting the armature and observe showing of target.

(11) Check for damage to the relay coils caused by prolonged high currents by smelling or squeezing the coil with the fingers. A burned smell or spongy or brittle insulation indicate thermal damage.

(12) Components of the relay that touch when the relay is in a "normal" position and part as the relay "operates" should be cleaned. These parts may become dirty and prevent the relay from operating properly on relatively low overloads.

(13) Pitted or burned contacts should be cleaned with a burnishing tool. Never use a solvent on these contacts or touch with fingers as the residue left on the contacts may cause improper operation.

15.5.7 Electrical Testing

(1) Test Methods.

(a) Testing with the relay disconnected from the power and trip circuits (most popular for field testing).

(b) Testing across the secondary of the current transformer with primary de-energized (secondary injection).

(i) This method may be used whenever it is possible to de-energize the power circuit being tested.

(ii) This test includes checking the operation of the circuit breaker, the presence of energy to trip the circuit breaker, etc.

(iii) Testing is done by introducing the test current at the secondary terminals of the instrument transformer. This test checks the relay connections as well as the relay.

(c) Primary circuit tests; high current, low voltage (primary injection).

(i) Complete system is checked including the current transformers.

(ii) The primary bus must be de-energized and exposed to perform this type of test.

(iii) This method requires that caution and safety practices be closely observed as test connections are made on the primary conductors.

(iv) Test simultaneously checks current transformer ratio, secondary wiring, polarity, relay operation, and identity of each phase on the switchboard.

(v) This test is valuable for initially checking bus differentials where many current transformer secondaries are paralleled at the relay.

(2) Test Connections.

(a) Relays with drawout construction.

(i) Whenever practical, the relay should be tested in place, disconnected from the power and trip circuits. The use of the proper test plug is convenient.

(ii) If the relay cannot be tested in place, carefully remove it from the case and proceed to set up at a convenient location. Make sure that the relay is level. If possible, a spare case should be used to hold the relay during the test.

(b) Relays permanently fixed to panel.

(i) Usually these relays have test facilities installed to disconnect the relay from the power and trip circuits and to permit access to the relay circuits from the front of the board.

(ii) If such facilities do not exist, it is necessary to test from rear of panel.

(3) Typical Test for an Overcurrent Relay.

(a) Insulation resistance (do *not* perform on solid-state relays).

(b) Zero check.

(c) Induction disk pickup.

(d) Time-current characteristics.

(e) Target and seal-in operation.

(f) Instantaneous pickup.

The importance of these tests and the procedures to be followed in conducting them are given below.

(a) Insulation resistance. This test is used to ensure the adequacy of the relay insulation.

(i) Apply insulation test voltage from each terminal to ground (500 V dc minimum).

(ii) Apply insulation test voltage across *open* trip circuit contacts.

(iii) Any finite reading should be investigated.

(b) Zero check. This test is to determine that the relay contacts close when the time dial is set at zero. For this test, a continuity light is used. Consult manufacturer's instruction leaflet or relay schematic to identify relay trip circuit contact terminals.

(i) With the continuity light connected across the relay induction disk trip circuit contact terminals, manually turn time dial until the continuity light glows.

(ii) Note reading of time dial.

(iii) If reading is not zero, consult manufacturer's leaflet for proper adjustment so that the time dial reads "zero" when contacts are made.

(iv) In some relays the position of the backstop may be changed on the shaft. In others, the position of stationary contacts may be changed. Some relays have no adjustment.

(c) Induction disk pickup test. This test is to determine the minimum operating current of the relay, that is, the minimum current needed to close the relay induction disk trip circuit contacts for any particular tap setting.

Consult manufacturer's instruction leaflet or relay schematic to identify operating coil terminals and the trip circuit contact terminals.

(i) The pickup value should be equal to tap value ± 3%.

(ii) Connect the induction disk operating coil terminals of the relay to the source of test current.

(iii) Connect a continuity light to relay induction disk trip circuit contacts.

(iv) Energize the relay operating coil with 150–200% of relay tap value so that the relay induction disk trip circuit contacts will close rapidly as indicated by glow of continuity light. Decrease the current until the light flickers.

(v) Record this value of current as "pickup." Usually pickup is adjusted by changing tension of the spiral spring that is employed to return the relay trip circuit contacts to their deenergized position. Increase spring tension to increase pickup. The relay manufacturer's instruction booklet should be consulted to assure the proper procedure.

(d) Time-current characteristics. A timing check should be made to see that the relay closes its contacts within a specified time for a given abnormal value of current. Normally this test is run with the relay tap in its designated position.

It is suggested that a test current of four times pickup be used. However, in any given specific plant, management personnel will indicate given test points and acceptable test results.

Consult manufacturer's instruction leaflet or relay schematic to identify operating coil terminals and the relay trip circuit contact terminals.

(i) The operating coil of the relay induction disk unit should be connected to the source of the test current. Adjust current to test value and then de-energize the relay.

(ii) Connect relay induction disk trip circuit contacts to a timer.

(iii) Energize relay to determine time of operation at the chosen value of test current.

(iv) Record the value of the test current and time. Compare with manufacturer's specified values. If the relay is too fast, either increase time dial setting (up to one half division) or adjust damping magnet. If relay is too slow, either decrease time dial setting or adjust damping magnet. Consult manufacturer's leaflet for method of making adjustment.

(e) Target (operations indicator) and seal-in operation. The purpose of the target is to indicate which relay tripped the circuit breaker. The seal-in provides a low-resistance trip circuit in parallel with the relay's induction disk trip circuit contacts and spiral spring.

There are three types of target units in common use:
(i)   Combined target and seal-in unit.
(ii) Separately operated target and separately operated seal-in.
(iii) Mechanical target.
(iv) Relay seal-in (contactor switch).

Induction disk relays have the combined target (operation indicator) and seal-in unit. Relays of older design may have a separate target. Some relays, such as the instantaneous overcurrent, have mechanical indicators.

(i) Combined target and seal-in test procedures.

(a) Close relay induction disk unit trip circuit contacts.

(b) Connect a dc supply to the relay induction disk trip circuit contact terminals. Note the tap setting of the target unit.

(c) Gradually increase the magnitude of the dc supply until the target shows. Value should be less than the tap value.

(d) Open the relay induction disk trip circuit contacts. The current through the dc circuit should remain unchanged if the seal-in unit is operating properly.

(e) When installing the relay cover, make sure the target reset operates correctly.

(ii) Separately operated target. The same test procedure as the first three steps for combined target and seal-in apply.

(iii) Mechanical-operation targets. These targets show as soon as the relay trip circuit contacts close. No separate electrical check is required.

(iv) Relay seal-in (contactor switch). Some relays are equipped with a seal-in (contactor switch) independent of the target. Note seal-in rating from the relay nameplate and proceed with testing as for combined target and seal-in. Be sure locknut on seal-in unit is tight.

(f) Instantaneous Unit Pickup. Consult manufacturer's instruction leaflet or relay schematic to identify instantaneous unit operating coil terminals and instantaneous unit trip circuit contact terminals.

(i) Connect relay instantaneous unit operation coil to a source of test current.

(ii) Connect relay instantaneous unit trip circuit contact terminal to a continuity light.

(iii) By alternately pulsing and adjusting the test current, determine the minimum amount that just causes the instantaneous unit to operate.

(iv) Note ammeter reading as instantaneous unit picks up. If pickup value differs from specified value, adjust and repeat the test. If target operation is not mechanical, test the operations indicator as described previously.

The preceding test procedures are given only as an example. There are many hundreds of different kinds of protective relays, and this chapter cannot go into detailed test procedures for each and every one. However, the relay manufacturer's instruction books normally contain data pertaining to maintenance and calibration tests.

15.5.8 Solid-State Relays. Since many solid-state relays have built-in operational test provisions, these tests are simple to make. Calibration tests are made in the conventional manner.

Generally, no maintenance is required in the usual sense of adjusting, cleaning, or lubricating.

15.6 Medium-Voltage Circuit Breakers

15.6.1 Medium-Voltage Circuit Breakers Using Series Trip Coils. The maintenance and testing of these devices follow the same format as for low-voltage power circuit breakers, except that the trip current is properly introduced through the primary of the external current transformer. Tripping by injection current through the secondary current transformer terminals is an acceptable alternate if isolation of the current transformer primary is not possible.

15.6.2 Maintenance and Inspection Procedures. It is recommended that a maintenance program be established that will provide for a periodic inspection of circuit breakers.

15.6.3 Air Magnetic Power Circuit Breakers Rated 4.16–13.8 kV. Maintenance tasks listed are suggested minimum guidelines. Consult specific manufacturer's instruction books for detailed information.

NOTE: Before inspecting or performing any maintenance on either the breaker or the mechanism, be sure the breaker is in the *open* position, is disconnected from all electrical sources, and is removed from the cubicle. Both the closing and the opening springs should be discharged or blocked mechanically before any maintenance is done.

(1) Record number of operations and perform a general visual inspection of the breaker. Report any unusual signs or problems.

(2) Using test coupler, operate breaker electrically. Check operation of all electrical relays, solenoid switches, motors, control switches, and indicating devices.

(3) Visual inspection.

  (a) Remove box barriers.

  (b) Wipe clean of smoke deposit and dust all insulating parts, including the bushings and the inside of the box barrier. Use a clean, dry, lint-free cloth; a vacuum cleaner would be helpful.

  (c) Inspect condition of bushing primary disconnect stabs on finger cluster.

  (d) Inspect condition of bushing insulation; should be clean, dry, smooth, hard, unmarred.

  (e) Check breaker and operating mechanism carefully for loose nuts, bolts, or retaining rings, and ensure that mechanical linkage is secure.

(f) Inspect insulation and outside of arc chutes for holes or breaks; small cracks are normal.

(g) Inspect magnetic "blow-out" coils (if used) for damage.

(h) Inspect all current-carrying parts for evidence of overheating.

(4) Functional inspection.

(a) Sandpaper throat area of arc chutes until thoroughly clean.

(b) Ensure that arc chutes are clear of contamination, and have no significant damage on grids or ceramics. If ceramic or fins are broken, replace arc chutes.

(c) Ensure that all brazed, soldered, or bolted electrical connections are tight.

(d) Inspect contacts of control relays for wear, and clean as necessary.

(e) Check actuator relays, charging motor and secondary disconnect for damage, evidence of overheating, or insulation breakdown.

(f) Check that all wiring connections are tight and for any possible damage to the insulation. Replace any wire that has worn insulation.

(g) On one design of stored-energy breakers, operate the breaker slowly, using a ⅝ in ratchet. By using the spring blocking device, check for binding or friction, and correct if necessary. Make sure contacts can be opened or closed fully.

(h) Inspect the arcing contact for uneven wear or damage. Replace badly worn contacts. Measure the arcing contact wipe, using ohmmeter. Make adjustment if necessary. Refer to appropriate instruction book. Do not grease arcing contacts.

(i) Inspect primary contacts for burns or pitting (caution loaded spring). Wipe contacts with clean cloth. Replace badly burned or pitted contacts. Rough or galled contacts should be smoothed with a crocus cloth or filed lightly.

(j) Inspect primary disconnect studs for arcing or burning. Clean and lightly grease contacts per manufacturers' instructions.

(k) Check primary contact gap and wipe. Make adjustments per appropriate instruction book. Grease contacts with GE D50H47 or equivalent and operate breaker several times.

(l) Check operation and clearance of trip latch. Wipe trip armature travel and release latch per appropriate instruction book. Replace worn parts.

(m) Inspect all bearings, cams, rollers, latches, buffer blocks for wear. Teflon coated sleeve bearings do not require lubrication. All other sleeve bearings, rollers, and needle bearings should be lubricated with SAE 20 or 30 machine oil. All ground surfaces coated with dark molybdenum disulphide do not require lubrication. Lubricate all other ground surfaces such as latches, rollers, or props with D50H15 grease or equivalent.

(n) Install box barriers.

(o) Measure insulation resistance of each bushing terminal-to-ground and phase-to-phase. Record readings along with temperature and humidity. Some users perform these tests immediately after removal from service.

(p) Hi-pot breaker bushing insulation per appropriate instruction book (optional).

(q) Check closed breaker contact resistance (optional).

(r) Perform power factor test (optional).

(s) Perform corona test (optional).

(t) Using test box, operate breaker both electrically and manually. Check all interlocks.

(u) Raise and lower breaker in cabinet. Watch for proper operation of the positive interlock trip-free mechanism. (Breaker should trip "in" not in "full up" or "test" position.)

(v) Remove breaker from cubicle and check primary disconnect wipe; refer to appropriate instruction book.

(w) Insert breaker into cubicle; ready for energization.

15.6.4 Vacuum Power Circuit Breakers Rated 4.16–13.8 kV. Maintenance tasks listed are suggested minimum guidelines. Consult specific manufacturer's instruction books for detailed information.

NOTE: Before any maintenance work is performed, make certain that all control circuits are de-energized and that the breaker is removed from the metal-clad unit. Do not work on the breaker or mechanism while in the closed position without taking precautions to prevent accidental tripping. Do not work on the breaker while the closing springs are charged unless they are secured in that position by the gag pin.

(1) Record number of breaker operations and perform a general visual inspection of the circuit breaker. Report any unusual signs or problems.

(2) Using the test coupler, operate the breaker electrically. Check operation of all electrical relays, solenoids, motors, control switches, and indicating devices.

(3) Visual inspection.

(a) Perform a visual inspection of the breaker and remove dust and contaminants from the interrupter housing, insulation, and mechanism. Check damage to the breaker exterior, condition of primary and secondary disconnects, evidence of overheating or tracking, and erosion of contacts.

(b) Inspect interrupters and operating mechanism for loose nuts, bolts, and damaged parts. All cam, latch, and roller surfaces should be inspected for damage or excessive wear.

(c) With the breaker open and the close spring discharged, check the trip coil plunger and close coil plunger move freely.

(4) Functional inspection.

(a) Check interrupter and mechanism adjustments (pull rod, trip latch clearance, contact wipe and erosion indication, over-travel stop gap, control switches).

(b) Lubricate breaker operating mechanism.

(c) Operate breaker slowly to be sure there is no binding or friction and the movable contact of the interrupter can move to the fully opened and fully closed positions.

(d) Inspect all wiring connections and insulation.

(e) Using test cabinet, operate breaker and observe that breaker charges opens and closes correctly, trips freely, and indicators show position correctly.

(f) Check primary circuit integrity by means of a 2500 V megohmmeter; breaker in closed position phase-to-phase and phase-to-ground.

(g) Check control circuit integrity with a 500/1000 V megohmmeter; do not apply to motor leads to prevent damage to winding insulation.

(h) Perform a vacuum integrity test on each interrupter.

(i) Perform a low-resistance contact test using a micro-ohmmeter.

15.7 Fuses and Switches. Maintenance of fuses is limited by nature of the device to an inspection to ensure that the proper size fuse is installed, that it shows no signs of deterioration, and that the enclosure is clean and the connections are tight. The size and type of fuses should comply with those specified by the responsible department.

The fuse clips used for 1–600 A fuses should be checked for alignment and tightness. The conductor terminations on both the switch and fuse terminals should also be checked for tightness. Additionally, the switch should be operated several times to ensure that the toggle mechanism is operable.

Manufacturer's instruction bulletins should be consulted for testing and maintenance of fused disconnects, bolted-pressure switches or fused power circuit devices.

15.8 Auxiliary Devices. Periodically, the substation battery should be checked for proper water level, voltage, and specific gravity of each cell. Terminals should be cleaned and connections tightened. The battery charger should be checked frequently to make sure that the charging rate is correct and that the charger is actually in operation. Control circuitry and auxiliary relays can be checked periodically through an operations test. The operations test consists of closing the protective relay trip circuit contacts manually and assuring that the proper circuit breaker will open. This test will check continuity of all control circuitry and the ability of the battery to trip the circuit breaker. It will also check operation of the circuit breaker. Since this test requires that the power circuit be de-energized, it must be scheduled at the convenience of operating personnel.

15.9 References. The following publications shall be used in conjunction with this chapter.

[1] ANSI/NFPA 70-1984, National Electrical Code, Section 230-95c.

[2] ANSI/NFPA 70B-1983, Recommended Practice for Electrical Equipment Maintenance.

[3] ANSI/UL 1053-1982, Safety Standard for Ground-Fault Sensing and Relaying Equipment, Section 29.

[4] NEMA AB2-1984, Procedures for Verifying Field Inspection and Performance Verification of Molded Case Circuit Breakers.

[5] NEMA PB2.1-1979, Instructions for Safe Handling, Installation, Operation and Maintenance of Deadfront Switchboards Rated 600 Volts or Less.

[6] NEMA PB2.2-1983, Application Guide for Ground-Fault Protective Devices for Equipment.

[7] GILL, A. S. *Electrical Equipment Testing and Maintenance.* Prentice Hall, Reston, VA.

[8] IEEE-PES Power System Relaying Committee. *Sine-Wave Distortions in Power Systems and the Impact on Protective Relaying,* IEEE-PES Report 84TH0115-6 PWR, 1984.

[9] MORROW, L. C., *Maintenance Engineering Handbook,* 2nd ed. New York: McGraw-Hill, 1966, Section 7, Chapter 11.

[10] NETA Acceptance Testing and Maintenance Specifications. National Electrical Testing Association, Dayton, OH, 1982.

Index

A

B

C

D